Applied Mathematical Sciences

EDITORS

Fritz John
Courant Institute of
Mathematical Sciences
New York University
New York, N.Y. 10012

Lawrence Sirovich
Division of
Applied Mathematics
Brown University
Providence, R.I. 02912

Joseph P. LaSalle
Division of
Applied Mathematics
Lefschetz Center
for Dynamical Systems
Providence, R.I. 02912

ADVISORS

H. Cabannes University of Paris-VI

J.K. Hale Brown University

J. Keller Stanford University

J. Marsden Univ. of California at at Berkeley

G.B. Whitan California Inst. of Technology

EDITORIAL STATEMENT

The mathematization of all sciences, the fading of traditional scientific boundaries, the impact of computer technology, the growing importance of mathematical-computer modelling and the necessity of scientific planning all create the need both in education and research for books that are introductory to and abreast of these developments.

The purpose of this series is to provide such books, suitable for the user of mathematics, the mathematician interested in applications, and the student scientist. In particular, this series will provide an outlet for material less formally presented and more anticipatory of needs than finished texts or monographs, yet of immediate interest because of the novelty of its treatment of an application or of mathematics being applied or lying close to applications.

The aim of the series is, through rapid publication in an attractive but inexpensive format, to make material of current interest widely accessible. This implies the absence of excessive generality and abstraction, and unrealistic idealization, but with quality of exposition as a goal.

Many of the books will originate out of and will stimulate the development of new undergraduate and graduate courses in the applications of mathematics. Some of the books will present introductions to new areas of research, new applications and act as signposts for new directions in the mathematical sciences. This series will often serve as an intermediate stage of the publication of material which, through exposure here, will be further developed and refined. These will appear in conventional format and in hard cover.

MANUSCRIPTS

The Editors welcome all inquiries regarding the submission of manuscripts for the series. Final preparation of all manuscripts will take place in the editorial offices of the series in the Division of Applied Mathematics, Brown University, Providence, Rhode Island.

SPRINGER-VERLAG NEW YORK INC., 175 Fifth Avenue, New York, N.Y. 10010

Printed in U.S.A.

Applied Mathematical Sciences | Volume 33

Ulf Grenander

Regular Structures

Lectures in Pattern Theory
Volume III

Springer-Verlag
New York Heidelberg Berlin

Ulf Grenander
L. Herbert Ballou University Professor
Division of Applied Mathematics
Brown University
Providence, Rhode Island 02912

AMS Classification 68G10

Library of Congress Cataloging in Publication Data

Grenander, Ulf.
Lectures in pattern theory.

(Applied mathematical sciences; v.18, 24, 33)
Includes bibliographies and indexes.
Contents: v. 1. Pattern synthesis—v. 2. Pattern analysis—v. 3. Regular structures.
1. Pattern perception—Collected works. I. Title.
II. Series: Applied mathematical sciences (Springer-Verlag New York Inc.); v. 18 [etc.]
QA1.A647 Vol. 18, etc. [Q327] 510s 76-210
ISBN 0-387-90174-4 (v. 1) [001.53'4] AACR2

Printed in the United States of America

9 8 7 6 5 4 3 2 1

ISBN 0-387-90560-X Springer-Verlag New York Heidelberg Berlin
ISBN 0-540-90560-X Springer-Verlag Berlin Heidelberg New York

PREFACE

Most of the material in this book has been presented in lectures at Brown University, either in courses taught in the Division of Applied Mathematics or in the author's Research Seminar in Pattern Theory. I would like to thank the several members of the Division of Applied Mathematics that have participated in the discussions and in particular W. Freiberger, S. Geman, C.-R. Hwang, D. McClure and P. Thrift.

I would also like to thank F. John, J. P. LaSalle, and L. Sirovich for accepting the manuscript for the Series Applied Mathematical Sciences published by Springer-Verlag.

The research reported here has been supported by the National Science Foundation, Office of Naval Research and the Air Force Office of Scientific Research. I am grateful for the active interest and help given in various ways by Dr. Eamon Barrett, Dr. Kent Curtis, Dr. Robert Grafton and Dr. I. Shimi of these agencies.

I also thank C.-R. Hwang and P. Thrift for help with proofreading.

I am indebted to Mrs. E. Fonseca for her careful preparation of the manuscript, to Miss E. Addison for helping me with the many diagrams, and to Mrs. K. MacDougall for the final typing of the manuscript.

Ulf Grenander
Providence, Rhode Island
October 1980

TABLE OF CONTENTS

INTRODUCTION

This is the third and final volume of the Lectures in Pattern Theory. Its two first chapters describe the science-theoretic principles on which pattern theory rests. Chapter 3 is devoted to the algebraic study of regularity while Chapter 5 contains new results in metric pattern theory. Some brief remarks on topological image algebras can be found in Chapter 4.

Two chapters deal with pattern synthesis: Chapter 6 on scientific hypothesis formation and Chapter 7 on social domination structures. In Chapter 8 we study taxonomic patterns, both their synthesis and analysis, while in the last chapter we investigate a pattern processor for doing semantic abduction.

The material contained in the three volumes has been presented in historical rather than logical order. A reader approaching pattern theory for the first time is advised to do it in the following order,

Introduction to regular structures	Chapters 1 and 2 of Volume III.

Pattern synthesis	Chapters 1,2,3 of Volume I Chapters 3,4,5 of Volume III Chapter 4 of Volume I Chapters 6,7 of Volume III
Pattern analysis	Chapters 1,2,3,4,5 of Volume II Chapter 8 of Volume III
Pattern processors	Chapters 6,7 of Volume II Chapter 9 of Volume III

Most of the content is due to the author and the members of the Research Seminar in Pattern Theory at Brown University. With a few exceptions it has not appeared in print before.

Space does not permit the inclusion of all the new results. So for example have we not included the analysis of star-shaped patterns and of spectroscopic patterns, nor the study of growth patterns based on contact transformations. The method of sieves, developed for pattern inference, will be presented in the author's forthcoming book, "Abstract Inference". A separate publication will also appear containing mathematical software that we have written for the computational experiments that have played an important role during the growth of pattern theory.

Lord Kenneth Clark once described the publication of lectures as "a well-known form of literary suicide". One can certainly argue against publishing lecture notes since they are likely to contain obscurities and mistakes and be too fragmented to offer a complete view of the subject.

In spite of this we decided to publish these Notes rather than to wait for a polished and complete presentation. As mentioned in the Introduction to Volume I a more definitive version will appear eventually. In the meantime these three volumes with all their imperfections will have to suffice.

CHAPTER 1
PATTERNS: FROM CHAOS TO ORDER

1.1. The search for regularity

The search for regularity is a dominant theme in man's attempt to understand the world around him. Any such attempt is based on an assumption, tacitly made or explicit, that phenomena in nature and in the man-made world are governed by laws that result in order and structure.

Or to quote Hume in his Treatise of Human Understanding, Book I, Sect. VI: "If reason determined us, it would proceed upon that principle, *that instances, of which we have had no experience, must resemble those, of which we have had experience, and that the course of nature continues always uniformly the same.*" This principle underlies the incomplete inductive reasoning used in science as well as in everyday life.

Indeed, it is hard to see how anything could be really understood in a completely chaotic world, where events followed each other in an arbitrary fashion, where chaos reigned and no rules restricted what could occur. It would be impossible to plan for the future, even to take action to make the individual or the species survive in such a frightening and mysterious environment.

Already in pre-scientific times man must have tried to find regularities that he could rely on in his everyday life or that would give him a feeling of security in a hostile world. Or, quoting from Frazer's Chapter LXIX of "The Golden Bough",

> "In magic, man depends on his own strength to meet the difficulties and dangers that beset him from every side. He believes in a certain established order of nature on which he can surely count, and which he can manipulate for his own ends. When he discovers his mistake, when he recognizes sadly that both the order of nature which he had assumed and the control which he had believed himself to exercise over it were purely imaginary, he ceases to rely on his own intelligence and his own unaided efforts, and throws himself humbly on the mercy of certain great invisible beings behind the veil of nature, to whom he now ascribes all those far-reaching powers which he once arrogated to himself" ...

Magic is superseded by a religious belief in gods -

> "But as time goes on this explanation in its turn proves to be unsatisfactory. For it assumes that the succession of natural events is not determined by immutable laws, but is to some extent variable and irregular, and this assumption is not borne out by closer observation. On the contrary, the more we scrutinize that succession the more we are struck by the rigid uniformity, the punctual precision with which, wherever we can follow them, the operators of nature are carried on."

Most sciences pass through an early stage of collecting isolated data, assembling curious objects or facts. Already taxonomic attempts to classify objects or facts represent a tendency towards generality and "immutable laws" and "rigid uniformity". At a somewhat later stage, usually overlapping with the earlier one, one strives for the explicit formulation of general principles. The scientist's rule is not just to discover or invent such principles, but it is at least as much concerned with the logical analysis of them and to deduce consequences. It depends upon the consequences and their relation to the observed world how successful the scientist has been in describing the regularities.

Viewed from our own time and in a more abstract setting such attempts could be formalized as *formal systems:* certain basic statements or procedures and rules how to apply them in order to explain certain phenomena. For example, statement A implies B, another statement C implies A or, formally

$$\begin{cases} A \rightarrow B \\ C \rightarrow A \end{cases} \tag{1.1}$$

In pre-Galilean mechanics A could be "object 1 is heavier than object 2", B="object 1 falls faster than object 2", and C="objects 1 and 2 have the same volume, the first is made of lead and the other of iron".

For a given set of basic statements (1.1) the richness of the results of applying rules will depend upon how sophisticated are the syllogisms to be used. If the usual rules of logic are applied one gets as consequences of the statements in (1.1) if B does not occur A cannot hold, if C is true then B must hold, etc:

$$\begin{cases} \sim B \rightarrow \sim A \\ C \rightarrow B \\ \ldots \end{cases} \tag{1.2}$$

In order that a system describing regularity deserve its name it must have some permanence in time and space. If it only applies to a particular time and a particular place it is a *datum*, an isolated observation, but not a law of nature. Therefore one must insist that the statements should be true in some generality.

When we speak of laws, order, patterns, we are concerned with more than isolated facts. Laws deal with several alternatives, interesting laws with a great number of alternatives. We therefore have to adopt an *ensemble* attitude: the pattern should refer to an ensemble of possible cases. In such an ensemble order is viewed as the uniform validity of certain properties. This is still rather vague but will become more precise when we examine a number of regular structures in Section 1.2.

The symbols used (A,B,...) are irrelevant, we could equally well have employed other abbreviations for the statements. We could express this by saying that we are thinking of a particular interpretation of the formal statement (1.1) and the interpretation is fixed while the formalization of it remains arbitrary to some extent. One and the same regular structure could be expressed through many formal systems, mutually equivalent. As long as the formulas mean the same we have no reason to prefer one before the other unless we bring in other criteria based on notions such as simplicity and convenience.

From a formal point of view we need not distinguish between statements like $A \to B$ and syllogisms like $(x \to y) \to (\sim y \to \sim x)$. In the interpretation used above the first one was based on *empirical knowledge* while the second one was an *analytical truth*. Formally they can both be viewed as laws or axioms that we can combine together to arrive at other, derived statements. The number of derived statements can be large, even infinite.

To bring out more clearly the conceptual structure of this kind of regularity, let us consider another case, a fragment of Newtonian mechanics for point masses. We would then have statements like

$$\left\{\begin{array}{ll} m_1 & = \ldots; m_2 = \ldots; \ \ldots \ x_1 = \ldots, y_1 = \ldots; \ \ldots \\ F_{gr} & = k\,\dfrac{m_1 m_2}{r^2}\ ; \\ k & = \ldots; \\ r & = \sqrt{(x_1-x_2)^2+(y_1-y_2)^2+(z_1-z_2)^2} \\ m\ddot{x} & = F^x; m\ddot{y} = F^y, \ldots; \\ F_{12} & = -F_{21}; \\ F_{fr}^x & = -f\dot{x}; \\ f & = \ldots \end{array}\right. \tag{1.3}$$

together with the other statements representing calculus and syllogisms. Combining statements together in a "meaningful" manner we can derive other statements and describe, analyze, and predict the behavior of mechanical phenomena. In other words, we can express the regularities of such phenomena.

In (1.3) the natural invariances are the invariances with respect to Galilean transformations.

$$\left\{\begin{array}{l} t = t' \\ x = x' + at \\ y = y' + bt \\ z = z' + ct \end{array}\right. \tag{1.4}$$

as well as scale changes for units of length, mass, and time. For the latter the induced changes have to be made for the mechanical constants depending upon their dimension. As before the names (symbols) used for labelling quantities can also be changed as long as it is done consistently.

Starting from (1.3) we arrive at one conclusion after another. For example one proves that in the absence of forces a mass point moves in uniform motion, a very direct consequence. Or, introducing second order concepts, such as energy and momentum, one proves conservation laws under certain conditions. A derivation can be viewed as a *sequence* of the original statements appearing in (1.3), or of the mathematical-logical auxiliary statements that are needed. Of course, to make sense, this sequence cannot be arbitrary, but *its successive elements must follow each other according to the rules specified.*

Again, from a formal point of view this sequence is just a formula, constructed from certain sub-formulas according to the manipulative rules. The *meaning we attribute to the formula* comes from the particular interpretation we have in mind, in this case mass point dynamics. The correctness of the formula in the present case is deducibility of the conclusion from the initial assumptions, or rather from the subset of the initial assumptions needed to carry out the chain of reasoning. The steps of reasoning "inside" the formula are irrelevant as long as they are correct. It is clear that several formulae may have the same meaning, and it is possible, although perhaps less obvious, that one formula (sequence) may have several interpretations when viewed in different contexts.

The regular structure of Newtonian mechanics is certainly one of the deepest in the natural science. It has a rich texture, unrivalled in its elegance and power. It may be instructive to consider a much simpler example, which brings out some of the logical features characterizing regular structures in a way that is easy to follow, unencumbered by technical reasoning.

Consider an infinite sequence of natural numbers $x_1, x_2, x_3, x_4, \ldots$, for example the sequence $1,3,5,7,\ldots$, the odd positive integers. This sequence, let us name it x, is a single object so that it may seem to be contradictory to the ensemble attitude to look for patterns. It is not necessary to think of the sequence x in restrictive terms ("not divisible by two") but instead generate it by recursion. We then start with the sequence with a single element $x' = (x_1) = (1)$, and apply repeatedly the recursion

$$\begin{cases} x^{n+1} = x^n \text{ concatenated with } x_{n+1}, \text{ where} \\ x_{n+1} = x_n^n + 2 \end{cases} \tag{1.5}$$

We can think of this as a sequence of applications of rule (1.5) plus the initial condition that $x_1 = (1)$. The elements of this sequence are identical except of course that they accept as inputs different values and, hence, also produce different outputs. This is the operation of the sequence, the interpretation or meaning of it is the sequence of values produced.

The sequence of applications of rule (1.5) and the resulting numerical sequence are closely related to each other, but it would be a serious mistake to treat them as identical.

We shall return to this question in Sections 1.2 and in 2.3.

If the constant 2 in (1.5) is changed to some other natural number, and if the initial condition is altered, we get other arithmetic series. Similarly we can modify (1.5) to get arithmetic series of higher order, geometric series, Fibonacci numbers, etc.

The sequence of applications of (1.5) is denumerably infinite in contrast to the examples discussed before. One should not attribute much importance to this difference, however.

Neither do we insist on the particular formal way of writing rules like (1.5). Indeed one could equally well use a programming language format. In APL for example we could write it as

$$X \leftarrow X,\ 2 + \text{¯}1 \uparrow X \tag{1.6}$$

initialized by the statement $X \leftarrow ,1$ and embedded in a loop executing (1.6) repeatedly. In other programming languages we would get other, usually less attractive, expressions depending upon what computational modules are available and how they are handled syntactically in the particular language.

In principle it would not matter what language (including mathematical notation) we use as long as it is powerful enough. This is true but misleading. In computer programming one could code everything in binary but most users of computers prefer a higher level language, and with good reason. Similarly, in mathematics, the choice of notation is important in that it can focus attention on the decisive aspects of a problem.

In the study of regularity we also need a systematic procedure, *a formalism*, to describe and analyze patterns.

It should be general enough to be applicable to the many varieties of patterns that will be encountered, but also flexible so that we can use it with convenience and be supported by its conceptual framework. The formalism will not solve the problems for us but help us to express them concisely, emphasizing the common features of seemingly different patterns.

The examples of regular structures mentioned above can guide us toward the design of a pattern formalism. They differed in their meaning: two of them described deductive processes and one arithmetic calculations. Formally, however, they shared the property that they *constructed objects by combining given ones* following certain rules of construction, and this will be one of the *leitmotifs* in our study.

The rules will put limits to the arbitrariness of the constructions: the more stringent the rules are, the more rigid will the resulting patterns be, farther away from chaos. In this connection the reader is reminded of Kolmogorov's notion of complexity of computation and the resulting definition of randomness. The reader is referred to Solomonoff (1964), Martin-Löf (1966).

Leaving out the technical aspects, which are somewhat complicated, the reasoning goes like this. Given an abstract computational set up, in terms of general machines or algorithms, consider long binary sequences $x = x_1, x_2, \ldots, x_n$; $x_i = 0$ or 1, n large. If x is a given sequence let p be a program, coded also as a binary sequence expressing the use of machine instructions in p, such that p computes x. The length $\ell(p)$ of p has a lower bound, say $K(x)$, which we interpret as $+\infty$ in the case when no program computes x.

Now the *complexity measure* K(x) will in general depend upon the machine used. It was shown, however, by Kolmogorov that this dependence is not crucial, when n tends to infinity, and this makes it possible to define a related complexity measure uniquely, without reference to a particular machine. A random sequence is then, a bit vaguely, one of maximal complexity: a long program is needed to compute it.

We have been looking at the other extreme when the object, which could be a numerical sequence but does not have to be one, can indeed be described by a concise program, for example the Fibonacci sequence. Hence our study seems to deal with situations which are diametrically opposed to randomness.

This is not quite accurate, however, and probabilistic ideas will play an important role in the pattern theoretic development. This is best illustrated by an example.

Say that we study geometric patterns where the objects are sets representing biological shapes, and where the sets are limited by given rules. Perhaps they are ellipsoidal, or convex, or defined in terms of more general geometric constructs. Such patterns could be quite rigid and possible to be described concisely.

It is a different matter when we ask what happens when the objects are observed and measured. Then it will depend upon what instrumentation is available to the observer: his *view of the object* can be quite different, conceptually as well as quantitatively, from the object itself. Usually the regularity is loosened up, the constraints limiting the shape need no longer apply strictly.

But this means that we will need more complex, perhaps much longer, descriptions of the view of the object than for the object itself, and we are led to employ probabilistic descriptions. This will be done by introducing probability measures over the possible values of objects: we shall have to study probabilities over *sample spaces that are often quite different from the ones usually considered in probability theory and statistics.*

The very rigid patterns are themselves of great interest, for example in terms of the generative power of different logical structures, or decidability questions related to recognition of patterns. In spite of this, such questions will receive little attention in our study, most of which will deal with regular structures in the middle of the spectrum - not completely random nor highly rigid.

As mentioned, randomness will be used to describe really observed patterns as distinguished from the hypothetical ones. But it will also be needed to describe *how likely are the different hypothetical ones*. Again we will have to develop some mathematical tools to handle the questions that arise. This can be seen as an extension of attempts to analyze probability measures on such sample spaces as groups and semigroups, topological vector spaces, algebras, and so on. We shall attempt to express these probabilities in terms of the underlying topological and algebraic properties of the patterns but it would be premature to go into any details of how this can be done. Instead we shall return to this important question in Chapters 3-5, see Notes A.

Returning to the observer's view of some regular phenomenon one should not think of it as just a disturbance caused by random noise in the technical sense of this term. That would be to underestimate the mathematical difficulties encountered when one tries to understand the relation between theoretical regularity and observable regularity. The *effect of the instrumentation* can profoundly effect the nature of the phenomenon. So, for example, can the dimensionality be changed when three-dimensional objects are viewed by a monocular instrument and appear two-dimensional, or when spherical objects (with four parameters) are transformed into more general convex sets (perhaps with an infinite number of parameters). The imperfection of the observations causes distortions that have to be understood and analyzed, which is one of the main difficulties in the mathematical study of regularity.

In the few examples that we have mentioned the regularity appeared as the systematic arrangements of simple objects according to given rules, and interpreted as viewed by an observer. We were not primarily concerned with a single law, a single regularity, but with systems or ensembles. Therefore we shall speak of *regular structures, patterns,* to emphasize the ensemble aspect.

So far, so good. But how can we express and characterize regular structures formally, that is in mathematical terms? This is no easy task and we need more insight into the nature of patterns before deciding on a formal framework. To get it, let us consider a number of special cases illustrating how the notion of regularity has been used in the historical development of the sciences, as well as in humanistic studies and technology.

Although the list of examples is fairly long, it represents only a miniscule portion of the ideas of regularity that have been put forward. This is obvious. What is less obvious a priori is that they have a lot in common, that similar ideas have been used to characterize regularity in subjects that might appear as completely unrelated. This will be no surprise to anyone believing in the unity of science, and it will certainly help us in our task.

Here a qualification is needed. The examples in the list in Section 1.2 have been selected with a certain bias, as the reader will notice. This will limit the range of applicability of the formalism to be constructed and we must keep this reservation in mind when using it.

1.2. Some regular structures

To learn about the general we shall study the particular. We shall begin with one orderly event that will have made primitive man wonder and reflect: the motion of the sun, the moon, and the planets around the earth and against the stars.

It must have been noticed early in most cultures that the stars could be seen as fixed, perhaps attached to some invisible sphere, and rotating around an axis through the Polar Star. The motion of the Sun and the Moon could also have been described by circular motions with the earth in the center. Indeed, this fits in well with Plato's belief in circular motion as the only ideal one, but the problem was, of course, to reconcile this belief with the seemingly irregular, back-and-forth motion of the other planets. Observing for example Jupiter, and plotting the observed successive position against a star chart, one finds that the planet occasionally reverses its motion: retrogression.

To account for such disturbing anomalies the classical astronomers had to modify a purely Pythagorean universe with a few spheres inside each other to what was to become the model described by Ptolemy in his Almagest. The idea was to preserve the circular motion as the basic assumption but combine such motions into compound ones. A circle is made relative to another one so that a point of the first one will move along an epicycle. In this way the resulting motion will sometimes appear as retrogression explaining the anomaly.

A Ptolmaic universe could look like the picture in Figure 2.1, see Notes A, where the earth is in the center, the moon and the sun rotate around it in slightly eccentric circles, and the five planets follow epicycles. This magnificently conceived system of the universe enabled the astronomers to numerical predictions of some accuracy. As astronomical observations became more accurate the Ptolemaic model had to be refined to reconcile it with data, and this was done by adding more circles. A late version of the model had 39 circles.

Ptolemy sums up his view in the Almagest, by saying that his aim has been to show that all phenomena in the sky are produced by uniform circular motions. He had set himself the task of proving that the apparent irregularities of the planets can be explained by such motions, that only such motions are appropriate to the divine nature of the universe. This was then the ultimate aim of mathematical science based on philosophy.

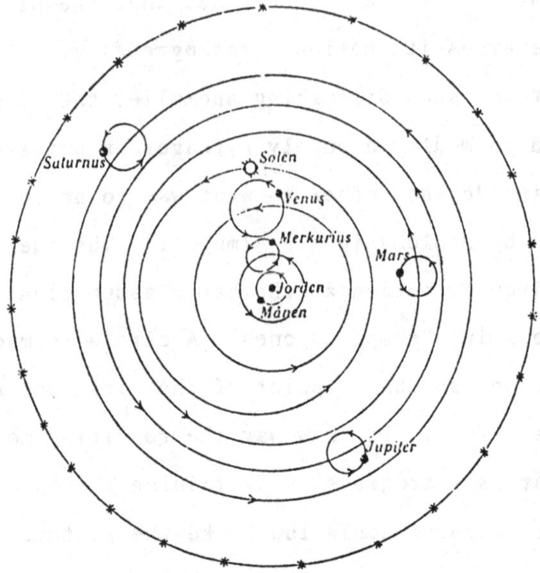

Figure 2.1

As these models developed from the simple one in Figure 2.1 to more complex ones, the basic idea is obviously to combine certain given motions - circular motions - with one-another. A uniform circular motion is determined by the plane in which it is carried out, its center in the plane, its radius and its angular velocity; it has seven parameters. On the other hand all circular motions are related to each other by simple transformations of space and time so that one could, perhaps, speak of *the* circular motion as the building block used to describe these celestial regularities.

When a new circle is added to the system its center is positioned on the periphery of another one. The appearance of the resulting system is what an earthbound observer can see as time goes on. His view of the planetary system is related to but conceptually distinct from the system itself as well as from the logical construction that serves him as a model. These distinctions may appear as scholastic hair-splitting but we shall see soon that they will return again and again, in changing forms. They will lead us to introduce certain formalized concepts that will play an important role in our mathematical study of regularity.

The circular motions could be said to constitute the *atoms* of the epicycle models of the inverse. According to the atomistic view, going back to Democritus and Epicurus, all matter is constructed of atoms which are themselves indivisible. Atoms are combined with others to form substances, mixtures, and compounds. Much later this was expressed quantitatively in terms of fixed proportions governing the composition of the weights forming a compound. Water should contain hydrogen and oxygen in the ratio 1:8, and similarly for other compounds - Proust's law.

Dalton described this in terms of atoms and their relative weights. Compounds were classified as simple - just one type of atom - binary, with two atoms, ternary, quaternary, and so on. Introducing pictorial symbols for the substances he used simple *diagrams* to describe the compounds as he illustrated in Figure 2.2 from Dalton (1808), p. 219.

From p. 143 ibid. we quote - "*the ultimate particles of all homogeneous bodies are perfectly alike in weight, figure, etc*. In other words, every particle of water is like every other particle of water; every particle of hydrogen is like every other particle of hydrogen, etc." The atoms of any given type differ in location but not in their intrinsic properties. If two objects, say both made of iron, are

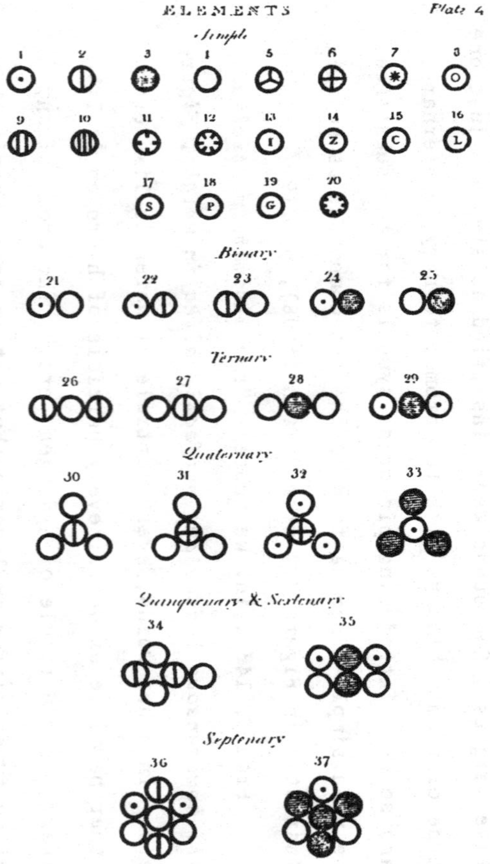

EXPLANATION OF THE PLATES.

PLATE IV. This plate contains the arbitrary marks or signs chosen to represent the several chemical elements or ultimate particles.

Fig.			Fig.		
1	Hydrog. its rel. weight	1	11	Strontites	46
2	Azote,	5	12	Barytes	68
3	Carbone or charcoal,	5	13	Iron	38
4	Oxygen,	7	14	Zinc	56
5	Phosphorus,	9	15	Copper	56
6	Sulphur,	13	16	Lead	95
7	Magnesia,	20	17	Silver	100
8	Lime,	23	18	Platina	100
9	Soda,	28	19	Gold	140
10	Potash,	42	20	Mercury	167

21. An atom of water or steam, composed of 1 of oxygen and 1 of hydrogen, retained in physical contact by a strong affinity, and supposed to be surrounded by a common atmosphere of heat; its relative weight = - - - - - - 8
22. An atom of ammonia, composed of 1 of azote and 1 of hydrogen - - - - - - - - - - 6
23. An atom of nitrous gas, composed of 1 of azote and 1 of oxygen - - - - - - - - - 12
24. An atom of olefiant gas, composed of 1 of carbone and 1 of hydrogen - - - - - - - - - 6
25 An atom of carbonic oxide composed of 1 of carbone and 1 of oxygen - - - - - - - - 12
26. An atom of nitrous oxide, 2 azote + 1 oxygen - 17
27. An atom of nitric acid, 1 azote + 2 oxygen - - 19
28. An atom of carbonic acid, 1 carbone + 2 oxygen 19
29. An atom of carburetted hydrogen, 1 carbone + 2 hydrogen - - - - - - - - - - - - 7
30. An atom of oxynitric acid, 1 azote + 3 oxygen 26
31. An atom of sulphuric acid, 1 sulphur + 3 oxygen 34
32. An atom of sulphuretted hydrogen, 1 sulphur + 3 hydrogen - - - - - - - - - - - - 16
33. An atom of alcohol, 3 carbone + 1 hydrogen - 16
34. An atom of nitrous acid, 1 nitric acid + 1 nitrous gas - - - - - - - - - - - - - - 31
35. An atom of acetous acid, 2 carbone + 2 water - 26
36. An atom of nitrate of ammonia, 1 nitric acid + 1 ammonia + 1 water - - - - - - - - 33
37. An atom of sugar, 1 alcohol + 1 carbonic acid - 35

Figure 2.2

different and appear different then, according to this view,
it is because the respective configurations differ, not because of any difference in the nature of their atoms.

The pictures in Figure 2.2 represent the combinations
of atoms with diagrammatic clarity, and led the way to the
modern system of chemical notation introduced by Berzelius.
Laws governing the composition of substances are given ex-
pressive form through this notation which made it possible
to represent the constitution of molecules concisely.

The regularities of molecular composition are brought
out further by the use of structural formulae indicating the
valence of each atom. The valence expresses the number of
bonds one atom forms in certain chemical combinations. So,
for example, can methane with the empirical formula CH_4 be
written by the structural formula, see (a),

$$
\text{(a)} \quad
\begin{array}{c}
\text{H} \\
| \\
\text{H}-\text{C}-\text{H} \\
| \\
\text{H}
\end{array}
\qquad\qquad
\text{(b)} \quad
\begin{array}{c}
\text{H} \\
| \\
\text{C} \\
\text{H-C} \diagup \quad \diagdown \text{C-H} \\
\text{H-C} \diagdown \quad \diagup \text{C-H} \\
\text{C} \\
| \\
\text{H}
\end{array}
\qquad\qquad (2.1)
$$

where the valence of the hydrogen atom is one and of the car-
bon atom four. In (b) we see the benzene structural formula
with the same valences but differently arranged atoms.

In the structural formulae we meet for the first time an
instance where the *topology of the bond structure* is brought
out explicitly. Going back, we see that the topology of the
combination of "atoms" occurred also in our earlier examples,
but not as clearly. In the epicycle models, for example, the
various circles are affixed to one another in a certain order

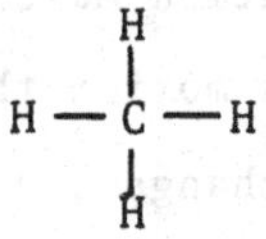

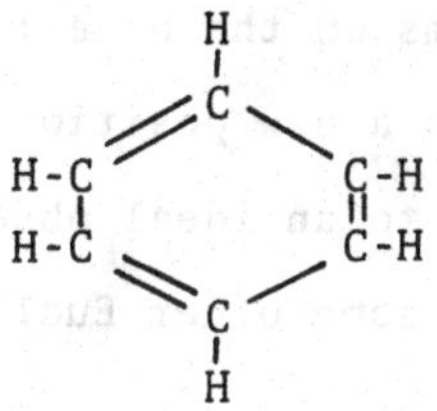

and this can be viewed as the topology of the arrangement; see Figure 2.1. From now on we shall consciously watch out for the topology of the combinations of primitive objects.

Let these objects still remain physical atoms. The most regular arrangement of them that we can imagine is of course a crystal. The simple rock salt crystal shown schematically in Figure 2.2 has sodium atoms (black) and chlorine atoms (white) arranged in a cubic lattice. If all sodium atoms are considered "identical", and similarly the chlorine atoms, then the structure will not change if it is translated a certain amount in each of three orthogonal directions. This assumes that the crystal is infinite in all directions.

We see here how the notion "same" or "identical" which we first applied only to the individual atoms is now applied to the set of all the sodium atoms on the one hand and all the chlorine atoms on the other: the translation moves the entire crystal to a new position. After this motion the crystal appears, to an ideal observer, as unchanged, and the same is true for some other Euclidean motions.

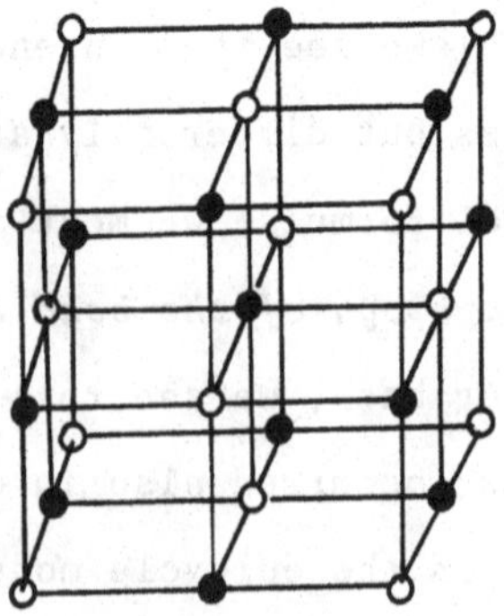

Figure 2.3

The crystal exhibits *symmetry*, a classical notion from geometry, that has played a dominant role in physics and chemistry at least since the 19th century. Formalized, symmetry means that a structure is invariant with respect to a group of transformations. The latter may operate in space or time and the idea of symmetry becomes precise only after it has been specified.

The successful classification of crystal structures in terms of the 230 space groups showed the power of the idea of symmetry. In the general study of regularity we can expect symmetry to play an important role. At the same time it should be mentioned that in the following, symmetry will be a consequence of order - it is a derived property rather than a primary one.

The ideal observer that can "see" exactly where all the atoms are situated is of course completely hypothetical. It was only when von Laue introduced X-ray methods into crystallography that detailed knowledge of complex crystal structures was made available. Using electromagnetic radiation with a wavelength short enough compared to the length scale of the atomic lattice, von Laue "illuminated" the structure and observed the effect. This effect, say as it appeared on a photographic plate, is related to the crystal type in a complicated manner. Therefore the *real observer* obtains a picture that requires a deep analysis, decoding, in order for one to get an answer about the arrangement of atoms.

This difficulty is typical of many patterns encountered in the sciences. The relation between the ideal structure, governed by rigid laws, and what can actually be observed, can be highly complex, making the mathematical analysis difficult. This relation must be taken into account in any theory of patterns.

Before leaving this case we should mention a suggestion of A. L. Mackay. He has argued the case for what he calls a generalized crystallography which would deal with regular structures more complex than crystals. The reader is referred to Mackay (1974, 1975).

The discrete crystal was explored by using the propagation of certain wave phenomena. Turning this around, let us look at the properties of waves in a continuous medium and how one describes wave-like patterns.

Say that we are studying waves in a homogeneous medium governed by the wave equation

$$Lu = \frac{1}{c^2}\frac{\partial^2 u}{\partial t^2}; \quad u = u(x;t). \tag{2.2}$$

Here L is a second order elliptic differential operator, say with constant coefficients, c is a material constant and u is a scalar quantity such as pressure. The coordinate x can be in R (a plucked string), in R^2 (a sounding membrane), or in R^3 (a vibrating column of air). The elementary approach to (2.2) goes like this. Try to "separate variables"

$$u(x,t) = a(x)b(t), \tag{2.3}$$

so that (2.2) gives

$$La(x)\cdot b(t) = \frac{1}{c^2}\, a(x)\, \frac{\partial^2 b(t)}{\partial t^2} \tag{2.4}$$

or, with some constant λ,

$$La = \lambda a, \quad \ddot{b} = c^2\lambda b. \tag{2.5}$$

Then b will be a combination of $\exp(\pm\sqrt{\lambda}ct)$, or in the most interesting case $\lambda < 0$ a combination of $\cos(c\sqrt{-\lambda}t)$ and $\sin(c\sqrt{-\lambda}t)$. This means that we have to solve for the eigenfunctions of L taking into account appropriate boundary conditions. Denote them by $\phi_\nu(x)$ corresponding to eigenvalues λ_ν, $\nu = 1,2,\ldots$

The next step is to try a series expansion

$$u(x,t) = \sum_\nu A_\nu \phi_\nu(x) \cos(tc\sqrt{-\lambda} + \alpha_\nu) \tag{2.6}$$

with amplitudes A_ν and phase angles α_ν, assuming optimistically that the series will converge in some meaningful sense. In order that this method should work L must of course have a pure point spectrum situated on the negative half axis.

Leaving aside the technical details, what concerns us here is the way the time dependence in (2.6) has been reduced to one written as a linear combination of trigonometric functions. The latter have frequencies of the form $2\pi/c\sqrt{-\lambda}$; these are the *harmonics of the oscillating medium*. For many systems of this type (2.6) describes all solutions although this statement needs some qualifications to be true. Then the continuous system has been analyzed in terms of a discrete set of harmonic motions. This is of course what gives the instruments in traditional music their characteristic sound in contrast to, for example, electronic music.

Starting from certain simple functions, here, for example, the trigonometric functions $\cos(tc\sqrt{-\lambda})$ and $\sin(tc\sqrt{-\lambda})$, one generates a large ensemble of temporal patterns from them. This generation involves two sorts of calculations.

The first one simply takes one of the functions, say $\cos \mu t$, and *changes the amplitude* by multiplying by a scalar to $A \cos \mu t$. This is not a drastic change since one could say that all the functions $A \cos \mu t$, fixed μ but arbitrary A, *sound the same*. It is only the intensity that differs, not the quality.

The other calculation consists in adding all the functions obtained in the first step. This means *superposition* of many functions into a single one, and expresses the way the listener's ear operates, or rather a conceivable way of operating. In the absence of disturbing noises, this latter calculation expresses an ideal observer in terms of additivity.

Historically these mathematical ideas go back to Fourier although he presented them in a different physical context, the analytical study of the propagation of heat. His statement that arbitrary functions can be expanded in trigonometric series must have seemed paradoxical to his contemporary colleagues. Now, when we are used to dealing with orthogonal series in general, and when the notion of convergence has been relaxed in the spirit of functional analysis, the idea of completeness (for function systems) is a familiar one.

Given some L_2-complete orthonormal system $\{\phi_\nu(x)\}$, say with $x \in [0,1]$, and an L_2 function f we can *describe* f by a sequence $(c_1, c_2, c_3, \ldots)$ of Fourier coefficients

$$c_\nu = \int_0^1 \phi_\nu(x) f(x)\,dx. \tag{2.7}$$

Knowledge of the sequence determines f uniquely a.e. Nothing would change in principle if we had doubly, triply, or arbitrarily indexed function systems. We would just have to make sure that the array of constants also carried information about the value of the index associated with each entry.

Say that we tried to represent a function f, for example from $L_2(0,1)$, by an array with entries c_α, where α labeled the elements in the function system. Then in order to get a uniquely determined representation we must of course demand that all the α's used in array are distinct: an exclusion relation between the α-values. Otherwise we would have linear dependence with resulting ambiguities.

In this way we represent functions by numerical arrays, the shape of which can be chosen for our convenience and where we use only distinct values of α. We have been mentioning L_2 but other function spaces could be used, including the simpler, finite dimensional ones with, for example, polynomials up to a certain order. In each case we have a regular structure well known to us from functional analysis.

Most of the regular structures we have considered so far have in common that they are already expressed in mathematical language and come either from mathematics itself or from the most mathematical of sciences: physics and astronomy. Before we turn to patterns from other sciences and from the humanities, let us summarize what we have found so far.

First, it is clear that these patterns have all been formed from certain primitive objects - they are *atomistic* in nature. The objects themselves vary from case to case in their mathematical properties, as does the notion of "*sameness*" which expresses that two objects are essentially the same although not necessarily identical.

Second, these objects are linked together in a specified way, sometimes expressed through a diagram describing the topology of the linkage. Rules determine whether bonds are valid or not in the set of links established. The structures can be said to be *combinatory*.

Third, combinations of objects formed according to the rules, are given an *interpretation* by introducing an ideal observer. The interpretation, or meaning, of the combination depends upon the observer's view of it, i.e., how he experiences it.

Finally, *real patterns* can be conceptually different from the ideal ones just mentioned, and the relation between the two should be defined precisely.

With these four observations in mind let us continue our examination of patterns encountered in the natural sciences. How does one describe regularities in biology, for example in botany? What are the logical principles behind the classification schemes in systematic botany? These questions go far back in time (see Notes D) and we shall take a brief look at the ideas of Linnaeus and Adanson.

Divisio et Denominatio - division and naming - this constitutes the main task of the biological taxonomist, according to Carolus Linnaeus, one of the founding fathers of systematic botany. Like his contemporaries, the young Linnaeus considered plants and other forms of life as organized into species, families and so on, as laid down by the Creator in his plan of the universe. The species were looked upon as immutable, given once and for all. It is true, though, that Linnaeus modified this static view in his later life, influenced by his observations on hybrids, and allowed for the possibility of change. Anyway, the botanist should try to classify his plants into groups of related individuals. The crucial word in this statement is "related"; it is not obvious what this should mean or how it should be interpreted to enable us to glance into the divine plan of creation.

Admitting that this sublime goal may not be within reach of a mortal, Linnaeus suggested a less ambitious system of classification, his famous sexual system. In this he divided plants into 24 classes:

1.	Monandria:	1 stamen
2.	Diandria:	2 stamens
3.	Triandria:	3 stamens
⋮		
11.	Dodecandria:	12 stamens
12.	Isosandria:	more than 12 stamens, attached to the calyx
13.	Polyandria:	more than 12 stamens, attached to the receptacle
⋮		
16.	Monadelphia:	stamens in 1 bundle
17.	Diadelphia:	stamens in 2 bundles
18.	Polyadelphia:	stamens in several bundles
⋮		
24.	Cryptogamia:	concealed flowers

These 24 classes were subdivided into orders depending mainly upon the number of styles in the pistil, with the corresponding names Monogynia, Digynia, and so on.

Linnaeus stressed that his was an *artificial system*, not based on intrinsic relationships between plants but rather on a few directly available and observable features, especially properties of the organs of reproduction. Whether or not this system of classification really went to the heart of the matter, it certainly proved its usefulness as have few other systems in natural history.

The basic idea in a classification scheme of this type is to split the classes successively according to the values of certain features, for example the number of stamens. The resulting classes, orders, etc. are *related to each by inclusion* relations and, when no overlap can occur in the splitting, this results in the *topology of a tree or of a set of disjoint trees, a forest.*

It is also relevant to this general discussion to see how the botanist defines or establishes a species. He could take a particular plant and designate it as the *type* of the species. Higher groupings in the hierarchy, such as genus and family, are typified through the specimen in terms of lower order groupings already defined.

The specimen or some reproduction of it, say a drawing or photograph, is used to *represent* a group. But this is not meant literally; it is not the exact number of leaves, the length of its stem, or the precise shape of its root system that is relevant. Instead, the botanist uses his experience to pick out certain features that he considers significant. It could be the topological arrangement of the flowers or the qualitative shape of the leaves. This just allows for the fact that biological variation affects certain features in a way that is judged irrelevant for classification while other features, often of qualitative nature, seem to be more *permanent* or invariant.

The efficiency of such a classification scheme depends upon the choice of features used for the splittings. These features can be viewed as the primitives that generate the partition of an idealized botanical universe. A combination of features, arranged into subtrees according to rules that have to be specified, can then be interpreted as a genus, a species, etc.

According to this view the subsets one arrives at in the hierarchy are calculated using the primitives as computational modules. Plants that do not differ for any feature used in the system are classified as the same, or perhaps with a better word, as similar, although they may differ in other respects. The relation between the ideal notions in the system of classification and actual individuals is likely to be imprecise and has to be described, perhaps, only in statistical terms.

This system and its later variations is phenetic in the sense that it is based only on the properties of the phenotypes. It has been argued that a more natural system should start from phylogenetic relations, to the extent that such knowledge is available. Knowledge about the way in which plants have evolved could then be exploited for classification purposes. Plants would be treated as "related" if their evolutionary developments have much in common.

This is not the right place to argue for or against the plausibility of any particular taxonomic system. We are concerned instead with the methodology of classification, and we should mention an alternative view put forward by Adanson in the latter half of the 18th century. Instead of using isolated features, more or less arbitrarily selected, the systematic botanist should build his taxonomy using *many features* and base the decision on whether two entities are close to each other or not on the correlation between their many characters. The subsets of the partition - the taxa - would then be richer in information content but it can happen that the taxa partially overlap.

Numerical taxonomy is a modern descendant of Adansonian thinking and employs statistical techniques to establish the taxa, see e.g. Sokal-Sneath (1963). Expressed more abstractly one could say that one starts with a high-dimensional sample space and wishes to *decompose an empirically available probability measure on the space into a mixture of sub-measures.* The supports of the sub-measures may overlap, hopefully not very much.

One technique that has received a good deal of attention in numerical taxonomy is principal components, by which one tries to single out subspaces, preferably low dimensional ones, such that most of the probability mass is close to one of them. This is done by eigen-analysis of covariance matrices. Hence we are brought back, somewhat surprisingly, to *eigenvectors viewed as the primitive objects,* just as eigenfunctions appeared in the study of vibrating media.

In taxonomy we can therefore expect to meet certain measures as primitives, connected together by topological relations. These relations, expressible by graphs, in the simplest case trees, describe the hierarchy of the regular structure. On the other hand, we may prefer to approximate the measures by ones having linear manifolds as support and we then deal with eigenvectors and the eigenspaces spanned by them.

The general principles developed by botanical taxonomists may well carry over to other fields, but for complex and diffuse phenomena they will be difficult to apply. As an example we mention classification of diseases in medicine.

There it is obvious that classification is used not just to organize a large body of knowledge; grouping phenomena under a certain label also serves the purpose of associating certain actions or treatments with this label. Admittedly, it is difficult to express the diagnostic process through an explicit and well-defined decision tree or recognition algorithm. There even seems to be a widespread emotionally laden reluctance to admit the feasibility of even partially formalizing any diagnostic process in terms of an algorithm. In spite of this pessimistic view, many attempts are being made in this direction for particular, usually small, groups of diseases. A good example is given in Peterson *et al.* (1966) where a decision tree is given for the Stein-Leventhal syndrome (see Fig. 2.4). A glance at it will show a strong similarity to the floras of systematic botany. Just as before we can let the primitives be the actions indicated in the lower part of the figure.

Note that these patterns are not intended as formalizations of diseases, which might have appeared as the natural thing to do. The very notion of a disease is so nebulous, however, that to deal with it in terms of recognition algorithms may just give the discussion a superficial and spurious exactness. An exception to this would be when the diseases concerned can be defined by associating them with one or several clearly delimited causes. Otherwise we will get involved in an ontological discussion of the very existence of diseases. If this is so it is preferable to talk about syndromes, symptom complexes, or about decisions (action complexes), as in the above example. To evaluate the

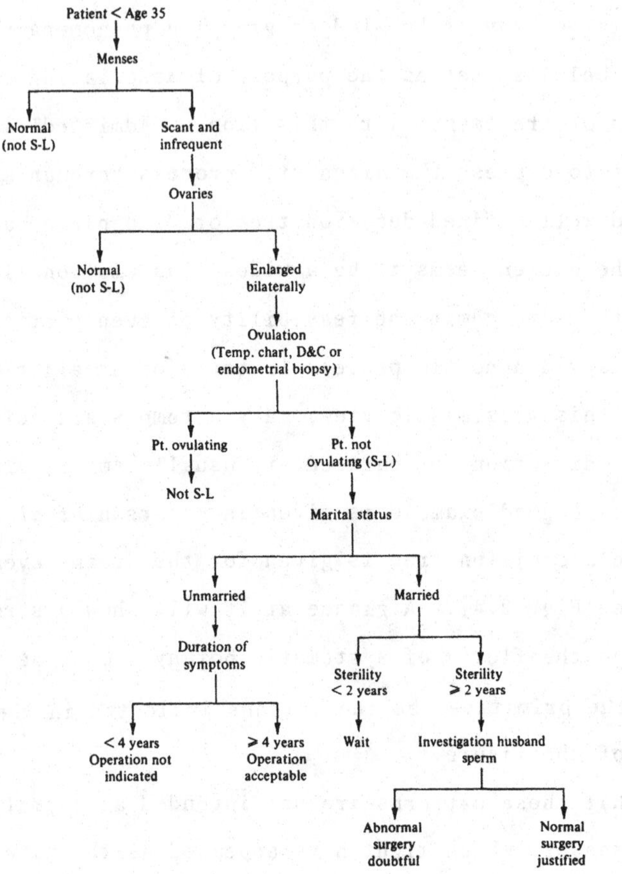

Figure 2.4

decision algorithm it would have to be compared with the diagnostic performance of the physician and related to the results of follow-up studies. A preliminary but illuminating discussion of this can be found in Peterson *et. al.* (1966).

It should be pointed out that although Figure 2.4 contains an implicit definition of a syndrome, it is more natural to look at it as a decision function. Its main role is to recognize a syndrome and take appropriate action. In this respect it typifies a good deal of the work done in *pattern recognition* where the emphasis is usually on the recognition of patterns and the construction of algorithms for doing this. The existence and definition of what the patterns are, receives little attention, most of the time, with certain exceptions to be noted later on in this section.

The most studied biological patterns are of course those associated with shape and the development of shape, both for the individual and during evolution. The overwhelming majority of these studies is descriptive, employing no mathematical tools, or, possibly mathematics as represented by curve fitting. Anatomy is by its very nature analytical in the sense that it proceeds by segmenting the organism into subsets which are themselves broken down into smaller subsets and so on. The body is made up of trunk, head, limbs, etc. and the limbs are segmented into more or less rigid parts, and so on.

The description of the segments on various levels of the hierarchy is made verbally, possibly with the addition of statistically determined numerical statements. It is tempting to try to formalize the segmentation analysis mathematically, and such attempts have indeed been made (see picture grammars below). At this point let us only mention the celebrated work by d'Arcy Thompson: *On Growth and Form*. One of the central themes of his work is to look for *general principles that force regularity upon the biological organisms*.

The most obvious of such principles is that of the symmetries that apply to living organisms with respect to certain axes, for example the dorsoventral one, or with respect to certain planes, for example the medial plane. More interesting are principles that try to account more fully for shape formation: minimum potential energy, minimum surface area and other variations on the isoperimetric problem, and maximum strength for given mass. It appears that many biologists view ambitious attempts like d'Arcy Thompson's with considerable scepticism, and perhaps rightly so. Nevertheless, whatever is the applicability to the real world, the idea that such principles induce patterns is a fascinating one which deserves study by mathematicians (see Notes B).

Patterns in biology can, however, also be approached from another angle, almost orthogonal to the above. Instead of asking what the regularities that characterize life are, and how one goes about their mathematical description and analysis, one could formulate the problem as follows: for a given animal, or group of animals, what patterns can it perceive and how is perception organized?

Most of this type (and there is an abundance of it) deals with visual perception. Here, *the starting point is the neural network* carrying out the processing, and one tries to correlate visual stimuli with properties of the individual neurons and the architecture of the network. Only two highly influential papers will be mentioned here: Hubel-Wiesel (1965) and Lettvin *et. al.* (1959), containing some startling observations. A parallel development in psychology has focused on the learning of patterns under laboratory conditions, to isolate effects, phenomena, and anomalies.

Assuming that the world surrounding an individual animal is ordered by some regular structure, one could examine idealized neural networks mathematically and ask what properties do we have to postulate in order that the network recognize and learn certain patterns. Such formal *pattern processors* studies are likely, at least at present, to tell us more about the mathematics of regularity than about the real world. One such pattern theoretic study has been described in Volume 2, Chapter 6.

On a less fundamental level, one can observe patterns in human behavior under stereotyped conditions, as in motion studies. To carry out a given manual task the human operator will use his hands, and possibly feet, in a sequence of motions. In the spirit of this section it would be natural to look for primitives, elementary motions, and ask how they are combined into *motion patterns*. This is exactly the object of enquiry in motion studies, see e.g. Larkin (1969). The rules that allow us to combine certain elementary motions, but not others, would express the continuity of motions, put limitations on the power needed for transitions between them, and so on.

In this connection one should mention also the patterns that can result from motion patterns. Say that an artisan is making some object. He has at his disposal certain tools, some material to start from, and, most importantly, a set of hand motions. The latter represent his skill. Given these, what are the possible shapes he can create? How difficult are they, compared to earlier ones, how time-consuming?

Such questions also deal with pattern processors: the problem is how the motion patterns induce spatial patterns and how technical constraints and changes influence the forms produced. Admittedly, the atomistic view of hand motions may be inappropriate, too restrictive, for a full understanding of such manufacturing processes. Whether this is true or not can only be decided by a detailed study of particular cases (see Notes C).

With the exception of the last example we have only discussed patterns appearing in nature. What about the man-made world: are the patterns in it of essentially different character from those studied in the natural sciences?

The natural languages exhibit regularities that are obvious but at the same time elusive. To a reader aware of modern developments in linguistics it is well-known that the language patterns governed by morphology and syntax can be viewed in generative terms. Indeed, modern grammarians have emphasized that grammar is rule-driven: behind the surface structure of language, which is confusing in its complex appearance, there is a set of rules and transformations. The rules and transformations may be many but their number is finite. We shall take a brief look at an example illustrating this, but let us first cite a historical note.

In what may be the earliest detailed grammar of any Indo-European language, the Sanskrit grammarian Panini described word formation in terms of verbal roots. The exact data are not known, but the grammar certainly existed some hundred years B.C. Panini's work has been called "one of the greatest monuments of human intelligence". The reason why it should be mentioned here is the generative attitude in which words are obtained from simple roots by *applying certain rules in a certain order*. The regularities in word formation are hence explained as due to primitives consisting of rules and roots. The rules are ingeniously chosen leading to extreme economy in the grammar. Or, quoting Robins (1967), p. 144: "This quest for economy was evidently part of the context of early Indian grammatical composition; a commentator remarked that the saving of half the length of a short vowel in framing a grammatical rule meant as much to a grammarian as the birth of a son."

The requirement that the rules used to analyze patterns should be simple and form a concise system is a recurrent, indeed indispensible, idea in the study of regularity. The rule system should not contain any redundant statements, *the individual rules should be logically independent*. On the other hand, *the system should be complete* so that it is powerful enough to account for the patterns that occur.

Recent work on generative grammars includes precise formalizations of phrase structure grammars. This is done on various levels of generative power: finite state language, context free languages, etc. The basic rule is the same: from a finite set $\mathcal{G}$ of rules and auxiliary concepts an infinite set L of sentences is derived. These are the grammatical sentences, $L = L(\mathcal{G})$ corresponding to the grammar $\mathcal{G}$.

For example the sentence "the dog chased the little boy" can be thought of as generated as in Figure 2.5. In it the

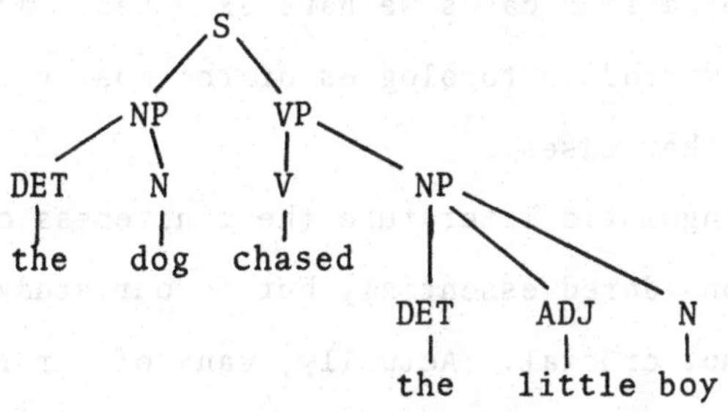

Figure 2.5

starting symbol S is successively expanded into strings of syntactic entities NP (noun phrase), VP (verb phrase), DET (determines), N (noun), V (verb), ADJ (adjective). First a rule S → NP,VP is applied and each of the NP,VP entities is rewritten separately. The first using a rule NP → DET,N, the second using the rule VP → V,NP. Then NP is rewritten, but this time with an optional adjective into DET,ADJ,N. Finally lexical items are related: "the" from DET, "dog" and "boy" from N, etc.

Leaving aside the technicalities, and they are many and intricate, it is clear that we here operate with *rewriting rules* as primitives, either such as NP → DET,N which rewrite one syntactic entity into others, or such as ADJ → "little", which results in lexical items. At the moment we need only concern ourselves with the operative principle behind schemes like Figure 2.5. *Combining primitives so that they fit with each other*, we obtain a diagram, the interpretation of which is the base line of lexical items viewed as a grammatical sentence. The diagram is really a *formula* that computes the sentence, very much like the formulas in (1.2) or like the arithmetical expressions generating sequences in the same section. The fact that the topology of the arrangement in Figure 2.5 is that of a tree does not distinguish it in principle from the earlier cases we have examined. Actually we shall have to introduce topologies of the most varying kinds to deal with other cases.

In the linguistic literature the finiteness of formal grammars is considered essential, but in our study of regularity it is not crucial. Actually, many of our regular structures belong to analysis and involve continua: *they are not necessarily finitistic.*

The problem of describing ideal language differs of course from that of analyzing usage - the distinction between linguistic competence and linguistic performance. To deal with regularities in usage it is natural to introduce probabilistic - or metric - aspects, to make it possible to talk in precise terms about what is common and what is unusual. It is not quite obvious, however, how to introduce the "noisiness"; should it be simply in terms of frequencies of lexical items, or is it possible also to describe syntactic variability in probabilistic terms?

Say now that we are dealing with language patterns that can be understood in terms of a given formal grammar and that the language is over the set X of words. The definition of distributional class expresses whether two words $x,y \in X$ allow the same context. More precisely, for any strings u and v of words $u,v \in X^*$ the concatenated sentences uxv and uyv should be grammatical or non-grammatical together, in order that we can say that x and y belong to the same distributional class. The corresponding equivalence classes express a certain *invariance*, in that for any word in a grammatical sentence we may substitute any other y from the same class without destroying the grammaticality. This is just a tautology using the direct definition of distributional class. Often, however, the distributional classes are determined, or partially determined, by non-syntactic considerations, especially semantics. The syntactic "sameness" of words may be given to some extent a priori, which makes it of greater interest. Considerations like this have been explored by Kulagina and others, and the interested reader should consult Marcus (1967), in particular Chapters I, III which are aimed at highly inflected languages.

An interesting spin-off from formal grammars has been to *pattern recognition.* With some over-simplification, pattern recognition can be said to consist of two distinct areas of research. In the first, the object of study is the algorithms, software, and hardware needed to implement the recognition process. A considerable portion of this is of statistical nature and has borrowed well-known statistical techniques and concepts, and applied them to recognition problems. In our study such problems will arise but they will not be in the center of our investigation.

The second area is of greater interest to us in this context, since it aims at the patterns themselves, their nature and definition. It is not limited to real language patterns - on the contrary most of the applications are not linguistic. We shall discuss it in this context, nevertheless, since it was inspired by linguistic thinking. The basic idea is to use the formal grammars introduced by theoretical linguists, or variations of them, to *generate patterns,* and the approach is often called *the syntactic or linguistic method in pattern recognition*, see Fu (1974) for a full presentation, with a substantial bibliography.

One of the pioneers in the syntactic approach was Narasimhan, who dealt with *tracks observed in bubble and spark chambers* in an early paper, 1964, see Bibliography.

In rough outline, the procedure goes as follows. Pictures are taken from the bubble chamber (actually this is a three-dimensional pattern) and preprocessed, which is not a simple step in itself. The points that form the track are identified and joined by continuity arguments. Singular points (end points and branch points) are isolated. Numerical attributes, such as curvature, are calculated and the analysis proceeds stepwise toward a structural description of the picture. Or, to quote Narasimhan, "The aim of any adequate recognition procedure should not be merely to arrive at a "yes," "no," "don't know" decision but to produce a structured description of the input picture. I have argued elsewhere at some length that no processing model could hope to accomplish this in a satisfactory way unless it has built into it, in some sense, a generative grammar for the class of patterns it is set up to analyze."

After the first steps in the analysis mentioned above, a decision is made on the separate parts that constitute the picture. An arc may be classified as a segment of a straight line or of a spiral. Or the algorithm may look for subpictures with given qualitative and/or quantitative properties. In this way one hopes to be led to a physically meaningful analysis.

Actually, a language specially designed for processing pictorial patterns has been developed by Miller and Shaw (1968). It is, at least in spirit, close to what Narasimhan suggested, but the detailed construction of the grammar differs. The primitive elements or subpictures each have two specially indicated points, the head (H) and the tail (T) and primitives are only concatenated via these points. Classes of primitive objects are given a priori, designated by name and possibly through some attributes, numerical or not. Subpictures may be labeled through superscripts, as S^1. There exists a null element λ, consisting simply of a tail and a head located at the same place. Denote two primitives by P_1 and P_2 and introduce the binary operations

$$\begin{cases} P_1 + P_2: & \text{head of } P_1 \text{ to tail of } P_2 \\ P_1 - P_2: & \text{head of } P_1 \text{ to head of } P_2 \\ P_1 \times P_2: & \text{tail of } P_1 \text{ to tail of } P_2 \\ P_1 * P_2: & \text{head to head, tail to tail.} \end{cases} \tag{2.8}$$

On a single primitive we may operate as follows:

$$\begin{cases} \sim P: & \text{switches head and tail} \\ P: & \text{blanks out all points} \\ TP: & \text{transforms } P \text{ through an affine transformation } T. \end{cases} \tag{2.9}$$

The syntax is given through the following rules generating sentences S.

$$\begin{cases} 1. & p \text{ is a primitive class.} \\ 2. & \theta \to +|\times|-|*|\sim \\ 3. & S \to p|(S\ S)|(\sim S)|TS|S^1. \end{cases} \tag{2.10}$$

As an example of how this works, Miller and Shaw analyze the picture in Figure 2.6 as follows. It is supposed to illustrate the pictures obtained starting with a negative particle. This particle may scatter from a positive particle, or decay into a neutral and one negative particle, or pass through the picture. The primitives are t_+ (positive track), t_- (negative track), t_n (neutral track). If T_r stands for an r-track with all subsequent events and where $r = +, -,$ or n, the rules could be formulated as

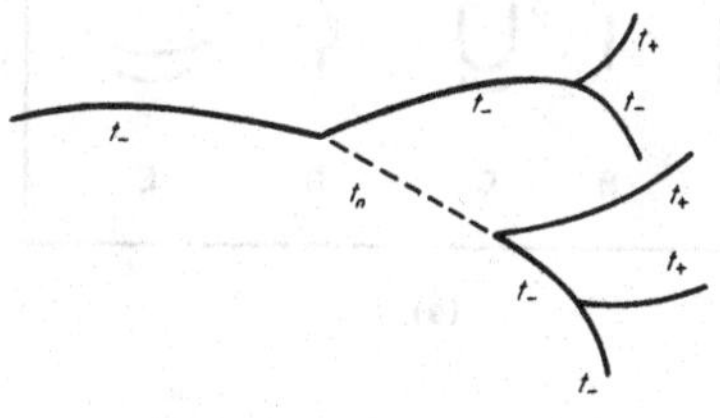

$$S \rightarrow (t_- + ((t_- + (t_+ \times t_-)) \times (t_n + (t_+ \times (t_- + (t_+ \times t_-))))))$$

Figure 2.6

$$\begin{cases} S \rightarrow T_- \\ T_+ \rightarrow t_+ \\ T_- \rightarrow t_- \mid (t_-+(T_-\times T_+)) \mid (t_-+(T_-\times T_n)) \\ T_n \rightarrow t_n \mid (t_n+T_+\times T_-) \end{cases} \tag{2.11}$$

giving rise to the grammatical analysis

$$S \rightarrow (t_-+(t_-+(t_+\ t_-))\times(t_n+(t_+\ (t_-+(t_+\ t_-)))))) \tag{2.12}$$

of Fig. 2.6.

In this connection we should mention another early idea in this spirit suggested by Ledley (1964) for the analysis of photographs of *chromosomes*. The picture is covered by a 700 × 500 rectangular lattice. In each of these 350,000 cells the blackness is expressed in an 8-leveled scale. To determine the contour of an object in the picture we decide on a threshold value for the blackness and join the point associated with this value by continuous arcs as far as this is possible. As usual, some form of preprocessing is necessary to compensate for the presence of optical noise. What is more characteristic of this procedure, however, is the

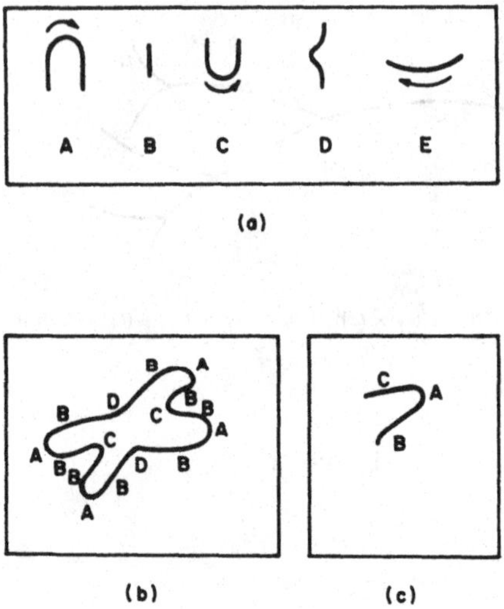

Figure 2.7

following syntactic step. The contour is separated into arcs and one tries to classify these into the five types in Fig. 2.7 denoted by A,B,C,D,E. Exact fit cannot be expected; it is a matter of approximate representation of the segments. Now a hierarchy of concepts is used such as "arm," "side," "pair of arms." Just to give an example, a syntactic rule may look like

$$\text{arm} = \text{B} - \text{arm or arm} - \text{B or A}. \tag{2.13}$$

This recursive rule tells us that an arm can have the form A,BA,AB,BAB, and so on. This will become clear when looking at the strongly idealized Fig. 2.7b, where the segmentation is indicated, and Fig. 2.7c where the arm BAB has been displayed isolated (these pictures are from Ledley's paper). Other syntactic rules function in a similar way.

The reader can find an up-to-date discussion of pattern recognition of biological organisms with many further references in Prewitt (1972). Before leaving the topic of language aspects of biological shape and change we must mention the *developmental systems* introduced by Lindenmayer. The basic building block is the cell, or more precisely, the cell together with its current state. The state may change in time as well as the number of cells. Assuming a finite number of possible states the development is governed by a finite number of productions, which may remind the reader of Markov algorithms.

Using an example from Lindenmayer (1975), say that the cell states form the "alphabet" {a,b,c,d,e,f,g,h,i,k} and introduce the productions

$$\begin{cases} a \to bc,\ b \to kd,\ c \to ek \\ d \to gb,\ e \to cf,\ f \to ik \\ g \to hi,\ h \to de,\ i \to k,\ k \to k \end{cases} \tag{2.14}$$

Starting from the initial state $\underline{a}$ one gets the development

$$\begin{array}{c} a \\ bc \\ kdek \\ kgbcfk \\ khikdekihk \\ kdekkgbcfkkdek \\ kgbcfkkhikdekihkkgbcfk \\ \text{etc.} \end{array} \tag{2.15}$$

by applying production rules successively to the individual letters of the alphabet. Note that here no context sensitive rules were used but such do appear in more complex systems of this nature.

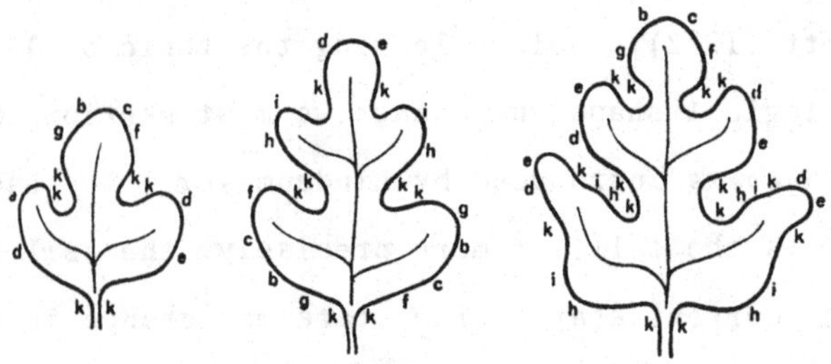

Figure 2.8

What makes this interesting is that the patterns obtained can in many cases be given attractive biological interpretations. For example (2.15) can be interpreted as shape formation for certain leaves if k means "no growth," see Figure 2.8, from Lindenmayer (1975). The developmental systems have attracted wide attention among theoretical biologists. Although they are certainly not identical with the syntactic approach in pattern recognition they share both the generative spirit and the finitistic attitude. Both produce linear strings but have been generalized to two- or higher-dimensional cases. See Culik-Lindenmayer (1976) for Lindenmayer systems, and Fu (1974), especially Appendices G and H for syntactic pattern recognition.

It is characteristic for the substantial literature on syntactic pattern recognition that most authors concentrate on the recognition algorithms and less on the mathematical properties of the patterns themselves. Related to this circumstance is the fact that the mathematical relation between the ideal patterns and real ones actually observed is seldom exposed to any penetrating analysis. It is then necessary to pre-process the real pattern, often by ad hoc filtering and smoothing techniques, before the recognition algorithm is applied.

Let us return to language in the strict sense of the word, but now with the emphasis on how it is presented: as speech, or in the form of handwriting, as literary works, etc. The alpha-numeric patterns, for example, have received much attention in the pattern recognition literature, resulting in very numerous but somewhat superficial studies. An exception is the pioneering work by Eden (1961) on *stylized handwriting*. Eden starts from the assumption that it is possible to *generate the handwritten characters* from a small number n of strokes. A few of them are shown in Fig. 2.9a. How large a value of n one should choose depends on what degree of approximation we aim at, considering the increase in complexity caused by a large value. Do we have to include the dot over the "i", the jot, and the bar across the "t", the tittle, for example? But these are details that we need not go into here. Instead we shall look at some more essential aspects of this way of generating the characters. The following three observations will be relevant.

First, all characters have been generated from a few basic patterns, the strokes. In this way we have restricted the set of possible patterns, or their representations, drastically.

Secondly, the strokes will not be combined arbitrarily but according to certain rules. We may ask that the strokes be connected continuously with each other, or that certain combinations be illegal. This regularity will also tend to

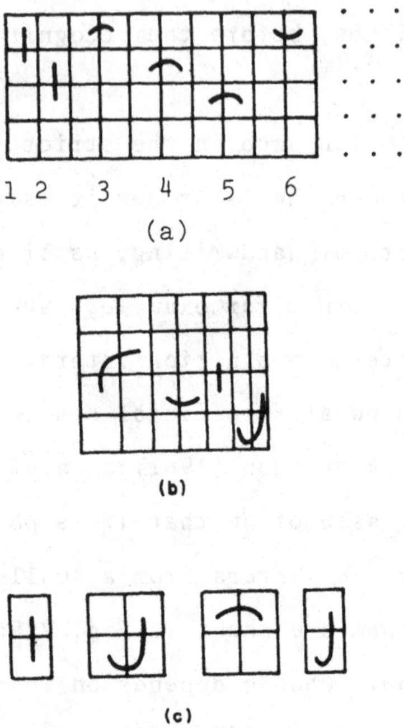

Figure 2.9

restrict the form of the resulting patterns. The letter "g" could be analyzed into the four segments of Figure 2.9b.

Thirdly, we can reduce the number of basic patterns even further. Look at the strokes 1 and 2. The second one could be considered the result of moving the first one downwards. In the same way, 6 can be considered as the mirror image of 3 with respect to a horizontal mirror. More generally, we start from the four segments of Fig. 2.9c and operate on them by a collection of transformations including certain translations, reflections, and rotations. All the n strokes are now obtained as the result of some operation applied to one of the four segments. In this way we have further reduced the number of building blocks.

Let us recapitulate the construction of the model. Starting with a collection of primitive units and a collection of transformations, we generate a set of simple figures (the strokes). Following given rules, these are then combined to give us the final result, the characters.

What makes Eden's paper important is not so much its utilitarian value for obtaining pattern recognition algorithms of stylized handwriting - this is somewhat doubtful - but the logic of his approach and the clarity with which it has brought out some fundamental aspects of pattern formation.

When we go to spoken, rather than written, language the speech patterns exhibit enormous complexity and richness. To introduce some order, phonologists try to decompose the speech patterns into units, the *phonemes*, which are supposed to carry most of the information in the pattern. A central idea has been to regard the phoneme as composed of certain distinctive features, and with these (binary) features indicating opposition to their absence. Such feature pairs could be voiced-voiceless, short-long, etc. The phoneme incorporates such properties and is an abstraction, not the actual sound.

But the phonemes do not exist in isolation, instead they are governed by relations and form a structure whose laws should be found by the phonologist. Troubetzkoy, one of the creators of modern phonology, expresses it thus:

"Un système phonologique n'est pas la somme mécanique de phonèmes isolés, mais un tout organique dont les phonèmes sont les membres et dont la structure est soumise à des lois."

In a seminal paper, Lieberman et al. (1967) showed that a human does not transmit and receive the separate entities that make up speech isolated from each other. Instead there seems to operate an intricate coding scheme resulting in a combined sound pattern with correlations between its parts.

The physical expression of speech as pressure variations in air can be viewed as a time series and be subjected to spectral analysis. Much work has been done in this direction, and it is of particular interest to us since it involves a type of regular structure that we have not yet discussed.

Over time intervals of several seconds it is clear that the time series could not be stationary. There is some evidence, however, that it is quasi-stationary over shorter time intervals, at least as a first approximation. If this is so, then the power spectrum over these time intervals is the natural mathematical tool to employ.

Calculating such spectra empirically it was observed long ago that they exhibit a very striking form, often with a high concentration of power at certain frequencies - the formants. Then a few formants may suffice to describe some of the sounds in steady state, approximately. The transition between quasi-stationary segments is more complicated.

Leaving aside the subject matter we could say that, mathematically, the above amounts to combining different stationary stochastic processes. They are restricted to disjoint and contiguous intervals and with boundary conditions at the junctures that have not been specified. This differs from the regular structures we have mentioned before, in that *the primitive objects here are stochastic processes, one level of abstraction higher than in the earlier cases.* Except for this difference, and it is a major one, the structure is still obtained in a combinatory manner from primitive objects.

The reader will have noticed that the regular structures we have just been considering have a great deal in common with those we encountered in the sciences. One difference is that in the sciences the formalization has often been carried further and the mathematical analyses may be deeper. This difference has little or nothing to do with the distinction between quantitative and qualitative methods: both can be expressed and studied mathematically.

This remark applies even more to the following cases where little mathematical formalization has been achieved. The first one concerns the remarkable study of the folk tale by Propp in 1927, an epoch making work that did not become widely known among Western scholars until the late 1950's. In it Propp attempted to construct a *morphology of folk tales* which could be used to analyze a tale into constituent parts of general nature. The primitives used for this purpose were especially the actions of *dramatis personae* and their functional role in the whole tale. Perhaps one could say, in a formalistic vein, that these functions were abstracted from their semantic content as much as possible. Anyway, the first goal would be to represent the tale as a sequence of primitives, *a formula,* which expresses the syntagmatic structure of the tale.

As typical primitives used by Propp we mention a few using Propp's notation (for a full list see Propp (1971), pp. 149-155)

ξ^1 "the villain receives information about the hero"

ξ^2 "the hero receives information about the villian"

.....

A^1 "kidnapping of a person"

A^2 "seizure of a magical agent or helper"

.....

A^{14} "murder"

.....

B^1 "call for help"

B^2 "dispatch"

.....

D^1 "test of hero"

D^2 "greeting, interrogation"

.....

E^1 "sustained ordeal"

E^2 "friendly response"

This is only a small subset but should give some ideas of the sort of primitives used. The result of a particular analysis is then a formula and the formulas corresponding to different folk tales can be compared for similarities and differences. While Propp's work is limited to the synchronic aspect his methodology points forward to diachronic studies.

The seemingly endless variation of folk tales is, if one follows Propp, reduced to a rigid pattern, or to quote him (p. 105): ..."it affirms our general thesis regarding the total uniformity in the construction of fairy tales."

Propp was a forerunner of the contemporary school of *narratologists* searching for regularities in narration. This movement with its center in Paris has also been influenced by the structuralists, especially by Lévi-Strauss who has himself critically examined Propp's work from the structuralists standpoint.

In Lévi-Strauss' study of the myths a basic principle is that what seems like chaos in the world of myths can be understood on a general level. They are made up of *constituent units*, but the meaning of the myth does not reside in the isolated elements. Instead it has to be sought *in the way these elements are combined*, see Lévi-Strauss (1955), sections 2.6ff. Transformations act upon these units, introducing paradigmatic schemes, but the transformations cannot be applied to the units in isolation but effect the entire system. The constituent units are relations or bundles of relations, and the analysis of the myth aims at these units explicitly and the way they are arranged into a structure.

Ideas like these certainly resemble those we have met earlier in this section but since they have not been formalized to the same extent we had better exercise some caution. The resemblance is striking but may not go very deep.

A more profound analogy can be found with the views expressed by Wittgenstein in his "Tractatus logico-philosophicus". Many of his aphorisms seem to indicate atomistic/combinatory/interpretative ideas quite similar to those we have discussed here. Since we shall return to this subject in Chapter 9 we shall not go into it in detail here.

We are no longer on very firm ground, but let us go ahead nevertheless, and consider the way intellectual doctrines can be analyzed into basic themata. Here "doctrine" is used in a wide sense and could be, for example, Hamiltonian mechanics, Thomistic theology, psychoanalysis, or statistical theory.

Holton (1973) has suggested that one should describe a theory by a three-dimensional metaphor. A theory usually has one component related to empirical matters and the "real world"; let us call it the x-component. It has another component of analytical nature based on logical calculi and employing (non-trivial) tautologies; call it the y-component. Of course these components need not be precisely defined.

Or, to quote Hume's celebrated aphorism: "... let us ask, does it contain any abstract reasoning concerning quantity or number? No. Does it contain any experimental reasoning concerning matter or fact or criteria? No. Commit it to the flames. For it can contain nothing but sophistry and illusion."

Holton argues that in addition to the x- and y-coordinates, a third one, a z-axis is required. This axis expresses *themata* or fundamental notions, terms, and ideas. He illustrates this by the physical concept of force. The x-component could be the measurement of effects caused by a force, the y-component could be vector calculus as applied to forces. The z-component is a principle of potency, *energy* in Aristotle, *vis* in Newton's Principia, and *Kraft* in the 19th century. This theme is that of the active principle.

Other examples of themata are "conservation principles", "atomistic discreteness", "wave-character", "evolution", "sexuality in children". The interested reader should consult Holton (ibid) especially pp. 47-68. It is tempting to ask whether one could formalize the way doctrines are constructed from themata in terms of regular structures similar to those that we have been considering.

In Henri Poincaré's famous lecture on his discovery of the Fuchsian functions he discussed mathematical discovery viewed as fruitful combination of ideas. He uses the suggestive metaphor *atoms of thought hooked together.*

Our problem is not just in identifying the atoms of thought - the themata - but also to find the rules that allow them to be hooked together.

Let us make the discussion more concrete by specializing to a particular doctrine, namely *statistical theory.* How would a thematic analysis of statistics appear? Can we single out any central themata?

Taking almost any elementary textbook in statistics one will find a large number of probabilistic models - statistical hypotheses - but the number of themata is much smaller. Certain themata occur frequently, for example the notion of a *random experiment,* or more narrowly, a *Bernoulli variable.* Let us write it generically $b(p)$ where p is the probability of success in the random experiment.

Another theme is *stochastic independence* which plays a fundamental role in probability theory and which is invoked in most statistical models although not always explicitly.

The *arithmetic functions* are also themes that are needed throughout the model building activity in statistics, as well as certain other functions.

Now we can start to build simple derived hypotheses, for example that of a binomial distribution

$$B(4,p) = b(p) + b(p) + b(p) + b(p) \qquad (2.16)$$

with the understanding that the four terms on the right should be interpreted as independent realizations. Similarly, if limit processes are used as themata, we can get the Poisson distribution $P(m)$ and the normal one $N(m,\sigma^2)$. We can introduce the chi-square distributions

$$\chi_3^2 = N^2(0,1) + N^2(0,1) + N^2(0,1) \tag{2.17}$$

and for F-distributions

$$F_{2,3} = \frac{3}{2}\frac{\chi_2^2}{\chi_3^2} \tag{2.18}$$

and so on. The standard statistical hypotheses can be obtained in this manner, and in general it can be seen that a *small number of themata is sufficient to explain the regularities in the hypothesis formation in elementary statistics.* We shall return to this topic in considerable detail in Chapter 6.

It is time for us to end this list of special cases of regular structures. It would be easy to add many more cases, but scarcely necessary to illustrate the point: *the unity of thought that underlies all of them.*

It was pointed out early that our selection of cases was biased, and the reader will see this more clearly now. All the examples were atomistic in nature: they started from primitives and combined them together producing an ensemble of structures. *We are not studying single laws, only combinatory regularity.* With this important restriction in mind, the examples show that we are led to some considerations that have occurred repeatedly:

Study the notion of "sameness" of primitives.

Regular structures have an internal topology.

When can primitives be hooked together?

What is the function attributed to a "formula"?

How does this function relate to observables?

These are questions that must be answered and we shall begin to express them in a precise formalism in Chapter 2.

1.3. The mathematical study of regularity

So far our discussion has been conceptual and we have employed mathematical formalisms only sporadically. We now turn our attention to the mathematical formalization of regular structures, those we have already mentioned as well as others. Such a formalization in precisely defined terms is a necessary preparation for the mathematical study of regular structures.

The many examples in the previous section will serve as a guide and help us to pinpoint the mathematical concepts that will be employed in our study. This is just as in other applications of mathematics where the subject matter problem leads to mathematical entities via a process of abstraction, elimination of the peripheral, and concentration, emphasis on the essential. What may differ is the role mathematics will play in constructing logical structures that should mirror regularities as they appear in reality.

It has become customary to view applied mathematics as a means to achieve successively better approximations to what can be observed. Starting from observations, perhaps rather primitive ones, the mathematician tries to think of a model that fits the data and is as simple as possible at the same time. The model is judged good if it also extrapolates to other situations, and, as better data become available, these are also consistent with the model. If they are not, then the model has to be changed, usually refined by adding another term, by relaxing an independence assumption, or other non-structural changes. In extreme cases the model will have to be scrapped entirely and the modelling would have to start again from the beginning.

During this process we operate on two levels, the analytical and the empirical. We use mathematical analysis, we compute numerically or non-numerically, and we employ induction and other principles of logic. Observations in the laboratory or in the field represent the only external information that we can legitimately use. This is reminiscent of the quotation from Hume in the last section.

According to the empiricist doctrine that dominates current scientific activities, at least in the English speaking world, this process is the only one that makes sense. A theory must be verifiable empirically, or, according to others, it should be such that it is possible a priori to falsify it by empirical tests. What cannot be observed does not belong to scientific enquiry.

We shall adopt a less positivistic and less pragmatic attitude towards the use of mathematics in model building. In the metaphor used by Holton we shall emphasize the y- and z-components, the analytical and thematic dimensions. The study of regular structures will be pursued with this in mind so that we will be more concerned with the analytical-mathematical aspects and with the general themes in regularity than with empirical examination and testing of the applicability of the mathematical models.

This implies that we do not have the right to claim that a particular, abstract regular structure adequately describes a certain observed phenomenon. We are aiming for theory, not for applications in the strict sense of this word. However, if the theoretical constructs are rich enough and deserve a study for their own sake, then it is likely that the results will also be of utilitarian value later on. This is what experience of applying mathematics has taught us again and again.

In the design of the pattern formalism to be discussed in Chapter 2 and in the study of the resulting structures we have been helped by analyzing regular structures appearing in a large number of special cases. Much of this is of speculative character and the relation to "real" patterns is often tenuous, as for example in Chapter 6 of Volume II or in Chapters 7 and 9 of the present volume. The highly speculative and non-empirical flavor will perhaps seem unappealing to a pragmatically inclined reader.

It has been said that a mathematical theory *is a collection of special cases held together by a uniform view*. The work on pattern theory reported in these volumes has been carried out in this spirit. When it was begun around 1965 there was only a logical skeleton - a way of thinking about patterns - but no special cases had been analyzed in any mathematical depth. The first priority was therefore to do this in a number of cases and the result of this has been given in the two previous volumes as well as in Chapters 5-9 in the present one. The logical glue that makes these cases hang together is examined in the first sections of each of Chapters 1-4 in Volume I, and in Chapter 3 of the present volume.

The different concepts that we have encountered and that will be formalized in Chapter 2 must be related to each other. We strive for a *unified theory* in the sense that the basic concepts are made to agree with each other in a natural way.

As an example that illustrates what we have in mind is the classical notion of a topological group. A topological group is not just a topological space on which a group operation is defined. In addition it is required that the group operation is continuous with respect to the topology given: the algebraic and the topological properties are related to each other.

Similarly, when we formalize such notions as "sameness", "configuration", "image", and so on, we shall make sure that they fit together. For example, a bond relation is related to similarities by our requirement that it be S-invariant, see Section 2.2, and many other instances will be encountered where we try to mold the theoretical requirements into a uniform theory.

CHAPTER 2
A PATTERN FORMALISM

2.1. The principle of atomism

We shall now examine the basic concepts and relations of pattern theory as expressed in a formalism whose terminology and notation will be discussed in this chapter.

The starting point is the *atomistic attitude*. In the examples presented in Section 1.2 the structures were generated from certain primitives whose nature varied from case to case. We speak of these primitives as the *generators*, denoting a generator generically by the symbol g, and the whole collection of them by G.

No assumption is made on the cardinality of G. It will occasionally be finite, but it can just as well be denumerably infinite or even have the power of some continuum.

Going back to the list of examples in the previous chapter let us look at some of the generators:

$$\begin{cases} \text{(i)} & \text{uniform circular motions in the plane} \\ \text{(ii)} & \text{arithmetic computational modules} \\ \text{(iii)} & \text{trigonometric functions} \\ \text{(iv)} & \text{rewriting rules} \\ \text{(v)} & \text{point atoms on a lattice.} \end{cases} \tag{1.1}$$

They may seem quite different but their role in the generation of regular structures is similar as will be more apparent in the next section.

Each generator is described by *attributes*. For (i) the attributes could be chosen as the coordinates of the center, the radius, and the angular velocity, and the phase angle. For (ii) a generator could be "ADD 5", MULTIPLY X AND Y", or "RECIPROCAL OF X", so that we can choose as attributes a label describing the type of operation together with constants as needed, e.g. the value 5 in the first case. The trigonometric functions, say in complex form $ze^{i\nu x}$, can be specified by an integer ν and a complex constant z. Rewriting rules can be specified by an ordered triplet of discrete symbols (i,j,x) meaning that variable i is rewritten into variable j while writing the lexical item x. This was for a finite state grammar; for context free ones we can use strings involving more than three symbols. For (v) we need one label indicating the type of atom as well as coordinates specifying its location.

The attribute vector $a = a(g)$ should of course describe the generators in sufficient detail so that generators are separated. In other words if $g_1 \neq g_2$ we should insist on $a(g_1) \neq a(g_2)$; the attribute specification is complete. If so, then we can identify the generator with the attribute vector.

In example (v) two atoms with the same label are identical except for location. As discussed in Section 1.2 we regard them as similar, meaning that they agree except for properties that are not crucial. One can be moved to where the other is and made to coincide with it. In other words, the two generators are the same except for a translation.

To formalize the notion of "sameness" we shall introduce the *similarity transformations*, generically denoted by the symbol s. The set of similarity transformations will be denoted by S. We shall often have reason to apply similarity transformations successively which leads us to require that S should be a semigroup. Actually, in most cases, we shall even have the group property for S, so that we shall assume that S forms a group of transformations, unless it is mentioned explicitly that only the semigroup property holds. We shall say that two generators $g_1, g_2 \in G$ are similar if there exists a similarity $s \in S$ such that $sg_1 = g_2$. Since S is a group this implies that similarity is an equivalence relation and partitions G into disjoint classes of similar generators.

Occasionally it will be seen to be natural to operate with more than one similarity group at the same time. If so, then one of them will often be a subgroup of the other. In case (v), for example, we may choose S as the Euclidean group in the plane or as the translation group.

If in case (i) we choose S as made up of the Euclidean group and uniform scale changes all generators with the same angular velocity are similar. If S also includes uniform scale changes of time then all generators are similar.

In case (iii) where generators are complex valued functions of a real variable $x \in [0,2\pi)$ let s represent multiplication with a non-zero complex number ζ. This means that the amplitude is multiplied by $|\zeta|$ and the phase angle is increased by $\arg \zeta$, but the frequency ν remains the same. The similarity classes are the (complex) trigonometric functions with given frequency.

For case (iv), for finite state languages, let S consist of permutations between the rewriting rules $i \to jx$ keeping both i and j fixed, but transforming x into some other lexical items. The corresponding notion of similarity played an important role in Chapter 7 of Volume 2.

In (ii) we could let S be a set of permutations of G such that any s leave the range and domain of the module unchanged.

Among the components of the attribute vector we shall single out one, denoted by $\alpha = \alpha(g)$, called the *generator index*. The role of this index is to express some fundamental characteristic of the generator so that we shall of course ask that α be S-invariant

$$\alpha(sg) = \alpha(g); \quad \forall s \in S, \quad \forall g \in G \tag{1.2}$$

The subsets

$$G^\alpha = \{g | \alpha(g) = \alpha\} \subset G \tag{1.3}$$

are called the *generator classes* and it is clear that the similarity classes form a partition of G which is at least as fine as that of the partition $\{G^\alpha\}$. The two partitions will coincide sometimes but need not always do so. In (v) we take G^{α} to be the atoms with the same label and the two partitions will coincide. In (iv) it is natural to choose the generator class as those rewriting rules that rewrite the same variable, and the partitions do not coincide.

Other components in the attribute vector are the *bonds* which play an important rule in the combinatory relations that we shall discuss in the next section. Anticipating this discussion we shall let a generator g have a number of bonds, the *arity* of g, denoted $\omega(g)$. The arity need not be the same for all g's, nor does it have to be finite.

The bonds may be of two sorts, in-bonds directed toward g, and out-bonds, directed away from g. Their numbers are denoted by $\omega_{in}(g)$, the *in-arity*, and by $\omega_{out}(g)$, the *out-arity*; see Notes A.

To keep track of the bonds for a given generator we need labels to identify them. This will be done in various ways but it is often convenient to have one subscript for the in-bonds and another for the out-bonds.

To each bond is associated a *bond value* taking values in some given space. The bond values will determine how generators can be joined together. The set of bonds, with their labels, but disregarding their bond values, constitute the *bond structure* of the generator. We shall always assume that the bond structure is invariant with respect to S, but g and sg can very well, and will in many cases, have different bond values.

Diagramatically a generator can be visualized as for example in Figure 1.1 when $\omega_{out}(g) = 3$ with the bond-values $\beta_1', \beta_2', \beta_3'$, and with $\omega_{in}(g) = 2$ with the bond-values β_1 and β_2.

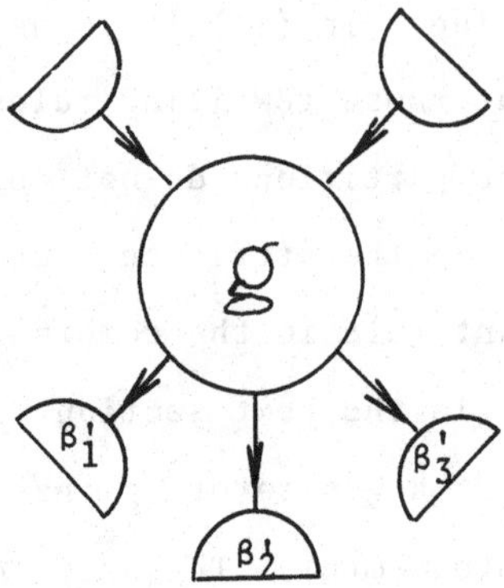

Figure 1.1

2.2. The combinatory principle

We shall now combine generators together by joining out-bonds to in-bonds between different generators g_1 and g_2 respectively; see Notes A. Such a combination will be called a *configuration* and it can be looked at as a formula as discussed in Section 1.2.

Consider the possibility of joining an out-bond, with bond value β_1 to an in-bond, with bond value β_2. The legality of such a juncture will be determined by the *bond relation* ρ. It is a function of β_1 and β_2 taking the values TRUE and FALSE and we write it as

$$\beta_1 \rho \beta_2 = \text{TRUE or FALSE} \qquad (2.1)$$

as the case may be.

As mentioned in Section 1.3 we aim for a unified theory whose constituent parts fit together naturally. Since similarity transformations represent changes that are non-essential and do not change generators drastically we are led to assume that the bond relation is S-invariant. In other words if $sg_1 = g_1'$ with β_1 carried into β_1' and β_2 carried into β_2' then we require that

$$\beta_1 \rho \beta_2 = \beta_1' \rho \beta_2'. \qquad (2.2)$$

Note that we allow only binary junctures of bonds: bonds are allowed to be joined pairwise. We never join three or more bonds together. This is a very restrictive assumption and in the early development of pattern theory it was believed that triple bonds etc. would be required. So far, however, in the many cases that have been studied we have not encountered a single case where this happened, so that the binary bonds will be the only ones allowed.

As examples let us look at (1.1) again. In (i) the in-bond would be the center of the circle and the out-bond its periphery. One way epicycles can be formed is by letting a center move on a periphery which would correspond to ρ being the "ELEMENT OF" relation.

In (ii) the bond values are subsets of spaces and in order to be able to carry out the computation of an arithmetic module we have to make sure that the incoming values lie in the domain of definition of the module. Hence ρ will be the "INCLUSION" relation between one range and another domain. One will have to make sure in each separate case that ρ is preserved by the similarities used.

For case "iii" the situation differs in an interesting way. Consider two configurations c_1 and c_2 made up of generators $g_1, g_2, \ldots g_n$ and $g_1', g_2', \ldots g_n'$ respectively. We have discussed briefly in 1.2 how such formulas c_1 and c_2 are interpreted and we shall return to it in more detail in the next section. We calculate the two functions obtained as trigonometric polynomials by adding generators pointwise. The meaning of c_1 and c_2 is given by the two functions so that c_1 can very well differ from c_2 although the functions coincide. In order to be able to identify the generators belonging to a given function of this type we should make sure that the formula (the configuration) does not include more than one generator from any generator class $\{ze^{i\nu x}\}$. Hence, if we make the bonds = $\alpha(g)$ and define ρ as the binary truth valued function "UNEQUAL" we achieve uniqueness. Then ρ is automatically S-invariant.

Note that the role of this bond relation is different from the earlier ones. In the previous cases the purpose was to make sure that the generators were naturally related to each other, say *adhesive* bond relations, in the present the purpose is to make sure that they are different from each other, *repelling* bonds.

In case (iv) let the in-bond value for a generator $i \rightarrow jx$ be i and the out-bond value be j. Then we can let ρ be the relation "EQUAL TO" which is of course S-invariant, and we get the sort of concatenation of rewriting rules we need.

Finally, in case (v) atoms are joined together if they are neighbors on some square lattice. The reader is referred to p. 170 of Volume 1 to see how this can be done. Again we encounter ρ as "EQUAL TO".

We must now discuss the topology of configurations. In 1.2 it was seen that they have an internal geometry which should now be formalized. To do this we shall use the notion of *connection type* denoted Σ. The connection type is a set of graphs, not necessarily finite, describing how bonds are joined together.

To make this concrete we give some examples in Figure 2.1 where generators are indicated by large circles and bonds by small semi-circles. Two connected bonds make up a small circle. In (a) all generators have arity = 2 with $\omega_{in}(g) = \omega_{out}(g) = 1$ and they connect into a linear chain. The whole configuration has one unconnected in-bond and one out-bond.

In (b) the generators have arity 3 with $\omega_{in}(g) = 2$, $\omega_{out}(g) = 1$. The topology is that of a bond two generators wide and of arbitrary length.

The connection type in (c) is of tree structure and all generators have in-arity one (see Notes C), but the out-arity can vary over G. The diagram shows a configuration with one unconnected in-bond and six out-bonds.

For (d) we have the topology often encountered when the generators are arithmetic modules but without loops. It induces a partial order: in order to execute a module all those modules must have terminated whose output goes into the given module. The configuration has in-arity two and out-arity three for the unconnected bonds.

The square lattice topology is shown in (e) where $\omega_{in}(g) = \omega_{out}(g) = 2$. The whole configuration has six unconnected in-bonds and the three unconnected out-bonds.

The two remaining examples are extreme but of considerable interest. In (f) the topology is completely vacuous: no bonds are connected so that no restrictions are imposed on the generators making up the configuration. In (g) the arity

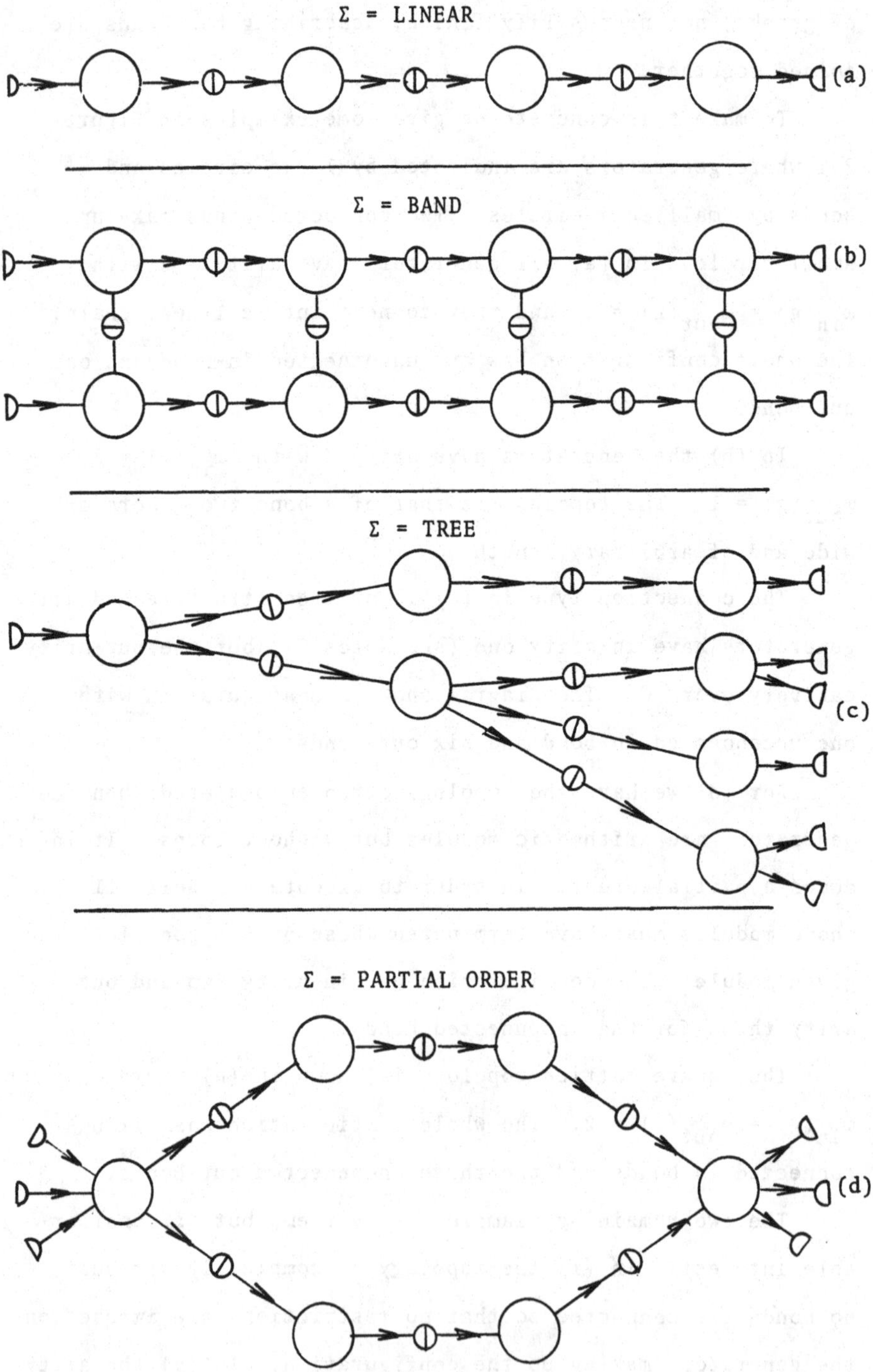

Figure 2.1

Σ = SQUARE LATTICE

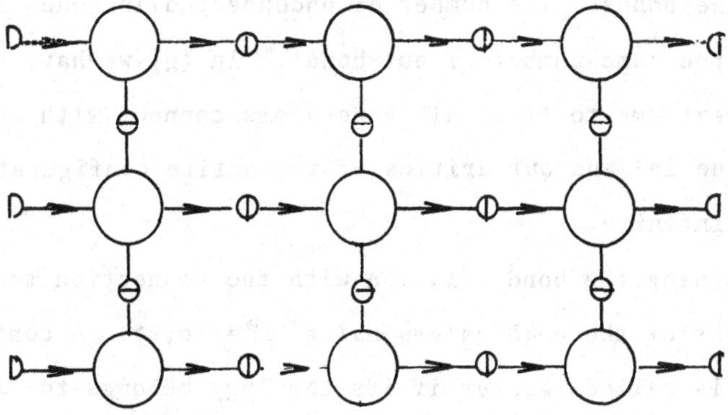

(e)

Σ = DISCRETE

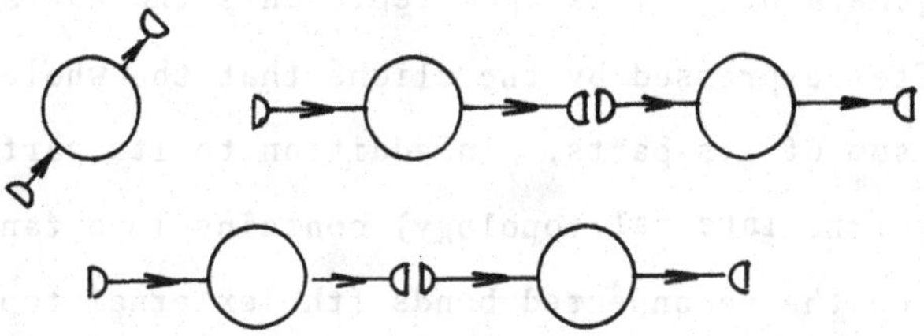

(f)

Σ = FULL

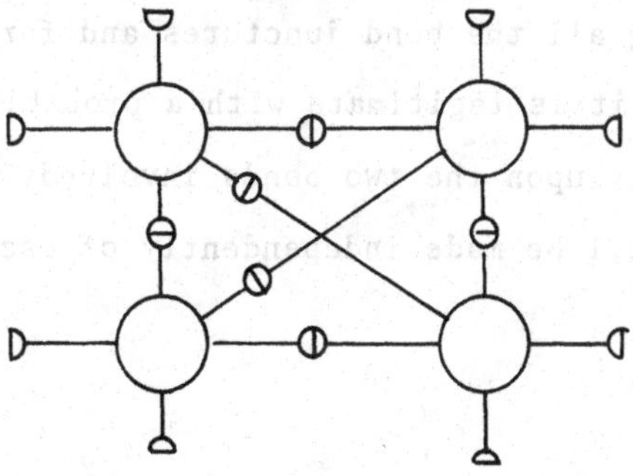

(g)

Figure 2.1 (cont'd.)

of any g is infinite but we can of course only indicate some of the bonds. The number of unconnected in-bonds is 5 and with the same number of out-bonds. In (g) we have the opposite extreme to (f): all generators connect with each other. The in- and out-arities of the entire configuration are both infinite.

Combining the bond relation with the connection type we shall speak of the *combinatory rules* $\mathcal{R} = \langle \rho, \Sigma \rangle$. A configuration c is called *regular* if its topology belongs to Σ and all its connected bonds satisfy ρ. *The set of all regular configurations* $\mathcal{C}(\mathcal{R})$, the *configuration space, is the set of regular structures that we shall study.*

Note that a regular configuration is not just a collection of generators. This fact represents the *holistic attitude* often expressed by the cliché that the whole is more than the sum of its parts. In addition to its parts the connectivity (the internal topology) contains important knowledge as do the unconnected bonds (the external topology).

An important extension of the bond relation is the notion of a *stochastic bond relation*. By this we mean that the regularity or non-regularity of a given configuration is determined by examing all the bond junctures and for each one decides whether it is legitimate with a probability $A(\beta_1, \beta_2)$ that depends only upon the two bonds involved. All the random decisions will be made independently of each other. See Notes B.

2.3. The principle of observability

The regular configurations are logical constructs that are introduced in order to give a precise definition to the various notions of regularity. They are not always directly observable; usually the observer can "see" only their "surface" but not their "inside".

The pattern formalism must therefore specify *how the regular configurations give rise to observables*. Let us assume that we have an *ideal observer* with accurate instrumentation so that for the moment we can neglect measurement errors etc. When this ideal observer examines a configuration c two cases are possible. Either he obtains complete information about c so that he can identify c uniquely from the set $\mathcal{L}(\mathcal{R})$. Otherwise he loses information so that he can only identify c to the extent that it belongs to some subset of $\mathcal{L}(\mathcal{R})$.

We shall formalize this by saying that there exists an equivalence relation R in $\mathcal{L}(\mathcal{R})$ such that $c_1, c_2 \in \mathcal{L}(\mathcal{R})$ appear identical to the ideal observer if and only if they are R-equivalent, which we shall write $c_1 R c_2$. The *identification rule* R is of course quite different in character from the combinatory rule $\mathcal{R}$.

The identification rule does not express imperfections of measuring devices such as noise; those will be discussed in the next section. Actually, such imperfections would not be well described by equivalence relations. *Instead it gives the interpretation of a regular configuration in terms of its functioning such as it appears to the ideal observer.*

The equivalence classes induced in $\mathcal{C}(\mathcal{R})$ by R are the entities that can be observed, at least in principle. They will be called the *images*, generically denoted by I, and they are elements in some space $\mathcal{I}$, the image algebra, that will receive more structure later.

If we think of configurations as formulas then the images are functions. They express the meaning of the formulas and for one function we may of course have several formulas. Images are *semantic* constructs.

To illustrate how R can be chosen let us look again at (1.1). In (i) the observer does not really see the circles; what he sees is the motion of certain points fixed to the circles as time develops. Although he may be able to deduce the parameters of the circles from his ideal observations it is clear that they are derived, not primary, observables.

The second case illustrates the difference between configuration and image more drastically. The interpretation of the configuration of computational modules is that of a function. The configuration is a formula or an algorithm or a computer program. The image is the function implemented by this configuration. A function has as part of its definition the domain which in the configuration is given as the unconnected in-bonds. Similarly the range can be obtained from the unconnected out-bonds. This makes it clear that the image must carry information about *unconnected (external) bonds in the configuration.*

This will make it possible to combine images with each other joining unconnected bonds to each other according to $\mathcal{R}$. Then $\mathcal{I}$ will receive an algebraic structure resulting from such combinations viewed modulo R. A reader familiar with universal algebras will see the resemblance to partial universal algebras.

In (iii) we let R simply be identification of functions obtained from adding the generators (trigonometric functions) appearing in the configuration. Then $\mathcal{T}$ happens to be just a vector space, finite- or infinite-dimensional as the case may be.

For case (iv) the configuration means the parsing (or derivation) of a grammatical phrase of lexical symbols. As is well known it can well happen that the parsing is not uniquely determined so that one image may contain several configurations. Again, we should include the first in-bond and the last out-bond of c in the image. In particular if these bonds are the initial and final states respectively we have a grammatical sentence as the image, otherwise a grammatical phrase.

In (v), the finite or infinite square lattice, we would naturally identify two configurations if they contain the same generators, atoms, and if they have the same unconnected bonds. The latter we do in order to be able to connect images to other images. In the configuration we also specify the internal bonds. Note that in the image the generators are not indexed by a double subscript; the generators need not even be uniquely determined. At the moment this is of no great importance but when we consider actually - not just ideally - observed images, then this fact will be seen to have an important consequence.

When we carry out the mathematical implementation of the R-relation and its consequences we should be careful and recall the attempt to build a unified theory, see Section 1.3.

The choice of R must fit both S and $\mathscr{R}$ so that the resulting set $\mathscr{T}$ will possess an algebraic structure appropriate for the case under consideration. We then speak of $\mathscr{T}$ as an image algebra. The details of how this can be done are given in Section 3.1 of Volume 1.

The role of pattern analysis is now beginning to emerge more clearly, at least when we assume an ideal observer. One is presented with the information contained in the image, no more and no less. One wishes to reconstruct the internal structure, the combinatory regularity, that underlies the image, or class of images, to which one has access. Borrowing a term from linguistics, but using it in a different sense, one could say that the pattern analyst tries to understand the deep structure having access only to the surface structure of the regular phenomenon in question. One is searching for this hidden truth, one could say, in the *hermeneutic* tradition.

Once we have the image algebra, the next natural level of abstraction is that of *pattern*. By a pattern or pattern class we mean a subset of $\mathscr{T}$ invariant under similarities. We must insist on S-invariance since we want pattern classes to exhibit homogeneous regularity. On the other hand, we do not restrict the cardinality of a pattern class. It can be equal to one (the patterns can be individual images), or it could be equal to $\mathscr{T}$ itself, containing all the images that obey the regularity rules as seen by the ideal observer. Of course, intermediate cases are possible, indeed they are the most interesting ones.

The construction configurations $\rightarrow$ image and image $\rightarrow$ pattern, both involve partitioning of sets. The two partitionings are, however, conceptually quite different, The first one is introduced to describe the loss of information due to limitations on observability. The second one involves no loss of information, but it brings together only images that differ in a non-essential aspect. The image represents a concept, a function in a very wide sense of the term. The pattern is a *generalization* of such a concept.

2.4. The principle of realism

The image algebra is a logical construct that describes regularities as they are seen by the ideal observer. In order to arrive at a realistic pattern formalism we must also specify in mathematical terms how these images are related to what is seen by a *real* observer.

This is a crucial, but unfortunately also difficult, step in building the mathematical theory of regularity. One could try to avoid this difficulty by assuming that the pattern analyst had available some technique of transforming observed objects into pure images belonging to $\mathcal{T}$. Such techniques could be approximations, filterings, or smoothings with the purpose of removing observational errors and imperfections while preserving regularity structure. This can certainly be done, sometimes, and in the pattern recognition literature one can find an abundance of such techniques.

The reason why we have chosen not to accept this approach is two-fold. First, on the utilitarian level, it may lead to less than efficient utilization of available information. Second, and this is our main reason, we want the theory to include the entire chain of generating real observables - including the last link involving the real observer.

What is, then, the nature of the relation between an image $I \in \mathcal{I}$ and what is actually observed? In qualitative terms, observation usually results in loss of information, introduction of errors, and loosening of regularity structure. The observed objects can therefore not be expected to exhibit the same sort of combinatory regularity as the images, except in certain instances.

To bring this out clearly we shall denote the observed object by $I^{\mathcal{D}}$, the *deformed image* and the set of the $I^{\mathcal{D}}$ by $\mathcal{I}^{\mathcal{D}}$. Although $\mathcal{I}^{\mathcal{D}}$ will usually be distinct from $\mathcal{I}$ it may have some other regularity structure. If so, it is usually weaker, less rigid, than that of $\mathcal{I}$.

Let us return to (1.1) again and mention some possible deformation mechanisms. What we really observe in (i) are the locations of certain distinguished points, perhaps the centers of the planets, the Moon and the Sun. Observations are of course made discretely in time, so that we have a finite number of readings in space-time. The fact that the readings are discrete rather than continuous causes some loss of information. Further, the readings will be subject to error. As usual part of the error will be described in deterministic terms: the systematic part. The remainder is the random part, the noisiness of the data.

Hence we have a mixture of deterministic and random deformations. It should be noticed that the images consist of certain trajectories in space-time. The deformed image is of logically different type, it consists of a finite vector. Although in this simple case the relation between the two is fairly straightforward it is nevertheless a fundamental distinction that should be kept in mind.

For example (ii), the image, a function, may be observed only for certain combinations of inputs, arguments. Then the deformed image will be a restriction of the original function. Whether it will contain sufficient information to determine the pure image will depend upon the complexity of the system of computational modules. There will also arise questions of decidability, whose answers also will depend upon the complexity mentioned.

In (iii) a very simple deformation could be just additive noise. If so we can immediately apply well known statistical methods for the pattern analysis, and the same will often be true when pure images are given as elements in topological vector spaces, say ℓ_2, L_2, C, etc., and deformations are just additive changes. If we also have errors on the x-axis the analysis will be harder.

Suppose for (iv) that changes are partly paradigmatic and partly syntagmatic. The paradigmatic deformations involve exchanges of lexical items for each other, including the possibility of deleting some entirely and also adding some others to the string. The syntagmatic ones represent grammatical errors so that the derivation as string or strings of syntactic variables is affected. The result need of course not be in I and we may have only a loose regularity structure over $\mathcal{T}^{\mathcal{D}}$. See Notes A.

Finally in (v) we may have access to readings of the (x,y) coordinates of the points. These readings will not cover the infinite lattice, of course. Another more fundamental effect is that unless we have markers on all the atoms observed, which would not be the case in a real set up, we will not have the reading labelled. We will not know to what ideal location a particular reading refers. This will be a serious obstacle to the analysis if the measurement errors are of the same order of magnitude as the inter-point distances.

These examples may suffice to show that we will have to expect unorthodox types of deformations to occur, with resulting difficulty in the ensuing analysis. It is also clear that the deformations are, with few exceptions, quite different from similarity transformations. The latter are not very drastic, involving little or no essential loss of information. They do not bring us outside $\mathcal{T}$. The deformations are more drastic and their influence makes for more serious mathematical difficulties.

Generically we shall denote a deformation mechanism by $\mathcal{D}$. It may involve random elements in which case $\mathcal{D}$ should specify the probability measures used. A particular deformation is denoted by d, $d \in \mathcal{D}$ and we write $I^{\mathcal{D}} = dI$.

Sometimes the strength of the deformation mechanism is specified by some parameter, say t, and we shall then assume that we have an additive semigroup

$$\mathcal{D}(t_1+t_2) = \mathcal{D}(t_2)\,\mathcal{D}(t_1); \quad t_1, t_2 \geq 0, \tag{4.1}$$

where $\mathcal{D}(0)$ stands for the identity mappings.

To make this part into a unified theory we shall have to consider how $\mathcal{D}$ is related to S, how G is transformed by $\mathcal{D}$, what is the combinatory structure of $\mathcal{T}^{\mathcal{D}}$, if any, and so on, but this has been studied elsewhere, see Notes B.

The pattern formalism that has been outlined will help us to approach regular structures from a certain point of view. It will not tell us how to choose generators, connection types, deformations, etc., and even less how to solve the analytical problems that arise. This has to be done separately for different special cases and the solution of these problems is at present emerging as a logically homogeneous mathematical theory. At the present stage of development of the theory it would be premature to claim that a fairly complete theory exists. This is not so, the existing theory is full of gaps and much remains to be done. However, a foundation has been laid.

CHAPTER 3
ALGEBRA OF REGULAR STRUCTURES

3.1. Generator coordinates

In Volume I we introduced the *objects of pattern theory:* generators, configurations, and images, as well as *pattern relations:* similarities, combinatory relations, and deformations. We are now going to deepen our understanding of these concepts with the emphasis on their algebraic properties.

Configuration spaces and image algebras have combinatory properties just as do other, more familiar, mathematical structures such as groups, vector spaces, algebras, and fields. These properties include algebraic ones that express how objects can be combined with each other, how similarities relate them, and how they are perceived ideally by an observer.

The *algebraization of pattern theory* is in its infancy, but it has been clear since its conception in the middle 60's that general algebraic concepts could be applied to elucidate the theoretical considerations. This was of course the reason why the author coined the term image algebra.

In this chapter we shall show how pattern theoretic ideas can be expressed in the language of universal algebra.

Combinations of configurations, or of images, are only possible if the result satisfies the regularity conditions. It is therefore clear that the basic operations are not *entire*, defined for all values of the arguments, but *partial*. This fact causes difficulties in the study of homomorphisms, congruence relations, etc., since these concepts have more than one natural definition - a difficulty that does not arise for semi-groups, groups, vector spaces and so on. Our frame of reference will therefore be *partial universal algebras*.

A reader may find some of the following analysis abstract and hair splitting, as, for example, the role of the empty configuration and image, or the labelling by coordinate systems of bonds and generators. Such distinctions cannot be avoided, however, in a rigorous treatment of pattern synthesis, and is necessary for dealing with the overwhelming variation and flexibility of regular structures.

As in all algebraic theories homomorphisms play an impartant role and we shall study them for many regular structures, both on the configuration level and for induced mappings of images. This will lead us to view configuration spaces and image algebras as categories, and to consider some functors for them. While this is undoubtedly a step in the right direction, it will be obvious that we have only scratched the surface below which a rich layer of algebraic structure remains to be tapped.

All the regular structures that we have considered are constructed from indivisibles - generators - which are combined together in an orderly manner to produce the appearance of regularity to the observer. *The regularity can be both globally and locally controlled: the connection type governs the first and the bond relation the second.*

Global regularity control will be discussed in considerable detail in the next section and we will not anticipate this discussion here. Instead we shall examine carefully the local regularity, via bond values, and bring up some questions that should be answered before we turn to global control of the configurations.

Let us recall the following from Chapters 1 and 2 in Volume I. The generator is the primitive carrier of information. A generator $g \in G$ possesses a set of information that can be expressed in terms of an attribute function $a(g)$. Among these attributes we mention some particularly important ones. The generator index $\alpha = \alpha(g)$ divides G into disjoint subsets G^α, forming a partition $\{G^\alpha\}$ invariant under transformations from the group S of similarities. The generator index serves to separate generators that are radically different from each other.

A generator may also contain a place holder that can carry an identifier whose role is to distinguish the generators that make up a configuration c. If all the generators in some c are distinct then this causes no problem and we can just use any labelling with some subscript i to denote the generators g_i, $i = 1,2,\ldots,n$, at least when $\#(c)$ is finite as will be assumed throughout this volume in order to simplify the discussion (the infinite arity case can also be dealt with). In the opposite case, when the regularity structure does not rule out identical g's in a configuration more care is needed. Then we have to decide whether they carry such place holders or not and the ensuing pattern synthesis will depend upon this decision. We shall return to this question in the next section in connection with coordinate systems for configurations.

At the moment we shall concentrate our attention on the attributes that control local regularity, viz. the bonds. To any g the bonds are given as

$$B(g) = [B_s, B_v]; \quad B_s(sg) = B_s(g), \forall s \in S; \tag{1.1}$$

where B_s expresses the bond structure and B_v the bond values.

In (1.1) B_s can be represented by a vector of length $\omega(g)$ whose entries are the *bond structure parameters*, taking values such as "in" and "out", "active" and "passive", simply natural numbers, or combinations of such values. It is necessary to label the bonds of g by some coordinate $i = 1,2,\ldots\omega(g)$ in order to be able to refer to them unambiguously, but the way we choose the coordinates is immaterial as long as we preserve information about the values of the bond parameters; this will be discussed later in this section. We will refer to them with the usage "the bond i" of g, where i is the *bond coordinate* for the bond in question.

The set B_v should be of the same cardinality as B_s and with the usage "the value of the bond i is β". The possible values of a β are the elements of some space V_b that may vary with b.

Two types of regularity will be studied and with a terminology that differs somewhat from *Pattern Synthesis*.

Definition 1.

(a) *Symmetric regularity* when all bonds are undirected and the bond relation ρ is symmetric.

(b) *Directed regularity* when all bonds have directions, either "in" or "out".

In case (a) we do not split up the arity of a generator into in-arity and out-arity. In case (b) in-bonds will only be connected to out-bonds and vice versa.

To illustrate how (1.1) is formed let us look at a few examples, starting with the simplest ones.

Example 1. A generator consists of a discrete "interval" $(k,k+1,k+2,\ldots\ell) = [k,\ell]$, $\ell-k+1$ integers with $\omega_{in}(g) = \omega_{out}(g) = 1$ and $B_s = (\text{in},\text{out})$. Also $B_v = (k,\ell)$; $V_b = \mathbb{Z}$.

Example 2. As above but g also contains the ordinates (real) $g_k, g_{k+1},\ldots g_\ell$ together with the old values.

Example 3. A generator consists of an (ordinary) interval $[a,b]$ with a and b integers and $\omega_{in}(g) = \omega_{out}(g) = 1$ and $B_s = (\text{in},\text{out})$. Also $B_v = (a,b)$; $V_b = \mathbb{R}$.

Example 4. As above but g also describes a continuous real function $g(x)$ still with $\omega_{in}(g) = \omega_{out}(g) = 1$ and $B_s = (\text{in},\text{out})$. Also $B_v = ((a,g(a)),(b,g(b)))$; $V_b = \mathbb{R}^2$.

Example 5. A generator consists of a rectangle with sides of length one, with corners at integral lattice points $(a,b),(a+1,b),(a,b+1),(a+1,b+1)$. Let $\omega(g) = 4$ with $B_s = (E,N,W,S)$ with $B_v = $ (side $(a+1,b)$ to $(a+1,b+1)$, etc. ...). V_b consists of elements in $\mathbb{Z}^4$ describing such sides. The bonds at E and N are out, those at W and S are in.

Example 6. As above but also with a continuous real function $g(x,y)$ given on the rectangle and bond structure as before with

$$B_v = ((\text{side } (a+1,b) \text{ to } (a+1,b+1), \text{ values of } g(\cdot,\cdot) \text{ on this side}), \text{ etc. } \ldots) \tag{1.2}$$

Note that V_b has as elements sides and functions restricted to those sides.

Example 7. As in (5) and (6) except that the rectangles are replaced by triangles as in Figure 1.1 with generator index = "upper" or "lower". A generator is either an upper triangle with corners $(a,b),(a+1,b+1),(a,b+1)$ and $B_s = (SE,N,W)$ with B_v the corresponding descriptions of the sides of the triangles, or a lower triangle in the same way. Here V_b represents sides. For α = upper the bonds at SE are out, at N and W they are in. For α = lower the bonds at NW are in and the others are outwards directed.

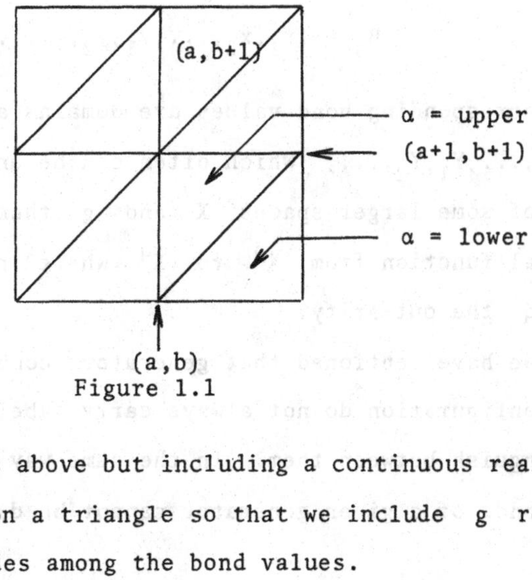

Figure 1.1

Example 8. Same as above but including a continuous real function $g(x,y)$ on a triangle so that we include g restricted to the sides among the bond values.

<u>Example 9</u>. Let g describe branches in a finite automaton $\omega_{in}(g) = \omega_{out}(g) = 1$ with $B_s = (\text{in},\text{out})$ and the bond values equal to the number of the state at the beginning and end respectively of the directed branch. Here $V_{in} = V_{out} =$ state space of the automaton.

<u>Example 10</u>. Let g describe a rewriting rule for a context free language with $\omega_{in} = 1$ and $\omega_{out} =$ arbitrary and finite. Here

$$B_s = (\text{in},\text{out } 1, \text{ out } 2,\ldots) \tag{1.3}$$

with the out bonds ordered as indicated. Also

$$B_v = (\text{symbol to be rewritten, resulting symbol string}) \tag{1.4}$$

again with the order indicated. Here $V_{in} = V_{out} =$ non-terminal symbol set V_N.

<u>Example 11</u>. Let g represent functions - say computing modules with $\omega_{in}(g) =$ number of arguments $x_1,x_2\ldots$ and $\omega_{out}(g) =$ number of results, so that we write

$$B_s = (x_1,x_2,\ldots,y_1,y_2,\ldots). \tag{1.5}$$

The corresponding bond values are domains and ranges $X_1,X_2,\ldots,Y_1,Y_2,\ldots,$ which often can be understood as subsets of some larger spaces X and g then defined as a partial function from X^p to X^q where p is the in-arity and q the out-arity.

We have mentioned that generators combined into a regular configuration do not always carry labels enabling us to distinguish between them. In the same way it can happen that the bonds of a given generator cannot be distinguished by us.

To make this clearer consider the generator diagrams in Figure 1.2. In (a) we illustrate Example 4 by the function sin 3x on the interval $[0,\pi/2]$. Here we have arity two and the question about being able to distinguish between the two

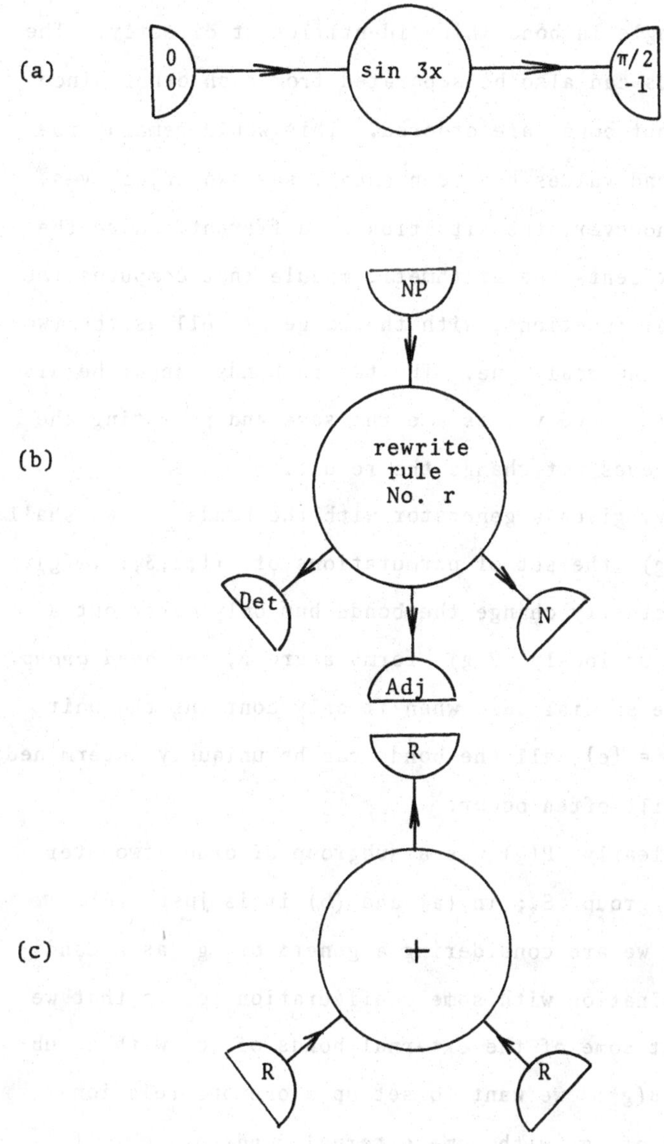

Figure 1.2

bonds is immediately answered: one is in and the other is out so that they are immediately identified.

In (b) the generator is a CF rewriting rule that rewrites NP = "noun phrase" into the string of syntactic variables, Det = "determines", Adj = "adjective", N = "noun". There is a single in-bond which identifies it directly. The three out-bonds can also be separated from each other since in this case out-bonds are ordered. This would remain true even if two bond values had been equal, say two adjectives.

In (c), however, the situation is different. Here the generator represents the arithmetic module that computes the sum of two real functions, with the range as well as the two domains being the real line. The two in-bonds cannot be distinguished: the bond values are the same and permuting the two arguments does not change the result.

Generally, given a generator with the bonds B we shall denote by P(g) the set of permutations of $\{1,2,3,\ldots\omega(g)\}$ that do not actually change the bonds but only carry out a relabelling. Obviously P(g) forms a group, the *bond group* of g. In the special case when it only contains the unit element, P(g) = {e}, all the bonds can be uniquely determined, a case that will often occur.

In (c) clearly P(g) is a subgroup of order two over the symmetric group S_3; in (a) and (b) it is just {e}.

Say that we are considering a generator g as a candidate for combination with some configuration c so that we try to connect some of the external bonds of c with a subset B' of B(g): we want to set up a one-one relation between bonds of g with some external bonds of c. If

$P(g) \neq \{e\}$ the labelling of the former is ambiguous which must be kept in mind when the connector is defined in terms of bond coordinates. We have just expressed the familiar principle that the choice of coordinate system should be irrelevant.

The labelling of the bonds is unique modulo $P(g)$. For two generators, g_1 and g_2 of the same arity, suppose we can establish a bijective mapping between their bonds so that the corresponding bonds (bond structure and bond values) are the same. Then g_1 and g_2 are *isomorphic as far as bonds are concerned*, or symbolically $B(g_1) \cong B(g_2)$, but they can of course differ drastically for other attributes.

As far as local regularity is concerned what determines whether a generator can, or cannot, be connected to a configuration is entirely determined by $B(g)$. This makes it natural to introduce:

Definition 2. *Two generators* g_1 *and* g_2 *are said to be locally congruent if* $B(g_1) \cong B(g_2)$; *they are isomorphic to each other as far as bonds are concerned.*

Clearly this relation is an equivalence so that it partitions G into classes of mutually (locally) congruent generators. It should be remarked, however, that this notion is not identical with congruence of configurations as used in Volume I: global regularity has been left aside for the moment.

In this connection let us remind the reader that all similarities preserve bond structure and generator index and map $G \to G$ bijectively. Consider the set of all bijective maps $G \to G$ leaving bonds (both structure and values) and generator index invariant. It is easily seen that they form a group: the *congruence similarities*. Any regular configuration remains regular when jittered by congruence similarities, see Volume I, p. 372, which is usually not true for general similarities.

Let us continue with two more examples:

Example 12. Any generator has four bonds with bond coordinates, e.g. E and N for out-bonds and W and S for in-bonds. The corresponding bond values are identical with the bond coordinates.

Since all four bond values are distinct we have $P(g) = \{e\}$ so that bonds can be distinguished from each other. With

$$\beta\rho\beta' = \begin{cases} \text{TRUE for } \beta = E \text{ and } \beta' = W, \text{ or } \beta = N \text{ and } \beta' = S \\ \text{FALSE else} \end{cases} \tag{1.6}$$

these generators can be used to build configurations topologically equivalent to subsets of the square lattice in the plane if we include an additional global regularity condition to be introduced later. The point of this example is that, compared to Example 5, in the latter examples the bond values are more informative: they contain information about location. Hence those generators can be interpreted to mean unit squares with given location, while in Example 12 they mean just unit squares. Such innocuous looking distinctions will have to be observed in many instances.

Example 13. Generators are closed half-planes in $\mathbb{R}^2$ with α = direction of the boundary. With $\omega(g) = +\infty$ and all bond values of g equal to α and with Σ = FULL the bond relation ρ = UNEQUAL we get descriptions of convex sets. The purpose of this regularity $\mathcal{R}$ = <FULL,UNEQUAL> is to avoid unnecessary generators in the description (see Case 9.1 in Volume I, Chapter 2) and the same $\mathcal{R}$ appears naturally in many other contexts for the same reason. To deemphasize simplicity of descriptions we can instead choose the wider regularity $\mathcal{R}$ = FREE.

We have mentioned Example 13 here, since it shows how to use local regularity (e.g. ρ = UNEQUAL) where we could instead have used global regularity by letting Σ require that only generators with different indices can be connected into regular configurations. The latter is possible since Σ is allowed to use information about α-values. Hence we sometimes have a choice of using local or global regularity to achieve a certain purpose.

More generally, to prepare for a detailed discussion of global regularity, consider a given generator g. Now let us disregard local conditions, so that bond values play no role momentarily. All the global information about g is then contained in the bond structure $B_s(g)$ and the generator index $\alpha(g)$. This information, symbolically written as the *generator skeleton*

$$\not{g} = \gamma = [B_s(g),\alpha(g)] \tag{1.7}$$

represents the *global connectivity* of g. All the values of the form (1.7) make up a set $\not{G} = \Gamma$ when g varies over G.

The global connectivity abstracts from all attributes but bond structure and index. Since we have blackened out all other information we shall diagrammatically show a $\not{g}$ as a major node, blackened circle (with α-value) together with bonds with coordinates. The global connectivity of the

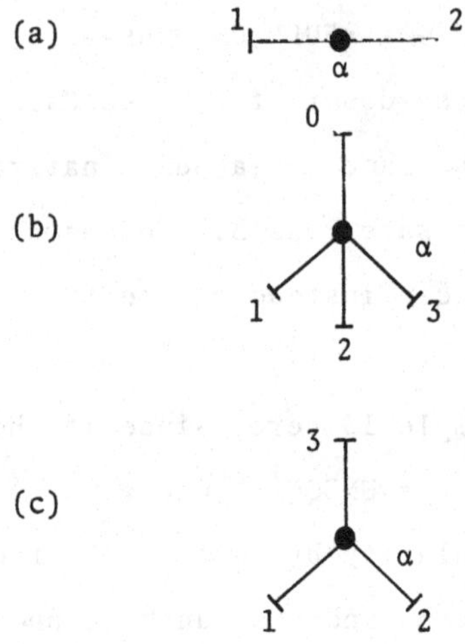

Figure 1.3

generators in Figure 1.2 would then appear as in Figure 1.3.

In this figure the choice of bond coordinates differs. In (a) and (b) bonds should be identifiable as described above, but the bond coordinates 1,2 can be permuted over $P(g)$. Classes of such *skeletons of generators* will be the starting point for discussing global regularity - connection types in Sections 2 and 3.

Just as the bonds of a generator are not always uniquely determined by bond structure and bond values, the bonds in a skeleton may be indistinguishable as the in-bonds in (c). We represent the group of permutations that do not actually change the bonds except for a relabelling by $\pi(\gamma)$. Note that $P(g)$ is a subgroup of $\pi(\gamma)$ for $\gamma = \not{g}$.

Given two families G_1 and G_2 of generators we can of course form the usual set operations $G_1 \cup G_2$, $G_1 \cap G_2$ etc. In addition to those we shall sometimes have occasion to apply other operations. One of these, the *product of generator families*, $G_1 \times G_2$; is formed as the set of all ordered pairs $g = (g_1, g_2)$, $g_1 \in G_1$, $g_2 \in G_2$ with the bond structure as the union

$$B_s(g) = B_s(g_1) \cup B_s(g_2) \tag{1.8}$$

and the corresponding bond values as the union

$$B_v(g) = B_v(g_1) \cup B_v(g_2) \tag{1.9}$$

with the natural relation between the new bond coordinates and the new bond values. Also $\alpha(g) = (\alpha(g_1), \alpha(g_2))$.

3.2. Configuration coordinates

We have used the analogy elsewhere that configurations play the same role in pattern theory as do the *formulae* in mathematical analysis, while images should be compared to *functions*. From this perspective configurations are *syntactic constructs* - although not limited to syntax in the formal languages - while images are *semantic concepts*. In the next section we shall see that this is not only an illuminating analogy but that images do indeed represent functions and hence meaning.

Consider a configuration skeleton γ resulting from the combination of the γ_k's. Introduce *configuration coordinates* (k,j) where k enumerates the generator skeletons and $j = 1,2,\ldots\omega(\gamma_k) = n_k$ are the generator coordinates of γ_k. A coordinate transformation for the configuration skeleton of the form $(k,j) \to (k',j')$ permutes the k-values: $(1,2,\ldots n) \to (i_1, i_2, \ldots i_n)$ as well as the generator coordinates $(1,2,\ldots n_k) \to (j_1, j_2, \ldots j_{n_k})$. Just as in the previous section (dealing with single generators) it can happen here that the relabelled configuration skeleton γ' coincides with the original γ, so that $\gamma_k = \gamma_k$, and so that the bond structure of the ℓth bond of γ_k equals the bond structure of the j_ℓth bond of γ_{i_k}.

The coordinate transformations for which this is true form a group, which will be denoted by $\pi(₵)$, where the symbol ₵ stands for the configuration skeleton. It plays the same role as $\pi(\gamma)$ for individual generator skeletons.

This way of labelling bonds is well suited for the general case. For particular configuration spaces it will sometimes be more convenient to label bonds not by a pair (k,j) but by specially designed coordinates; this will be illustrated in the examples that follow.

To be able to analyze global regularity we have systematically employed the notion of connection type Σ, see Volume I, Section 2.1. To fix ideas, say that the generators to be connected have skeletons (global connectivity) from a given set $₲ = \Gamma$. With an arbitrary collection of skeletons $\gamma_1, \gamma_2, \ldots \gamma_k \ldots \gamma_n$, some of which may coincide, form any combination by connecting bonds with no restriction except, of course, that <u>out</u> $\rightarrow$ <u>in</u> direction is observed if we deal with directed regularity. For symmetric regularity not even this restriction applies.

To simplify the discussion we have assumed that the generators have finite arity and that the combinations are of finite cardinality. This is no essential restriction; otherwise we would well-order generators and bonds and go ahead in the same way as below.

The set of all configuration skeletons that can be obtained arbitrarily, such as combinations, were only asked not to violate bond structure constraints, and will be denoted by ALL(Γ). In the important case of directed regularity this means that we can connect the bonds of the generator skeletons in any way we wish except that in-bonds can connect only to out-bonds.

Note that for ALL(Γ) we do not take bond values into account, so that the bond relation ρ need not be defined, nor do we ask for global regularity, so that the connection type Σ need not be defined.

The skeleton of the combination can be represented as bond structure + connections + generator indices

$$\not{c} = [B_s(c),\sigma(c),\alpha(c)] \tag{2.1}$$

where the bond structure is $B_s(c)$ and $\sigma(c)$ describes how the bonds, appropriately labelled, are connected together, and $\alpha(c)$ is an n-vector of indices of the generators appropriately labelled. Just as for single generators (and their skeletons) we express this information in terms of a coordinate system, and again the coordinates can not be uniquely determined if $\pi(\not{c}) \neq \{e\}$.

The connection type ALL(Γ) is usually too wide to describe the global regularities that we encounter and we will have to narrow it down to some subset generically denoted by Σ, or if we wish to emphasize the skeleton family, by $\Sigma(\Gamma)$. Let us give a few examples illustrated by the diagrams in Figure 2.1 where small slashes indicate bonds. For simplicity we assume the same α-value everywhere; otherwise these values should have been entered in the figure.

In (a) we have Σ = LINEAR and Γ is as in Figure 1.3(a). Because of the linear ordering a natural coordinate system is

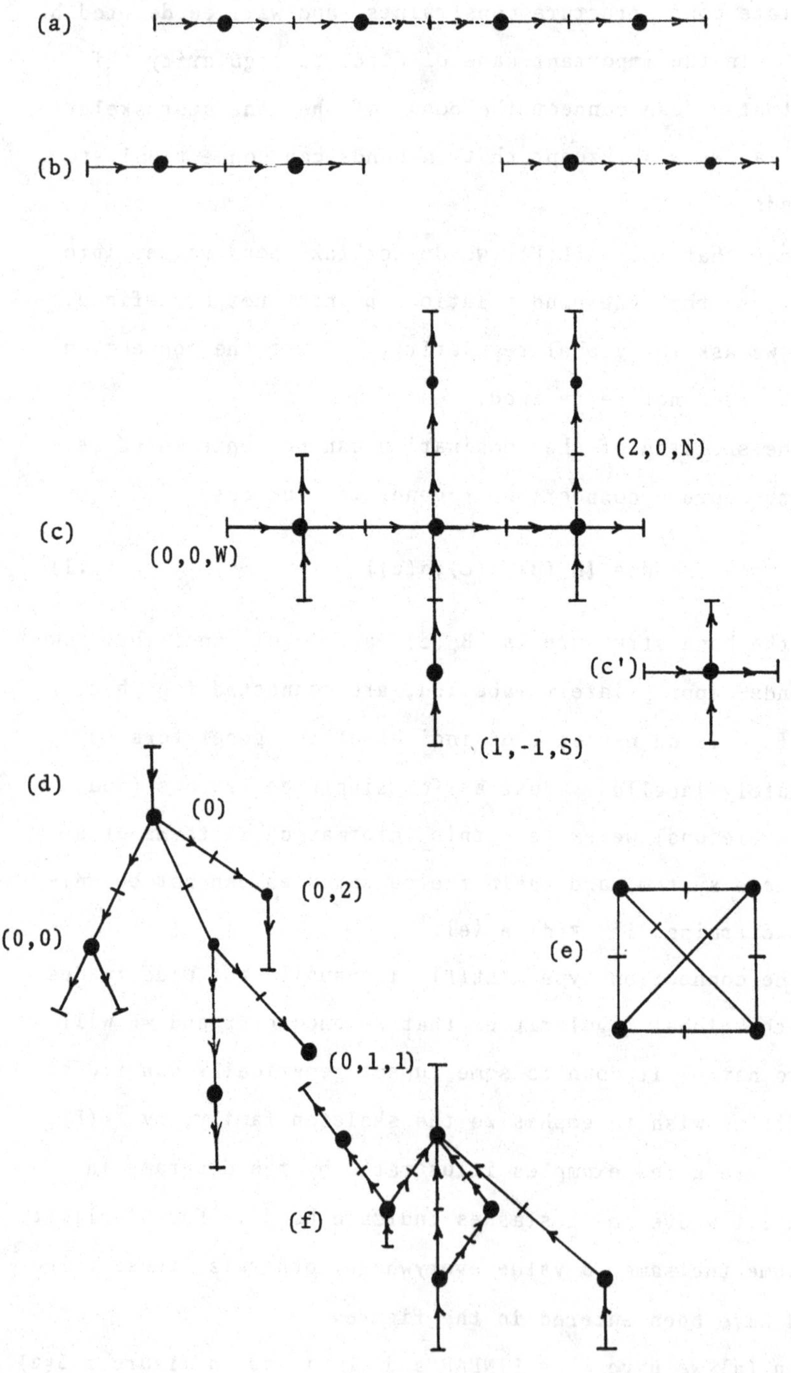

Figure 2.1

obtained by enumerating the generators by a subscript $i = 1,2,\ldots$ starting from the "left" and the bonds by (i,in) and (i,out) in the obvious manner.

Note that local conditions are not enough. We exclude for example a connection from the "right" - most bond in the figure to the "left" - most one.

If we do not insist on connected configurations so that unconnected components are allowed as in (b), then we can use coordinates as in (a) for each sub-component plus a marker for each sub-components. These markers are not distinguishable in the figure: $\pi(\not{c}) \neq \{e\}$. They would have been distinguishable for example if the sub-components had contained unequal numbers of generators (black dots). This connection type will be denoted $\mathcal{MIX}$(LINEAR) for reasons that will be clear later on.

We remind the reader that a connection type Σ is called monotonic, if for any skeleton it also contains any sub-skeleton obtained by deleting any of its generator nodes (together with their bonds) and opening any closed bonds arbitrarily. Obviously the connection type $\mathcal{MIX}$(LINEAR) is monotonic.

In (c) we have Σ = SQUARE LATTICE and $\not{c}$ is topologically equivalent to a subset of the plane lattice of pairs of integers; Γ is given as in (c'). It is important to realize that local conditions are not enough to describe this Σ: we do not allow any loops for example. One possible coordinate system could be constructed as follows. Consider all the "W"-most generator nodes and select the "S"-most of these. Give it the coordinates $(0,0)$ and continue by adding one to the first or second coordinate as we go along. The bond nodes are given coordinates as (0,0,W) or (2,0,N) as indicated.

Again if we allow sub-components not connected together, a connection type to be denoted by $\mathcal{MIX}$(SQUARE LATTICE), we also need a marker to denote the connected sub-components.

In (d) we illustrate Σ = TREE with Γ as in Figure 1.3(b). A reasonable coordinate system could be to assign the "upper-most" generator node by (0), the ones in the next layer by (0,0), (0,1), etc. and so on as indicated. Bonds could be enumerated by the coordinate of the major node from which it originates together with the coordinate of the bond in the single skeleton to which it belongs.

For Σ = FOREST = $\mathcal{MIX}$(TREE) the obvious modifications should be made.

In (e) Σ = FULL, so that all γ have infinite arity, we cannot show all the bonds. Here we deal with symmetric regularity. Unless we have distinct α-values associated with the major nodes they cannot be distinguished. Once they have been marked, however, internal bonds are determined by the two generator nodes involved for each bond.

In (f) we illustrate Σ = POSET with Γ as in Figure 1.3(c). Here we would associate the major nodes by values λ from some partially ordered set consistent with the connectivity of the graph. Once the major nodes have been determined (uniquely or not) the bonds can be marked by the two major nodes involved together with the coordinates of the single skeletons involved.

An extreme case is Σ = DISCRETE where only isolated generators are allowed in the regular configurations. It is in a sense the opposite extreme to Σ = FULL.

We are now ready to consider the *total regularity* $\mathcal{R} = <\Sigma,\rho>$ by introducing a bond relation ρ as in Volume I, Section 2.1 and this leads us to the space $\mathcal{C}(\mathcal{R})$ of regular configurations. To describe them we also need the bond values. The bonds of a configuration $c \in \mathcal{C}(\mathcal{R})$ can then be described by

$$B(c) = [B_s(c),\sigma(c),B_v(c)] \tag{2.2}$$

where $B_s(c)$ identifies bonds, with their structural parameters, in relation to their generators, $\sigma(c)$ describes their connections, and $B_v(c)$ gives the bond values associated with the bonds. To abstract from the connectivity we use the content of c

$$\text{content}(c) = \{g_1,g_2,\ldots g_n\} \tag{2.3}$$

where it should be remarked that some of the g_i may coincide so that content(c) is a set + frequencies of occurrence, a multiset. The frequency counts can of course be left out if $\mathcal{R}$ demands that generators in a regular configuration be distinct. Otherwise set structure alone is not sufficient.

When it comes to connecting c to other configurations local regularity is concerned only with the external bonds of c that are not already connected to any others. The external bond structure and bond values are denoted by

$$B^e(c) = [B_s^e(c),B_v^e(c)] \tag{2.4}$$

but it should be remembered that global regularity usually imposes conditions on $\sigma(c)$.

For two configuration spaces $\mathscr{C}_1$ and $\mathscr{C}_2$ we shall say that they are *globally equal* if the corresponding $\mathbb{C}_1 = \mathbb{C}_2$ and $\Sigma_1 = \Sigma_2$, in other words they exhibit the same global regularity.

Given two configuration spaces $\mathscr{C}_i = \langle G_i, S_i, \Sigma_i, \rho_i \rangle$, $i = 1,2$, we can form the union and other set operations directly. By the product $\mathscr{C}_1 \times \mathscr{C}_2$ we shall understand the new configuration space $\langle G_1 \times G_2, S_1 \times S_2, \Sigma_1 \times \Sigma_2, \rho \rangle$ where $\Sigma_1 \times \Sigma_2$ connects G_1-bonds according to Σ_1 and G_2-bonds according to Σ_2. The new bond relation ρ reduces to ρ_1 on G_1-bonds and to ρ_2 on G_2-bonds.

As an example of a product let Σ_1 = LINEAR, Σ_2 = CYCLIC, $\omega(g) = 2$ in both Σ_1 and Σ_2. In Figure 2.2 the $G_1 \times G_2$ generators are indexed by (i,j), $\underline{i}$ for the G_1-generators, and $\underline{j}$ for the G_2-generators that are involved. The two bonds of a G_1-generator are indexed ℓ, r for (left,right), those of a G_2-generator (u,d) for (up,down). There are also variations of this configuration, with a more "twisted" diagram, but with the same product regularity.

A related construction is *composition of connection types*. Given $\mathscr{C} = \langle G, S, \Sigma, \rho \rangle$ treat all its regular configurations as macro-generators, $G_1 = \mathscr{C}$, with bond structure and bond values given by $B^e(c)$. For a connection type Σ_1, defined in terms of $B^e_s(c)$ construct the composition by letting the global regularity $\Sigma_1(\Sigma)$ consist of all Σ_1-connections of $\mathscr{C}$-regular configurations $\mathscr{C}_1 = \langle \mathscr{C}, S, \Sigma_1, \rho \rangle$.

Regularity, as we see it, comes in two forms: local and global. It can happen that one or both are trivial. The configuration space $\mathscr{C}(\mathscr{R})$ is *locally free* if ρ is

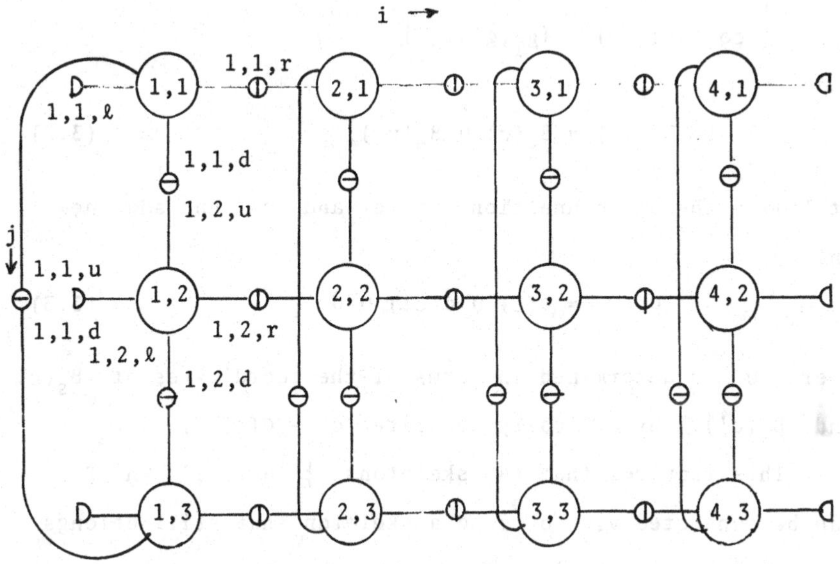

Figure 2.2

TRUE, and it is *globally free* if Σ is ALL. If $\mathscr{C}(\mathscr{R})$ is both locally and globally free we express this by writing $\mathscr{R}$ = FREE, with a change of terminology from Volume I.

3.3. Connectors

Regular configurations are combined into other regular configurations by connectors. In the previous volumes we have encountered many connectors operating in concrete configuration spaces.

In the abstract a connector σ, let us say with two arguments c and c' for simplicity, maps a set $C \subseteq \mathscr{C}(\mathscr{R}) \times \mathscr{C}(\mathscr{R})$ into $\mathscr{C}(\mathscr{R})$ in such a way that

$$\begin{cases} \text{content}(c'') = \{g_1,g_2,\ldots g_1',g_2',\ldots\} \\ \text{content}(c) = \{g_1,g_2,\ldots\} \\ \text{content}(c') = \{g_1',g_2',\ldots\} \end{cases} \tag{3.1}$$

and

$$B_s(c'') = B_s(c) \cup B_s(c'). \tag{3.2}$$

It leaves the old connections in c and c' and adds new ones

$$\sigma(c'') = \sigma(c) \cup \sigma(c') \cup \sigma \tag{3.3}$$

where σ is determined in terms of the coordinates of $B_s(c)$ and $B_s(c')$. Symbolically we write $c'' = \sigma(c,c')$.

This requires that the skeletons $¢$ and $¢'$ in Σ can be connected via σ into a skeleton that still belongs to Σ, global regularity of c'' and that the newly established bonds satisfy ρ, local regularity. Usually this is possible only for proper subsets C of $\mathscr{L}(\mathscr{R}) \times \mathscr{L}(\mathscr{R})$. Therefore $\sigma(c,c')$ is only a *partial function* on $\mathscr{L}(\mathscr{R}) \times \mathscr{L}(\mathscr{R})$.

Although the main role of connectors is on configuration spaces (and later on image algebras) it is natural to introduce them as *operations on configuration skeletons*.

It should be recalled that the skeletons carry information about the generator indices. Therefore connectors can be restricted by the values of the generator indices that belong to the nodes to be connected in the skeletons.

Given c and c', both in $\mathscr{L}(\mathscr{R})$, and with $(c,c') \in C$, the connector σ calculates what open bonds of c and c' should be connected. The specification of σ can be made in the configuration coordinates used in $\mathscr{L}(\mathscr{R})$, but if this is done we must remember that admissible coordinate transformations should not affect the result of σ. Hence, if the labelling of generators and bonds of c is changed by an element of the group $P(c)$, and similarly for c', then the calculated value of $\sigma(c,c') = c''$ should be the same, just with relabelled generators and bonds with an element from the group c''.

This just expresses that *the choice of coordinate system should be immaterial*. In concrete cases this is usually easy to verify directly, but in abstract situations some care is needed.

We mention a special connector that in spite of its trivial role will appear sometimes. This connector takes c_1, c_2 and combines them into a single configuration c without closing any bonds. It will be denoted as $c = \phi(c_1,c_2)$, since this connector changes no bonds at all, it is empty.

Let us mention a few examples.

Example 1. With G as in Example 1.4 and Σ = LINEAR, ρ = EQUAL we define $\sigma(c_1,c_2)$ as c_1 concatenated to the left of c_2. Global regularity always allows this but the local condition requires that the external out-bond value of c_1 coincides with the external in-bond value of c_2.

Similarly we can define another connector that concatenates c_1 to the right of c_2.

Example 2. Take G as in Example 1.9 except that $\Sigma = \mathcal{MJQ}$(LINEAR). We then have several ways of defining a left concatenation of c_1 with c_2, one of which goes as follows. The new configuration consists of the union of all the connected sub-components of c_1 and c_2 respectively, except that we have selected one from c_1 and one from c_2 and connected them with c_1 to the left if possible locally. The selections are computed by a rule belonging to the definition of σ, the result must as usual be unique modulo $P(c_1)$ and $P(c_2)$.

Example 3. With G as in Example 1.5 the connection type associated with topological equivalence to subsets of $\mathbb{N}^2$ and ρ as in (2.6). In the definition of σ an algorithm computes one generator in c_1 and one of its bonds, and similarly for c_2, and connects these two bonds if possible. Of course this can be impossible in several ways, for example if one of the bonds computed is already connected.

Another more complicated connector would compute several bonds for c_1 and for c_2 and connect them in a given manner.

Still another one takes c_1 and with $c_2 = \phi$, σ connects unconnected bonds of c_1 without offending against Σ or ρ.

We also have the trivial relation $\sigma(c,\phi) = c$ for any regular c.

Note that more generally when $c_2 = \phi$ we actually operate only on the bond structure of c_1 alone so that we write with some abuse of notation, $\sigma(c_1,c_2) = \sigma(c_1)$.

Example 4. With G as in Example 1.10, Σ = TREE, and ρ = EQUAL define $\sigma(c_1,c_2)$ as the tree obtained by connecting the in-bond of c_1 with the left-most out-bond of c_2 with the same bond value if there is any bond carrying the same value. See Notes C.

Example 5. If $\mathscr{R}$ = DISCRETE, then connectivity can be disregarded entirely so that configurations are just sets (with frequency counts) so that any σ only operates on contents so that if content(c_1) = $\{g_1,g_2,\ldots\}$, content(c_2) = $\{g_1',g_2',\ldots\}$ then $\sigma(c_1,c_2) = \{g_1,g_2,\ldots g_1',g_2'\ldots\}$.

In this way the connectors define binary partial functions with their arguments taking values in $\mathscr{L}(\mathscr{R})$ and producing values in $\mathscr{L}(\mathscr{R})$. The configuration space can therefore be regarded as the base set for a *partial universal algebra*, see Grätzer (1968), Chapter 2. If we also consider the identification rule R, which will be discussed in more detail in Section 8, we have an *algebraic system* in the sense of Mal'cev (1973), where R plays the role of a congruence relation since *all the basic operations are stable under* R. Indeed, for any $s \in S$ we have

$$c_1 \equiv c_2 (\text{mod } R) \Rightarrow sc_1 \equiv sc_2 (\text{mod } R). \tag{3.4}$$

Similarly for any connector σ we have if $c_1, c_2, \sigma(c_1,\sigma_2)$, c_1', c_2', $\sigma(c_1',c_2')$ all belong to $\mathscr{L}(\mathscr{R})$ and

$$\left\{\begin{array}{l} c_1 \equiv c_1' (\text{mod } R) \\ c_2 \equiv c_2' (\text{mod } R) \end{array}\right. \Rightarrow \sigma(c_1,c_2) \equiv \sigma(c_1',c_2') (\text{mod } R) \tag{3.5}$$

see Volume I, Section 3.1. In this way *we can consider the image algebra* $\mathscr{T}$ as the quotient algebra modulo R: $\mathscr{T} = \mathscr{L}(\mathscr{R})/R$; see Grätzer (1968), p. 35.

We can then go ahead and form *polynomials* over $\mathscr{L}(\mathscr{R})$, for example

$$sc,\ \sigma(c_1,c_2),\ \sigma_2[s_1\sigma_1(c_1,c_2),s_2c_3],\ldots \tag{3.6}$$

but it should be remarked that such polynomials are usually

not entire functions since the configurations involved do not always fit under $\mathcal{R}$ via the connectors $\sigma, \sigma_1, \sigma_2, \ldots$. The order of operations in the polynomial must be completely specified, for example using parentheses. In particular, to each group element $s \in S$ we have a unary entire function $s\colon \mathcal{C}(\mathcal{R}) \to \mathcal{C}(\mathcal{R})$ defined by $s\colon c \to sc$. For any polynomial given by repeated use of similarities and connectors its natural domain is the set of arguments for which the successive subexpressions in its expression are all regular.

3.4. Configuration homomorphisms

The homomorphisms defined in Volume I, Chapter 2, are sufficient for many situations, namely when the two configuration spaces have the same global regularity. Otherwise we need a more general definition.

To give this a general algebraic motivation let us consider a group X with the group operator written multiplicatively as xy, x and y in X. Let h be a homomorphism to another group X' so that (see Notes A)

$$h(x \cdot y) = h(x) \cdot h(y), \quad \forall x,y \in X. \tag{4.1}$$

Similarly if X and X' are rings with operators written as addition and multiplication let $h\colon X \to X'$ be a homomorphism so that

$$\begin{cases} h(x+y) = h(x) + h(y) \\ h(x \cdot y) = h(x) \cdot h(y),\ h(1) = h(1') \end{cases} \tag{4.2}$$

for all $x,y \in X$.

Now the terms "addition" and "multiplication" indicate general operations and one could ask what happens if we let the two binary operations in X' change roles with each other (but not in X) so that

$$\begin{cases} h(x+y) = h(x)\cdot h(y) \\ h(x\cdot y) = h(x)+h(y) \end{cases} \tag{4.3}$$

This does not make sense, however, since the two operations in a ring have to satisfy the distributive law

$$x(y+z) = (x\cdot y) + (x\cdot z) \tag{4.4}$$

and this makes them play different roles; they do not appear on equal footing. Hence the definition in (4.3) is not reasonable and is ruled out.

In a general configuration space with several binary (partial) operations $\sigma_1, \sigma_2, \ldots$ the situation is different. Here we do not always have any requirement like (4.4) relating the different σ's to each other. Of course conditional distributivity must hold, as will be discussed later, but this relates similarities to one of the σ's, not the σ's to each other. Therefore it can be natural to let a σ in one configuration space be related to a σ' in another configuration space in more than one way.

Consider a mapping h defined on $\mathscr{G} \cup S \cup \{\sigma\}$, where $\{\sigma\}$ stands for the set of connectors that we have decided to use. The values of h on S shall be in S', those on $\mathscr{G}$ shall be in $\mathscr{G}'$, and those on $\{\sigma\}$ shall be in $\{\sigma'\}$. The map restricted to the group S shall be homomorphic in the usual sense of the term so that

$$h(s_1 s_2) = h(s_1)h(s_2). \tag{4.5}$$

<u>Definition 1</u>. *Such a mapping is said to be a homomorphism if*

(i) $h(sc) = h(s)h(c)$, *for any* s *and regular* c

(ii) $h[\sigma(c_1,c_2)] = [h(\sigma)][h(c_1),h(c_2)]$ *for any regular* c_1,c_2, $\sigma(c_1,c_2)$.

<u>Remark</u>. In the special but important case where $\mathscr{C} = \mathscr{C}(\mathscr{R})$ and $\mathscr{C}' = \mathscr{C}(\mathscr{R}')$ have the same connection type Σ over the same generator skeleton Γ we shall always assume that $\{\sigma\} = \{\sigma'\}$ and $h(\sigma) = \sigma$, the connector is left unchanged, unless the opposite is stated explicitly.

Let us consider some examples of homomorphisms, starting with equal global regularity; different global regularity will be discussed later.

<u>Example 1</u>. Let μ be a covariant generator map $G \to G'$, so that μ permutes with similarities $\mu sg = s\mu g$, $S = S'$, so that $\alpha(\mu g) = \alpha(g)$ and such that μg has the same bonds (structure and values) as g. Typically this can occur when μg is a chosen prototype for g.

Define hc, $c = \sigma(g_1,g_2,\ldots g_n) \in \mathscr{C}(\mathscr{R})$ as $c' = \sigma(\mu g_1,\mu g_2,\ldots \mu g_n)$. Then c' is locally regular since bond values are preserved and globally regular since c' has the same bond structure as c and all generator indices are preserved by μ. Hence $c' \in \mathscr{C}(\mathscr{R}')$.

Define $hs = s$, $S = S'$, and $h\sigma = \sigma$. To show that h is a homomorphism it is enough to observe that

$$\begin{aligned} h(sc) &= \sigma(\mu sg_1,\mu sg_2,\ldots,\mu sg_n) \\ &= \sigma(s\mu g_1,s\mu g_2,\ldots,s\mu g_n) = s\sigma(\mu g_1,\mu g_2,\ldots,\mu g_n) \\ &= sh(c) \end{aligned} \tag{4.6}$$

so that (i) in (5) holds, and that $\sigma(c_1,c_2)$ is regular, $h[\sigma(c_1,c_2)] = \sigma[h(c_1),h(c_2)]$, since h just replaced generators.

This homomorphism, *substituting prototypes*, can be exemplified by finite state languages where branches $i \overset{x}{\rightarrow} j$ are replaced by prototypes $i \overset{\xi}{\rightarrow} j$.

Remark 1. Covariant generator maps can appear also when $S \neq S'$. We will then assume that the map $\mu:G \rightarrow G'$, $\Gamma = \Gamma'$, preserves bond structure as before but now it is also defined as a homomorphism $\mu:S \rightarrow S'$ such that $\mu(sg) = \mu(s)\mu(g)$.

Example 2. Let $\mathscr{R}$ = DISCRETE and G_o an S-invariant subset of G, $sG_o = G_o$, $\forall s$. Then the only σ is the empty connector ϕ, so that we need only define $h\phi = \phi$. Also put $hs = s$, $S = S'$. Configurations can be identified with their contents and we define hc by

$$\text{content}(hc) = \text{all } g_i\text{'s in } c \text{ except the ones belonging to } G_o. \tag{4.7}$$

We have with c containing the generators $g_1,g_2,\ldots g_n$

$$\begin{aligned} h(sc) &= \text{content}\{sg_\nu \mid sg_\nu \notin G_o\} \\ &= s\ \text{content}\{g_\nu \mid g_\nu \notin s^{-1}G_o\} = s\ \text{content}\{g_\nu \mid g_\nu \notin G_o\} \\ &= sh(c) \end{aligned} \tag{4.8}$$

so that (i) holds. Also

$$h[\sigma(c_1,c_2)] = \text{content }\{\text{all } g\text{'s in } c_\nu \text{ not in } G_o, \text{ all } g\text{'s in } c_2 \text{ not in } G_o\} = \sigma[h(c_1),h(c_2)] \tag{4.9}$$

so that (ii) is true.

This homomorphism, *elimination of generators*, occurs naturally for "descriptions" and makes them cruder.

Example 3. Still with $\mathscr{R}$ = DISCRETE, and $S = S'$, $hs = s$, let the generators form sets in a background space X and let G be closed under intersection. Let g_0 be S-invariant, $s'g_0 = g_0$, $\forall s \in S$ and define hc by its content

$$\{g_1 \cap g_0, g_2 \cap g_0, \ldots, g_n \cap g_0\} \tag{4.10}$$

is content(c) = $\{g_1, g_2, \ldots, g_n\}$. It is easily verified that the conditions in Definition 1 hold.

Example 3 can directly be extended to other set operations involving more than one fixed set g_0 but this will not be done here.

Example 4. Let Σ = LINEAR with G as in Example 2.4. Choosing G' as in Example 2.2 and with $S = S'$ as the translation group with integral translations over $\mathbb{R}$ with $hs = s$, define the generator map μ as the restriction of $g(x)$ from the interval $[a,b]$ to the "discrete interval" with the same end points. Applying μ to each generator in a regular configuration gives us a homomorphism h. Note that $h\sigma = \sigma$.

Indeed, $\rho' = \rho$ and with $\Sigma' = \Sigma$ = LINEAR it is clear that h maps $\mathscr{C}$ configurations into $\mathscr{C}'$ configurations: bond structure is preserved, $B_s^e(c)$ = (in,out) = $B_s^e(hc)$, and since local regularity is preserved we see that h maps $\mathscr{C} \to \mathscr{C}'$.

Applying a similarity means translating the support of g by an integer so that (i) holds. The relation in (ii) is obvious: both supports are reduced in the same way.

This is easily extended to more general situations of which we give just two.

Example 5. Let $\Sigma = \Sigma'$ be SQUARE LATTICE, ρ = EQUAL, and G as in Example 2. S = S' should be the translation group over $\mathbb{Z}^2$. Define G' as the restrictions of all $g \in G$ to the four corner points, same skeleton, but now with the bond values as 2-vectors $(g(a+1,b), g(a+1,b+1))$ for the bond E, $(g(a,b), g(a+1,b))$ for the bond S, etc. If we define h through the generator map which restricts g to the corner points it can be shown, just as in the previous case, that h is a homomorphism.

Example 6. Same but with triangulations as in Example 2.8. The homomorphism restricts to the corners of the triangles.

Example 7. Same $\mathscr{L}$ as in the last example. Let g in G' consist of linear functions, and define the generator map $g \to g'$ by letting g' be the function in G' coinciding with g at the three corners of the triangle forming the support of g.

This homomorphism representing *approximation by linear spline functions* can also be given some interesting variations, but this will be left to the reader's imagination, except for the following, rather special one.

Example 8. Let generators be of contrast type, $g: X \to Y$ as in Volume I, p. 16, and let $\{Y_\nu\}$ be a partition of the contrast (not the background space!) space Y with an element y_ν singled out in each subset Y_ν. The generators shall have $\omega(g) = +\infty$, all the bond values given by the support

β of g, and with the bond relation $\rho = \rho' =$ DISJOINT and $\Sigma = \Sigma' =$ FULL. The similarities in $S = S'$ should represent bijective mappings $Y \to Y$ preserving the $\{Y_\nu\}$ partition together with the y_ν-values, so that a similarity s: $Y_\nu \to Y_{\nu'}$, $y_\nu \to y_{\nu'}$, where $\{\nu\} \to \{\nu'\}$ is a permutation of the integers. As before we let $hs = s$, $h\sigma = \sigma$.

Define the generator map $\mu: G \to G'$ by letting $\mu g = g'$ be equal to y_ν wherever g takes values in Y_ν, $\forall\nu$, and with the same bonds as g. A regular $\mathcal{L}$ configuration is a partial function on X and the regularity expresses the condition that its definition via several generators is logically consistent. If c is consistent it follows that $hc = c'$, obtained by applying μ to each of its generators, and $\mathcal{L}'$ is regular. Bond structure is not affected by h.

Then the resulting map $h: \mathcal{L} \to \mathcal{L}'$ is a homomorphism. Indeed, (i) in Definition 1 is true since $\mu(sg)$ represents a partial function on X with its value shifted according to the similarity $\underline{s}$ and then collapsed into the y_ν-values, while $s(\mu g)$ does the same thing in the opposite order but with the same result. The relation in (ii) is obvious.

This h can be viewed as *approximation by step functions*.

Example 9. Consider a background space X with a measure m and let the generators be real or complex valued functions on X belonging to $L_2(X;m)$. The similarities represent multiplication by non-zero real or complex scalars. With $\mathcal{R} = \mathcal{R}' =$ DISCRETE and L a subspace of $L_2(X;m)$ define the generator map $\mu g = g'$ by projecting g down to L, $g' = P_L g$. Here G' consist of the functions

$$G' = \{g' | g' = P_L g,\ g \in G\}. \tag{4.11}$$

Since free configurations can be identified by their content, and since P_L commutes with multiplication by scalars, it is easily verified as above that h, obtained by applying μ to each generator of the configuration, is a homomorphism.

Example 10. Consider generators as in Example 1.11 with all X_ν and Y_ν as subsets of the same space X, with $\Sigma = \Sigma'$ as POSET structure, ρ = INCLUSION, and $S = S' = \{e\}$; see Volume I, Section 2.7. Let X_o be a subset of X such that any $g \in G$, restricted to $(X_1 \cap X_o) \times (X_2 \cap X_o) \times \ldots$ takes values in $(Y_1 \cap X_o) \times (Y_2 \cap X_o) \times \ldots$

Let $\mu g = g'$ be the restriction of domain and range as above and with the same bond structure but the new in-bonds $\beta'_\nu = X_\nu \cap X_o$ and out-bonds $Y_\nu \cap X_o$. If c is the σ-connection of $g_1, g_2, \ldots$ and is $\mathscr{L}$-regular, then the σ-connection c' of $g'_1, g'_2, \ldots$ is $\mathscr{L}'$-regular since

$$Y_\nu \subseteq X_\mu \Rightarrow Y_\nu \cap X_o \subseteq X_\mu \cap X_o \tag{4.12}$$

so that the extended map $h: \mathscr{L} \to \mathscr{L}'$. It is easy to show that h is a homomorphism.

Example 11. Let the generators be measures on some S-invariant σ-algebra $\mathscr{A}$ over the background space X. Let the sub-σ-algebra $\mathscr{A}'_o \subseteq \mathscr{A}_o$ be S'-invariant where S' is a subgroup of S, let $\Sigma = \Sigma'$ = FULL, $\omega(g) = +\infty$, $\forall g \in G$, and with all bond values of g = support(g). With $\rho = \rho'$ = DISJOINT consider the generator map that takes g into the restriction to $\mathscr{A}_o$ and extending this generator-wise to any regular configuration, it follows easily that the extension h is a homomorphism.

Here S and S' should represent bijections $X \to X$ preserving $\mathscr{A}$ and $\mathscr{A}_0$ but we could also let them include multiplication by scalars.

These examples involving *covariant generator maps*, more generally meaning $\mu(sg) = \mu(s)\mu(g)$ lead up to the following result. Note that μ now also represents an associated map $S \to S'$.

Theorem 1. *Consider two configuration spaces $\mathscr{C}$ and $\mathscr{C}'$ with the same global regularity $\Gamma = \Gamma'$, $\Sigma = \Sigma'$. Let μ be a covariant generator map $G \to G'$, $S \to S'$, such that if the bonds b_1 and b_2 of g_1 and g_2 connected via $\sigma(c)$ have values satisfying $\beta_1 \rho \beta_2$ then μg_1 and μg_2 have (homologous) bonds b_1 and b_2 with values satisfying $\beta_1' \rho' \beta_2'$. Define the configuration map h by*

$$h\sigma(g_1,\ldots g_n) = \sigma(\mu g_1,\ldots \mu g_n) \tag{4.13}$$

Then h is a homomorphism.

Conversely, if Σ is monotonic, then any surjective homomorphism $h\colon \mathscr{C} \to \mathscr{C}'$ with $h\sigma = \sigma$ can be considered as the extension of a covariant generator map $\mu\colon G \to G'$ preserving bond relations as above.

Proof: To see that the direct part of the statement is true we first note that h maps $\mathscr{C}$-regular configurations into $\mathscr{C}'$-regular ones. Indeed, the bond structure of hc is the same as that of c so that hc belongs to the same connection type and global regularity holds in $\mathscr{C}'$. We also have local regularity since all internal bonds established by $\sigma(c)$ and satisfying ρ will still hold for the generators in G' relative to the bond relation ρ'.

Now let $c = \sigma(g_1,\ldots g_n) \in \mathscr{C}$ and apply h as in (4.13) to c. Now form hsc, also leading to a regular configuration as we already know, namely to

$$hsc = h\sigma(sg_1,\ldots sg_n) = \sigma[\mu(sg_1),\ldots\mu(sg_n)]. \quad (4.14)$$

But μ is covariant so that (4.14) can be written as

$$\sigma[\mu(s)\mu(g_1),\ldots\mu(s)\mu(g_n)] = \mu(s)\sigma[\mu(g_1),\ldots\mu(g_n)] \quad (4.15)$$

which is (i) in Definition 1 if we put $h(s) = \mu(s)$.

To prove (ii), observe that if $c = \sigma(c_1,c_2)$, $c_1 = \sigma_1(g_1,g_2,\ldots)$, $c_2 = \sigma_2(g_1',g_2',\ldots)$, all three configurations $\mathscr{C}$-regular, then h(c) will consist of the generators $\mu g_1,\mu g_2,\ldots\mu g_1',\mu g_2'\ldots$. These new generators have the same bond structure as the old ones and are connected in the same way. Hence

$$hc = \sigma[\sigma_1(\mu g_1,\mu g_2,\ldots),\sigma_2(\mu g_1',\mu g_2',\ldots)] = \sigma(hc_1,hc_2) \quad (4.16)$$

which proves (ii).

To verify the converse part note that since Σ is monotonic it is automatically monatomic so that h induces a map $G \to G'$ by $\mu: g \to h(\{g\})$: recall that $h\sigma = \sigma$ so that $h[\{g\}]$ must be monatomic and represent a single generator in G'. This new generator must have the same bond structure as the old one; see Notes B. Also we must have (i) in Definition 1 so that $\mu(sg) = \mu(s)\mu(g)$, where we have written μs for hs. Then, for arbitrary s_1,s_2 in S and g in G,

$$\mu(s_1 s_2)\mu(g) = \mu(s_1)\mu(s_2)\mu(g) \quad (4.17)$$

and since $\mu G = G'$ (recall that h was assumed to be surjective), it follows that $\mu:S \to S'$ is homomorphic. Hence μ is a covariant generator map.

To see that h is the extension of this map μ note that any $\mathcal{L}$-regular configuration c can be obtained by successive connections of regular configurations with single generators (this follows from the monotonic property of the connection type) so that

$$\begin{cases} c = \sigma_1(\{g_1\},c_2) \\ c_2 = \sigma_2(\{g_2\},c_3) \\ c_3 = \sigma_3(\{g_3\},c_4) \\ \cdots \end{cases} \tag{4.18}$$

Applying (ii) in the homomorphism definition repeatedly we get finally

$$hc = \sigma_1(\mu g_1,hc_2) = \ldots \sigma(\mu g_1,\mu g_2,\ldots) \tag{4.19}$$

and (4.13) follows easily. Q.E.D.

This shows that if Σ is monotonic and if we have the same global regularity configuration then homomorphisms are just extension of a simple type of mapping of single generators. An example of different type is

<u>Example 12</u>. Let $\Sigma = \Sigma' =$ LINEAR with an even number of nodes, $S = S' = \{e\}$, $G = G' = \{g^1,g^2\}$, $\rho = \rho' =$ TRUE, and $\omega_{in}(g) = \omega_{out}(g) = 1$ with $B_s = (in,out)$. Define hc, with $c = (g_{11},g_{12},g_{21},g_{22},\ldots,g_{r1},g_{r2})$, as $hc = (g'_{11},g'_{12},g'_{21},g'_{22},\ldots)$ where $hs = s$, $h\sigma = \sigma$, and

$$g'_{i1} = g'_{i2} = g_{i1}, \tag{4.20}$$

see Figure 4.1.

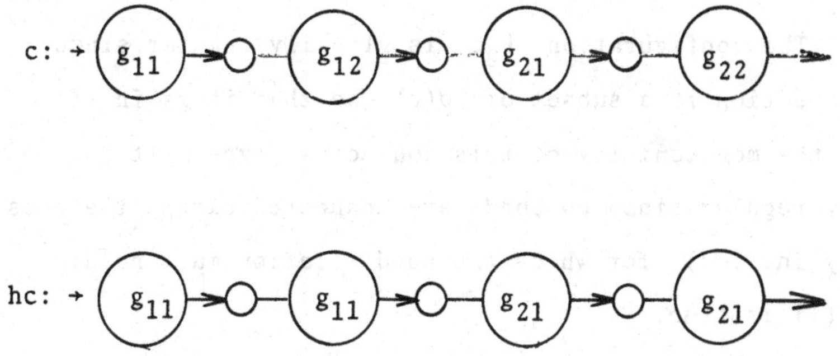

Figure 4.1

It is clear that h satisfies the conditions in Definition 1: it is a homomorphism. It is not an extension of a generator map since it involves "two generators at a time". Note that this Σ is not monotonic and that h is not surjective.

The notion of homomorphism can be used to extend the discussion in Grenander (1969), Sections 3 and 4, and we shall prove

<u>Theorem 2</u>. *Assuming* Σ *to be monotonic, let* B *be a subset of the generator index space* A, $B \subseteq A$. *For any* $c = \sigma(g_1, g_2, \ldots g_n)$ *define the projection* P_B *as the configuration with content*

$$\{g_i | \alpha(g_i) \in B\} \tag{4.21}$$

and connected with the subgraph of σ *connecting the* g_i*'s in* (4.21). *Then*

(i) P_B *maps* $\mathscr{L}(\mathscr{R})$ *into itself*

(ii) $P_{B_1} P_{B_2} = P_{B_1 \cap B_2}$

(iii) P_B *is a homomorphism.*

<u>Proof</u>: The configuration $P_B c$ is globally regular since its connection is a subset of $\sigma(c)$, so that it is in Σ due to the monotonicity of this connection type. It is locally regular since no bonds are connected except the ones already in $\sigma(c)$ for which the bond relation must hold. Hence (i) is true.

To see that (ii) holds it is clear that $P_{B_1} P_{B_2}$ is defined everywhere in $\mathscr{L}(\mathscr{R})$ as in $P_{B_1 \cap B_2}$. But both of them take a configuration c into one with content

$$\{g_i | \alpha(g_i) \in B_1 \cap B_2\}. \tag{4.22}$$

The two resulting configurations also have the same internal connections obtained by leaving out all the connections in $\sigma(c)$ for which the generators belong to any G^α with $\alpha \in B_1 \cap B_2$. Hence the two configurations are the same so that $P_{B_1} P_{B_2} = P_{B_1 \cap B_2}$.

Putting $B_1 = B_2 = B$ (ii) reduces to $P_B^2 = P_B$, so that P_B is idempotent.

Define the map $\underline{h}$ on $\mathscr{L} \cup S \cup \{\sigma\}$ as P_B on $\mathscr{L}$, as the identity maps on S, and let $h\sigma$ be the subset of connections in σ that do not involve any bonds with $\alpha \in B$ (see the discussion in Section 3). To know that $\underline{h}$ is a homomorphism we note that if c is regular with connections σ then sc has the same σ. Hence $sP_B c$ and $P_B sc$ have the same connections. Their content is also the same since the generator index in S-invariant. Therefore (i) holds in Definition 1.

To prove (ii) in Definition 1 it is clear that P_Bc has the same content as P_Bc_1 taken together with P_Bc_2; in both cases we delete all generators with generator index in the given set B. Also

$$P_B\sigma(c_1,c_2) = \sigma'(P_Bc_1,P_Bc_2) \tag{4.23}$$

where σ' is the subconnector of σ leaving out connections involving any generator g with $\alpha(g) \in B$. At the same time P_Bc_1 has lost those internal connections of c_1 that involve any g with $\alpha(g) \in B$, and similarly for P_Bc_2. Hence the $P_B\sigma(c_1,c_2)$ has the same content and connections as $\sigma(P_Bc_1,P_Bc_2)$ so that (ii) in the definition is satisfied (see Notes C) and P_B is indeed a homomorphism. Q.E.D.

A special case is when B contains a single element $\alpha, B = \{\alpha\}$. We then write simply P_α for the projection down to B.

3.5. Configuration categories

In this section the configuration homomorphisms shall be over configuration spaces with the same (fixed) global regularity, a connection type Σ over a (fixed) family of generator skeletons and with $h\sigma = \sigma$. Definition 4.1 allows for more general homomorphisms, but this will be left until later.

The following theorem sheds light on the algebraic structure of regular structures and will be exploited further later on.

Theorem 1. *For a fixed family* Γ *of generator skeletons and connection type* Σ *over* Γ *consider the set* $\underline{\underline{\text{Conf}}}(\Sigma)$

of all the corresponding configuration spaces together with the homomorphisms $\text{hom}(\mathcal{C}, \mathcal{C}')$. *Then* $\underline{\text{Conf}}(\Sigma)$ *forms a category.*

Proof: It is obvious that for any $\mathcal{C}$ over Σ the identity $\text{id}_{\mathcal{C}}$ is a homomorphism $\mathcal{C} \to \mathcal{C}$. Also with three configuration spaces $\mathcal{C}_1, \mathcal{C}_2, \mathcal{C}_3$ on Σ and with similarly groups S_1, S_2, S_3 consider two homomorphisms

$$f \in \text{hom}(\mathcal{C}_1, \mathcal{C}_2),\ g \in \text{hom}(\mathcal{C}_2, \mathcal{C}_3). \tag{5.1}$$

The mapping $h = gf: \mathcal{C}_1 \to \mathcal{C}_3$ is then also a homomorphism. First, h maps S_1 homomorphically into S_3. Second, for $c \in \mathcal{C}_2$, $s \in S_1$

$$\begin{aligned} h(sc) &= g[f(sc)] = g[f(s)f(c)] = g[f(s)](gf)c \\ &= h(s)h(c) \end{aligned} \tag{5.2}$$

so that (i) holds in Definition 4.1. Third, with c_1, c_2, $c = \sigma(c_1, c_2) \in \mathcal{C}_1$, then

$$fc = \sigma(fc_1, fc_2) \in \mathcal{C}_2, \tag{5.3}$$

so that

$$\begin{aligned} hc &= (gf)c = g(fc) = g\ (fc_1, fc_2) = \sigma[(gf)c,(gf)c_2] \\ &= \sigma(hc_1, hc_2) \end{aligned} \tag{5.4}$$

so that (ii) in the definition is satisfied. As for all concrete categories the composition operation is associative. Q.E.D.

This is a very general category of configuration spaces, for example $\underline{\text{Conf}}$(LINEAR), $\underline{\text{Conf}}$(TREE), $\underline{\text{Conf}}$(POSET), $\underline{\text{Conf}}$(SQUARE LATTICE), etc., but we shall often consider more restricted categories. This will be indicated by a qualifying adjective within the parentheses.

Let us now consider some functors on categories of configurations and start with a trivial one $\mathcal{F}$ denoted by $-\times \mathcal{C}_0 : \underline{\underline{\text{Conf}}}(\Sigma) \to \underline{\underline{\text{Conf}}}(\Sigma\times\Sigma_0)$. Here $\mathcal{C}_0$ is a fixed configuration space over the connection type Σ_0: $\mathcal{C}_0 = \langle G_0, S_0, \Sigma_0, \rho_0 \rangle$. To any object $\mathcal{C}$ in $\underline{\underline{\text{Conf}}}(\Sigma)$, $\mathcal{F}$ associates the configuration space $\mathcal{C}\times\mathcal{C}_0$. For two objects $\mathcal{C}_1, \mathcal{C}_2 \in \underline{\underline{\text{Conf}}}(\Sigma)$ and a homomorphism $h: \mathcal{C}_1 \to \mathcal{C}_2$, $\mathcal{F}$ associates the map $\mathcal{F}h = h \times id_{\mathcal{C}_0} : \mathcal{C}_1 \times \mathcal{C}_0 \to \mathcal{C}_2 \times \mathcal{C}_0$ and it is easily seen that these maps include the identity if $\mathcal{C}_1 = \mathcal{C}_2$ and that $\mathcal{F}(gh) = \mathcal{F}(g)\mathcal{F}(h)$ when applicable.

A more important functor is $\mathcal{MIX}(\Sigma)$ *which takes any* $\mathcal{C} = \langle G,S,\Sigma,\rho \rangle$ *into* $\langle G,S,\mathcal{MIX}\Sigma,\rho \rangle$, *where* $\mathcal{MIX}\Sigma$ *consists of all finite unconnected unions of connections from* Σ. For any two $\mathcal{C}_1$ and $\mathcal{C}_2$ over Σ and homomorphism $h: \mathcal{C}_1 \to \mathcal{C}_2$ the functor $\mathcal{MIX}$ associates the map $\mathcal{F}h$ which simply applies h to each unconnected configuration belonging to $\mathcal{C}_1$.

<u>Theorem 2</u>. *$\mathcal{MIX}$ is a functor.*

<u>Proof</u>: A general configuration belonging to $\mathcal{MIX}(\mathcal{C}_1)$ can be written as $c = \phi\{c_1, c_2, \ldots\}$ where each $c_i \in \mathcal{C}_1$ and no connections exist between different c_i. For fixed h we then get

$$(\mathcal{MIX}h)c = \{hc_1, hc_2, \ldots\} \tag{5.5}$$

again with no connections between the different hc_i. But (5.5) obviously belongs to $\mathcal{MIX}\mathcal{C}_2$ so that $\mathcal{MIX}h$ maps the object of the category $\underline{\underline{\text{Conf}}}(\Sigma)$ into objects of the category $\underline{\underline{\text{Conf}}}(\Sigma)$. Further, if $\mathcal{C}_1 = \mathcal{C}_2$ and $h = id_{\mathcal{C}_1}$ then

$(\mathcal{MIX}h) = \mathrm{id}_{\mathcal{MIX}\mathcal{C}_1}$. Finally, if $h \in \hom(\mathcal{C}_1, \mathcal{C}_2)$ and $g \in \hom(\mathcal{C}_2, \mathcal{C}_3)$ then with the above notation

$$\begin{aligned}\mathcal{MIX}(gh)c &= \{ghc_1, ghc_2, \ldots\} = \mathcal{MIX}g)\{hc_1, hc_2, \ldots\} \\ &= (\mathcal{MIX}g)(\mathcal{MIX}h)c\end{aligned} \tag{5.6}$$

and $\mathcal{MIX}$ is indeed a functor. Q.E.D.

If Σ = LINEAR then $\mathcal{MIX}$ gives us the category of configuration spaces whose global regularity means unconnected linear chains. If Σ = TREE applying the function $\mathcal{MIX}$ we get the configuration spaces with global regularity FOREST. If Σ = CYCLIC the $\mathcal{MIX}$ function gives us cycles and sets of cycles as the global regularity, and so on.

Note that if Σ is closed under $\mathcal{MIX}$, $\Sigma = \mathcal{MIX}\Sigma$, then nothing new results from applying $\mathcal{MIX}$, for example if Σ = FOREST.

Another functor is $\mathcal{ISOLATE}$ which takes Conf(MONATOMIC Σ) into Conf(DISCRETE) so that all connections are broken in any configuration. It takes any object $\langle G,S,\Sigma,\rho\rangle$ in the category Conf(MONATOMIC Σ) into $\langle G,S,\text{DISCRETE},\rho\rangle$ and any homomorphism $h\colon \mathcal{C}_1 \to \mathcal{C}_2$ into $H = \mathcal{ISOLATE}h$ by the relation

$$\begin{cases} c = \{g_1, g_2, \ldots\} \in \mathcal{ISOLATE}(\mathcal{C}_1) \\ Hc = \{hg_1, gh_2, \ldots\} \in \mathcal{ISOLATE}(\mathcal{C}_2) \end{cases} \tag{5.7}$$

where both configurations have isolated generators. Note that hg_i are defined since $\mathcal{C}_1$ is monatomic so that any monatomic configuration $\{g_i\}$ is regular.

The next functor reduces categories of configuration spaces over compositions $\Sigma_1(\Sigma)$ to the categories associated with Σ_1. More precisely the functor $\mathcal{MACRO}$ takes objects

$\mathscr{C} = \langle G,S,\Sigma_1(\Sigma),\rho\rangle$ where Σ_1 is monatomic, into objects $\mathscr{C}' = \langle \mathscr{C},S,\Sigma_1,\rho\rangle$, and homomorphisms $h: \mathscr{C}_1 \to \mathscr{C}_2$ into $H = \mathscr{MACRO}h$ defined by

$$H\{c_1,c_2,\ldots\} = \{hc_1,hc_2,\ldots\} \tag{5.8}$$

where the "external" connections between the c_i, viewed as generators, should be the same as between the hc_i. Note that the hc_i are defined since Σ_1 is monatomic. The global regularity according to $\mathscr{MACRO}\mathscr{C}_1$ of the right hand side of (5.8) is correct since the c_i were connected according to Σ_1 and we have the same bond relation for $\mathscr{C}$ and for $\mathscr{MACRO}\mathscr{C}$. Hence $\mathscr{MACRO}$ is a functor.

The operation that takes an arbitrary configuration space over $\Sigma = \Sigma(\Gamma)$ and extends it to the monotonic extension of Σ would also seem to be a functor. This is not so, however, since the homomorphisms in the given configuration spaces do not naturally extend to the ones for the monotonic extension of Σ. At any rate the author has not been able to construct such an extension.

There is clearly a need for a careful examination of these and other categories and functors of configuration spaces, but this will not be attempted here. In the next section we shall instead turn to *set operations on configuration spaces*.

3.6. Set operations in $\mathcal{C}(\mathcal{R})$

Consider the set $2^{\mathcal{C}}$ of all subsets of $\mathcal{C} = \mathcal{C}(\mathcal{R})$ and maps $\phi: 2^{\mathcal{C}} \to 2^{\mathcal{C}}$. If a set operator ϕ is *additive* in the sense that for $A = \{c_1, c_2, \ldots\}$ the result is $\phi A = \{\phi c_1, \phi c_2, \ldots\}$ then ϕ can be viewed as operating simply on each individual elements c_i of A. There is no interaction between the different c_i, so that we could treat each ϕc_i separately as another instance of configuration mappings, except for the fact that the value ϕc_i can be a set of configurations, not just a single configuration. Additive set operations of configurations are therefore rather special.

Here we shall study instead some operations that are not additive, and we start with *combine*. If $A = \{c_1, c_2, \ldots, c_n\}$, consider all regular configurations c that can be obtained as a combination from the given configurations as

$$c = \sigma(s_1 c_{i_1}, s_2 c_{i_2}, \ldots), \tag{6.1}$$

with arbitrary connector σ resulting in a regular configuration c, and where $1 \leq i_1 < i_2 < \ldots < i_n \leq n$ and arbitrary $s_i \in S$. The set of all c that can be obtained from A as in (i) is denoted $\underline{\text{combine}}(A) \subseteq \mathcal{C}$. Note that each c_i in A can occur at most once in (6.1), and also that any c_i from A is also contained in $\underline{\text{combine}}$ A: the set operation *combine* is *non-decreasing*, $\underline{\text{combine}}\ A \supseteq A$. If $A \subseteq B$ then it is clear from the definition of *combine* that $\underline{\text{combine}}\ A \subseteq \underline{\text{combine}}\ B$ so that the set operation *combine* is *monotonic* relative to the partial ordering of sets in $\mathcal{C}$ by inclusion.

Now we can iterate by applying *combine* to combine A, and so on for higher iterations. Since the resulting sets form a non-decreasing sequence their limit is well-defined. The limit will be denoted by span A.

Before studying the *span* operation let us rephrase the above construction in greater generality.

Theorem 1. *If* $\phi: 2^{\mathscr{C}} \to 2^{\mathscr{C}}$ *is a monotonic, non-decreasing set operation, define the iteration closure of* ϕ *as*

$$\phi^*(A) = \lim_{k\to\infty} \phi^k(A) = \bigcup_{k=1}^{\infty} \phi^k(A), \quad A \subseteq \mathscr{C}. \tag{6.2}$$

Then

(i) ϕ^* *is idempotent*, $\phi^*\phi^*(A) = \phi^*(A)$, *and*

(ii) $\phi^*(A \cup B) = \phi^*[(\phi^*A) \cup (\phi^*B)]$.

Proof: To prove (i) it is enough to note that

$$\phi^*\phi^*(A) = \lim_{k\to\infty} \phi^k[\lim_{\ell\to\infty} \phi^\ell(A)] = \lim_{k\to\infty}\lim_{\ell\to\infty} \phi^{k+\ell}(A) = \phi^*(A) \tag{6.3}$$

where we have used the definition in (6.2). Statement (ii) can be seen to follow from the inclusion

$$\phi^*(A \cup B) = \lim_{k\to\infty} \phi^k(A \cup B) \supseteq \lim_{k\to\infty} \phi^k(A) = \phi^*(A), \tag{6.4}$$

obtained from the monotonic character of ϕ, and similarly for B replacing A, so that

$$\phi^*(A \cup B) \supseteq (\phi^*A) \cup (\phi^*B). \tag{6.5}$$

Applying the monotonic operation ϕ^* to both sides we get from (i)

$$\phi^*(A \cup B) = \phi^*\phi^*(A \cup B) \supseteq \phi^*[(\phi^*A) \cup (\phi^*B)] \tag{6.6}$$

On the other hand, using the non-decreasing property of ϕ, and hence of ϕ^*,

$$A \cup B \subseteq (\phi^* A) \cup (\phi^* B) \tag{6.7}$$

so that

$$\phi^*(A \cup B) \subseteq \phi^*[(\phi^* A) \cup (\phi^* B)]. \tag{6.8}$$

Together with (6.6) this proves the assertion (ii). Q.E.D.

Theorem 2. *If the connection type* Σ *is monotonic, then for any subset* $A = (c_1, c_2, \ldots c_r) \subseteq \mathscr{L}(\mathscr{R})$ *the new set* span A *consists of all regular configurations that can be obtained as*

$$c = \sigma(s_1 c_{i_1}, s_2 c_{i_2}, \ldots) \in \mathscr{L} \tag{6.9}$$

where $i_1, i_2, \ldots$ *takes values between* 1 *and* r *and* s_i *are arbitrary similarities.*

Note: In (6.9) the subscripts are allowed to repeat themselves in contrast to in (6.1).

Proof: For any natural number k the set obtained by iterating combine k times applied to A consists of configurations that can be written as

$$c = \sigma\{\ldots s_3 \sigma_3 [\sigma_1(s_{11} c_{i_1}, s_{12} c_{i_2}, \ldots), \\ s_2 \sigma_2 (s_{21} c_{j_1}, s_{22} c_{j_2}, \ldots), \ldots], \ldots\}. \tag{6.10}$$

In (6.10) the depth of the nesting is at most k. Since similarities are distributive over connectors it is clear that such a c can be rewritten as in (6.9), and it follows that all configurations in span A can be expressed as in (6.9).

On the other hand, any regular c as in (6.9) can be rewritten as a connection of shifted subconfigurations, each without any repetition of subscripts. But Σ was monotonic so that any such subconfiguration has a connection belonging to Σ, global regularity holds. Further, local regularity holds automatically for the subconfiguration of a regular configuration, so that it follows that each of the subconfigurations is regular and belongs to combine A. Repeating this argument it follows that c can be obtained by iterating combine on A, and the proof is complete. Q.E.D.

Given $A = \{c_1, c_2, \ldots\} \in 2^{\mathscr{C}}$, again with monotonic Σ, let us consider the c's as macro-generators $g', G' = A$. Introduce the connection type Σ' over the skeletons $¢_1, ¢_2, \ldots$ as all σ's that connect c's so that global Σ regularity holds. One could say that Σ' equals Σ modulo A.

Consider now the new configuration space $\mathscr{C}' = \langle G', S, \Sigma', \rho\rangle$, where S and ρ are not changed. The map $G' \leftrightarrow A$ then estends automatically to a homomorphism $\mathscr{C}' \leftrightarrow$ span A which can be seen to be an isomorphism: $\mathscr{C}' \cong$ span A. Therefore span A has the same regularity structure as the configuration space $\mathscr{C}'$ built from the macro-generators.

3.7. Operations on images

As we have seen, a configuration space $\mathscr{C} = \langle G, S, \mathscr{R}\rangle$ together with an identification rule R and a set $\{\sigma\}$ of connectors forms a partial algebraic system. When the observer is looking at $\mathscr{C}$ he can distinguish only between the objects in the partition $\mathscr{C}/R$, i.e. the images; see Volume I, Chapter 3. Observing a regular configuration $c \in \mathscr{C}$, he can only see the image, the equivalence class, $I = [c]_R$ to which it belongs.

Although the observer may lose information about the internal bonds of the configurations that belong to I he will always know the external bonds, see Volume I, Definition 1.1(ii) in Chapter 3. The bond structure and bond values of I is given by these external bonds. To refer to them we need bond coordinates, and the choice of coordinate system is arbitrary to the extent we have discussed before.

An image u such that for any I and connector σ we have $\sigma(I,u) \in \mathcal{T} \Rightarrow \sigma(I,u) = I$ is called a (partial) *unit* of the image algebra. For example if ϕ is counted as a regular configuration, (see Notes A) and if Σ is monatomic, then $e = [\phi]_R$ is a unit.

If u is a unit then su is also a unit for any similarity s. Indeed, $\sigma(I,su) \in \mathcal{T} \Rightarrow \sigma(s^{-1}I,u) \in \mathcal{T} \Rightarrow$ $[\sigma(s^{-1}I,u) = s^{-1}I] \Rightarrow [\sigma(I,su) = I]$.

If Σ is monotonic and if u_1 and u_2 are units; then $u = \sigma_o(u_1,u_2) \in \mathcal{T} \Rightarrow \sigma_o(u_1,u_2)$ is also a unit. Indeed, for any given I and connector σ such that $I' = \sigma(I,u) \in \mathcal{T}$, form the sub-image $\sigma'(I,u_1)$ from I' by deleting u_2 and its bonds from $I' = \sigma(I,\sigma_o(u_1,u_2))$. Since the sub-image also belongs to $\mathcal{T}$ because of the monotonic property of the connection type, it follows that $\sigma'(I,u_1) = I$. But then I' can be written as $\sigma''(\sigma'(I,u_1),u_2) = \sigma''(I,u_2) = I$ which proves the assertion. Note that if $\sigma(u_1,u_2)$ is regular then $\sigma(u_1,u_2) = u_1 = u_2$ so that units can be combined only when equal.

<u>Remark</u>. The term "unit" is ambiguous, perhaps "identity" would have been better.

The discrete image algebras, $\mathcal{R}$= DISCRETE, are of course the simplest ones. Then two images I_1 and $I_2 \in \mathcal{I}$ can be combined only in one way and the result does not depend upon the order. It is then natural to write the combination as I_1+I_2 with commutative and associative addition. The similarities satisfy $s(I_1+I_2) = sI_1 + sI_2$, see (iii) in Definition 1.1 of Volume I, Chapter 3. In other words, *a discrete image algebra can be viewed as a commutative semi-group with the similarities acting distributively over addition.*

We now turn to the discussion of the identification rule R and make some remarks in addition to what was said in Chapter 3, Volume I about this topic.

The notion of congruence relation, which is obvious for a universal algebra, is more sensitive if some of the basic operations of the algebra are only partial which is the case of greatest interest for us. The reader is referred to Grätzer (1968), pp. 79-99. If we consider a basic operation σ connecting c_1 and c_2 into a regular configuration c, and c_1' and c_2' into a regular configuration c' then we have asked that $c_i R c_i'$, $i = 1,2,; \Rightarrow cRc'$. Hence our notion of identification rule corresponds to "congruence", not "strong congruence"; see Notes A. For the basic operations $c \to sc$ this distinction is unnecessary, since c is regular if and only if sc is regular, when the similarities form a group as assumed.

Just as we need more than one notion of homomorphisms, when operations are partial, we must be prepared to need more than one notion of identification rule, and we shall indeed occasionally use strong ones.

In Volumes I and II we confined our attention to a fixed identification rule R over $\mathcal{C}$. We shall now study what happens when *several observers are involved with different* R's over the same $\mathcal{C}(\mathcal{R})$.

Therefore, let us consider two observers, 1 and 2, with the identification rules R_1 and R_2 respectively. If $R_1 \leq R_2$, meaning that observer 1 has more powerful instrumentation, then the images seen by 1 are contained in the images seen by 2; they contain more detailed information than the latter ones.

On the other hand, R_1 and R_2 need not be comparable. If 1 and 2 pool their observational information they can make statements about images belonging to $\mathcal{I}_1 \wedge \mathcal{I}_2$ corresponding to the identification rule as the *meet* $R = R_1 \wedge R_2$ meaning that c_1 and c_2 are identified if and only if $c_1 R_i c_2$ for both $i = 1$ *and* 2. This is indeed a legitimate identification rule. Indeed, R is automatically an equivalence, and if cRc' then cR_1c' so that c and c' have the same external bonds. Also $cRc' \Rightarrow cR_ic' \Rightarrow (sc)R_i(sc')$ for $i = 1,2$, so that $(sc)R(sc')$. Finally, if $c = \sigma(c_1,c_2)$, $c' = \sigma(c_1',c_2')$, all regular, and if c_1Rc_1 and c_2Rc_2' then $c_1R_ic_1'$ and $c_2R_ic_2$ so that $cR_ic' \Rightarrow cRc'$ and Definition 1.1 in Volume I, Chapter 3, holds.

If 1 and 2 do not pool their information, what is their *common observed regularity?* The corresponding identification rule R must be an equivalence such that cR_ic' for either $i = 1$ or $i = 2$ imply cRc'. The smallest such equivalence is the *join* $R = R_1 \vee R_2$, which also satisfies Definition 1.1 (ibid.). Indeed, if cRc' there exists a chain $c = c_o, c_1, c_2, \ldots c_n = c'$ such that

$$c_{i-1}R_{k_i}c_i, \quad i = 1,2,\ldots n, \quad k_i = 1 \text{ or } 2 \tag{7.1}$$

see e.g. Grätzer (1968), p. 18. Then $B^e(c) = B^e(c_1) = B^e(c_2) = \ldots = B^e(c')$, so that c and c' have the same external bonds. Also $(sc_{i-1})R_{k_i}(sc_i)$ so that the chain $\{sc_i\}$ connects sc with sc', which implies $(sc)R(sc')$. To prove the fourth condition in the definition of identification rules with the same notation as above, c_1Rc_1' and c_2Rc_2' implies the existence of two chains $\{a_i\}$ and $\{b_i\}$ such that $a_o = c_1$, $a_n = c_1'$ and $b_o = c_2$, $b_m = c_2'$ with

$$\begin{cases} a_{i-1}R_{k_i}a_i, & i = 1,2,\ldots n, \quad k_i = 1 \text{ or } 2 \\ b_{i-1}R_{\ell_i}b_i, & i = 1,2,\ldots m, \quad \ell_i = 1 \text{ or } 2 \end{cases} \tag{7.2}$$

Then, successively using (7.2), we get

$$\begin{cases} \sigma(a_o,b_o)R_{k_1}\sigma(a_1,b_o) \\ \sigma(a_1,b_o)R_{k_2}\sigma(a_2,b_o) \\ \ldots\ \ldots \\ \sigma(a_{n-1},b_o)R_{k_n}\sigma(a_n,b_o) \\ \sigma(a_n,b_o)R_{\ell_1}\sigma(a_n,b_1) \\ \ldots\ \ldots \\ \sigma(a_n,b_{m-1})R_{\ell_m}\sigma(a_n,b_m) \end{cases} \tag{7.3}$$

so that we have a chain connecting $\sigma(a_o,b_o) = \sigma(c_1,c_1')$ with $\sigma(a_n,b_m) = \sigma(c_2,c_2')$ implying $\sigma(c_1,c_1')R\sigma(c_2,c_2')$ and the proof is complete. The corresponding image algebra will be denoted $\mathcal{T}_1 \vee \mathcal{T}_2$.

A degenerate case is when $R_1 \vee R_2 \equiv \text{TRUE}$ so that $\mathcal{T}_1 \vee \mathcal{T}_2$ contains only the two trivial images ϕ and $\mathcal{C}$. Then the two observers perceive no common regularity.

Combining what we have just proved we announce

Theorem 1. *Given the configuration space* $\mathcal{C}$ *the image algebras* $\mathcal{T} = \langle \mathcal{C}, R \rangle$ *form a complete lattice with meet* $\mathcal{T}_1 \wedge \mathcal{T}_2$ *and join* $\mathcal{T}_1 \vee \mathcal{T}_2$ *as above.*

Example 1. Let configurations be of contrast type so that $c: X \to Y$. For simplicity let $\mathcal{C}$ consist of all such functions; this is not essential and can be weakened considerably. Let R_1 and R_2 identify configurations that agree on the subsets X_1 and X_2 respectively.

Then an $\mathcal{T}_1$-image consists of all functions defined by $\mathcal{C}$ that take given values on X_1, and similarly for the $\mathcal{T}_2$-images. The images in $\mathcal{T}_1 \wedge \mathcal{T}_2$ consist of $\mathcal{C}$-functions agreeing on $X_1 \cup X_2$.

To see what $\mathcal{T}_1 \vee \mathcal{T}_2$-images mean, say first that R_1 and R_2 are comparable, for example by the inclusion $X_1 \subseteq X_2$. Then $\mathcal{T}_1 \vee \mathcal{T}_2$-images consist of all $\mathcal{C}$ functions with fixed values on X_1. However, if R_1 and R_2 are not comparable, the situation can be quite different. Then $\mathcal{T}_1 \vee \mathcal{T}_2$-images consist of functions agreeing on $X_1 \cap X_2$ so that if X_1 and X_2 are disjoint we have only the degenerate R-images ϕ and $\mathcal{C}$. The observers will not be able to see any common and non-trivial regularity.

In a universal algebra there is only one natural definition of homomorphisms but it is well-known that this is not so for partial universal algebras. We then have a choice and may have to operate with several homomorphism definitions as argued by Grätzer (1968), p. 81.

For the moment we shall use only one definition, and leave alternatives till later. It is a modification of the one used in Volume I, Section 2.11, and more generally in Grenander (1977d) where it was pointed out that even the definition given there is too restrictive and should be extended to cover cases where the two image algebras involved in the mapping have different global regularity: the two connection types allowed to be different. Now we shall attempt to give a generally applicable definition of homomorphisms between image algebras.

Suppose that we have constructed a configuration space

$$\mathscr{C}(\mathscr{R}) = \langle G,S,\mathscr{R}\rangle \tag{7.4}$$

with local regularity expressed by the bond relation ρ and global regularity introduced by the connection type Σ so that the total regularity is

$$\mathscr{R} = \langle \rho,\Sigma\rangle. \tag{7.5}$$

When we pass to an image algebra $\mathscr{T} = \mathscr{C}(\mathscr{R})/R$ that corresponds to an observer, we are no longer automatically able to "see" the components of (7.4) and (7.5), except for S, since the identification rule R may hide such information from the observer. The similarities in S are known to the observer, of course; otherwise he could not form copies sI of a given image I. When we abstract from the construction of the image algebra from generators etc. we shall speak of *abstract image algebras*.

To fix ideas let us consider a single binary connection σ partially defined over $\mathscr{T}$. This means that we neglect the workings of other connectors that may be defined on $\mathscr{T}$, but without affecting the image algebra itself. To emphasize that we view $\mathcal{F} \times \mathcal{F}$ as a (partial) domain of σ we shall speak of the image algebra $\mathcal{F}$ *via* σ. See Notes B.

In the following of this section we shall assume, to simplify the discussion, that $\mathcal{F}$ is *globally free* via σ in the sense that if the configuration skeletons $\not{c}_1, \not{c}_2 \in \Sigma$ then the skeleton of the connected configurations $\sigma(\not{c}_1, \not{c}_2) \in \Sigma$.

This condition on $\mathcal{F}$ seems to be satisfied fairly often when σ is chosen in a natural manner. This is in contrast to *locally free*, when ρ = TRUE, so that all bonds fit. That applies only in exceptional cases.

Hence, for a single binary connector σ, the algebraic structure of $\mathcal{F}$ can be expressed by a set of tables, the *similarity tables* and the *composition table*. The first ones express how copies are related to each other and could look like Table 1, where the first table describes $s_1 : I \to s_1 I$,

s_1-table	
I_1	I_1
I_2	I_2
I_3	I_3
⋮	⋮

s_2-table	
I_1	I_2
I_2	I_3
I_3	I_4
⋮	⋮

s_3-table	
I_1	I_3
I_2	I_4
I_3	I_5
⋮	⋮

. . . . .

Table 1

the second one describes $s_2 : I \to s_2 I$, etc. for all $s \in S$. The cardinalities of S and $\mathcal{F}$ need of course not be denumerable.

The composition table describes how a pair I_1, I_2 is combined through the connector σ and results in $I = \sigma(I_1, I_2)$ if this is defined or in N, meaning not defined. The composition table could look like Table 2. If we deal with more than one connector we need of course several composition tables.

σ	I_1	I_2	I_3	...
I_1	I_4	I_5	N	...
I_2	N	N	I_6	...
I_3	N	N	I_7	...
⋮	⋮	⋮	⋮	⋮

Table 2

The tables for tables and composition are far from arbitrary. The similarity tables are restricted by the fact that S is a group of transformations so that the tables describe permutations of images, finite or infinite. A composition table for σ is restricted by *conditional distributivity*, that is the relation

$$s\sigma(I_1, I_2) = \sigma(sI_1, sI_2) \tag{7.6}$$

holds when the combinations involved are defined; see Volume I, p. 95.

We do not always have *conditional commutativity*

$$\sigma(I_1, I_2) = \sigma(I_2, I_1) \tag{6.7}$$

(when one side is defined), indeed this is really an exceptional case. It is therefore easy to give examples of when conditional commutativity does not hold, for example when σ denotes the concatenation operation of functions on intervals on the real line.

Conditional associativity, that is

$$\sigma(\sigma(I_1,I_2),I_3) = \sigma(I_1,\sigma(I_2,I_3)) \tag{7.8}$$

(when one side is defined) requires a closer examination.

To verify that (7.8) holds we should verify that the set of closed bonds due to the two applications on the left of (7.8) is equal to the set of applications on the right, see Volume I, Theorem 1.1(ii) in Chapter 3. This condition is clearly sufficient. It need not be necessary: that depends upon R, but in a general context, with nothing assumed about R, we must verify this condition. For example if R means "IDENTITY" then an image is the same as a configuration and then the condition is necessary and sufficient.

As an example when conditional associativity does not hold let Γ consist of skeletons with one in bond and two recognizable out bonds and let Σ = TREE. Define the binary connector σ by the rule

σ: connect the in bond of I_1 to the second leftmost out bond of I_2 if it exists. (7.9)

Consider the three images in Figure 7.1, each one with $\omega_{in} = 1$ and $\omega_{out} = 2$. In (b) we see the image represented by the expression on the left of (7.5) and in (a) the one given by the right hand side of the relation. It is obvious

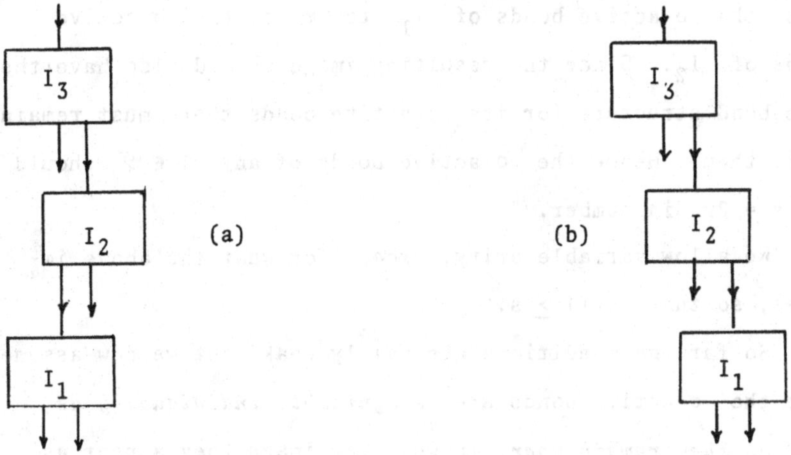

Figure 7.1

that the two sets of closed bonds are not equal in (a) and (b), so that conditional associativity does not hold generally in this example.

On the other hand let σ denote concatenation of I_2 to the right of I_1 viewed as functions on the real line as discussed several times in Volume I. For three images I_1, I_2, I_3 whose bond values fit in the usual manner, it is obvious that it does not matter if we first concatenate I_1 with I_2 and the result with I_3, or first concatenate I_2 and I_3 and the result to the right of I_1: this image algebra via concatenation is conditionally associative.

To shed light on the question when conditional associativity holds we shall deal with a situation that is not general but of sufficient scope to bring out clearly the heart of the matter. We shall deal with symmetric regularity.

Let us assume that $\mathcal{T}$ via σ is characterized by the fact that each image has certain *σ-active bonds* used exclusively for σ-connections. If σ has arity r, meaning that it can be expressed as an $r \times r$ matrix, it connects r of the σ-active bonds of I_1 to r of the σ-active bonds of I_2. Since the resulting image should also have the same bond structure for its σ-active bonds there must remain r of them. Hence the σ-active bonds of any $I \in \mathcal{T}$ should be $s = 2r$ in number.

We allow variable arity, except for what the above implies, so that $\omega(I) \geq s$.

So far the conditions are fairly weak, but we now assume that the σ-active bonds are *recognizable individually as long as they remain open,* in whatever image they appear as bonds of sub-images. Hence, numbering the σ-active (external) bonds of I by the labels $1,2,3,\ldots s$ and if I is a subimage of I' then the ith σ-active bond of I coincides with the ith σ-active bond of I' if the latter is open. This condition is more severe than may appear at first glance.

Theorem 2. *If $\mathcal{T}$ via σ is characterized by $s = 2r$ σ-active bonds, all individually recognizable, then $\mathcal{T}$ via σ is conditionally associative.*

Proof: Say that σ connects the active bonds $1,2,3,\ldots r$ of I with the active bonds $i_1, i_2, \ldots, i_r$ of I' in the order stated. This is no restriction since it can always be achieved by a fixed permutation of the labels used. None of the $i_1, \ldots, i_r$ can be equal to an integer from 1 to r, since if this happened, $i_\nu = k \in \{1,2,\ldots r\}$, the resulting image $I_0 = \sigma(I,I')$ would have no active bond labelled k. Similarly no $i_1, \ldots, i_r$ can be equal, since if $i_\nu = i_\mu$, $\nu \neq \mu$ then I_0 would have two σ-active bonds with the same label, hence unrecognizable from each other. Hence each $i \in \{1,2,\ldots,r\} = v$ is connected to exactly one of the σ-active bonds in $\{r+1,r+2,\ldots 2r\} = w$ in I'. Call this mapping p, it is a permutation of r objects.

Now consider three images I, I', I'' such that

$$I_o = \sigma[\sigma(I,I'),I''] \in \mathcal{T}. \tag{7.10}$$

Enumerate the σ-active bonds of the images as in the column vectors

$$\begin{pmatrix} v \\ w \end{pmatrix}, \begin{pmatrix} v' \\ w' \end{pmatrix}, \begin{pmatrix} v'' \\ w'' \end{pmatrix}. \tag{7.11}$$

Then $\sigma(I,I')$ connects the bonds v to pw' and the resulting image has the σ-active bonds

$$\begin{pmatrix} v' \\ w \end{pmatrix} \tag{7.12}$$

When this is connected to I'' via σ the bonds v' are connected with pw''.

On the other hand $\sigma(I',I'')$ connects the σ-active bonds v' to pw'', and when the resulting image is connected to I via σ, we have the σ-active bonds for the resulting image

$$\begin{pmatrix} v'' \\ w' \end{pmatrix} \tag{7.13}$$

and when it is connected to I we close the bonds v-pw'.

Hence in both cases we close the same bonds. Since $I_o \in \mathcal{T}$ then

$$I_1 = \sigma[I,\sigma(I',I'')] \in \mathcal{T} \tag{7.14}$$

and $I_o = I_1$ which completes the proof. Q.E.D.

To illustrate how this theorem applies consider some examples.

<u>Example 1</u>. For the <EQUAL,LINEAR>-regularity used as an example above we have $r = 1$ and it is clear that the two σ-active (σ = concatenation on the right) bonds are individually recognizable as described above. Hence σ should be conditionally associative as we already know.

<u>Example 2</u>. Consider <EQUAL,TREE>-regularity and let the connector σ connect the upper (single) bond of I_1 to the leftmost of the lower bonds of I_2. Again $r = 1$ and a moment's reflection shows that the σ-active bonds are recognizable. Hence this connector is also conditionally associative.

<u>Example 3</u>. Same as Example 2, except that σ now connects the upper bond of I_1 to the second leftmost bond of I_2. We already know that this connector is not conditionally associative and we can now see why not. Indeed, the property of being "second leftmost" depends upon how a subimage is connected to other parts of the entire image, so that the σ-active bonds are not individually recognizable and the theorem does not apply.

<u>Example 4</u>. Let $r = 2$, $\sigma(I) \geq 4$ with the σ-active bonds $i = 1,2,3,4$, to I', with $i = 1',2',3',4'$ by setting up the connections 1-$3'$ and 4-$2'$ so that $\sigma(I,I')$ has the -active bonds $1'2,3,4'$. For three images I,I',I'' the resulting combination appears in Figure 7.2. Some non-active bonds are indicated by dotted lines. The σ-active bonds are recognizable so that σ is conditionally associative.

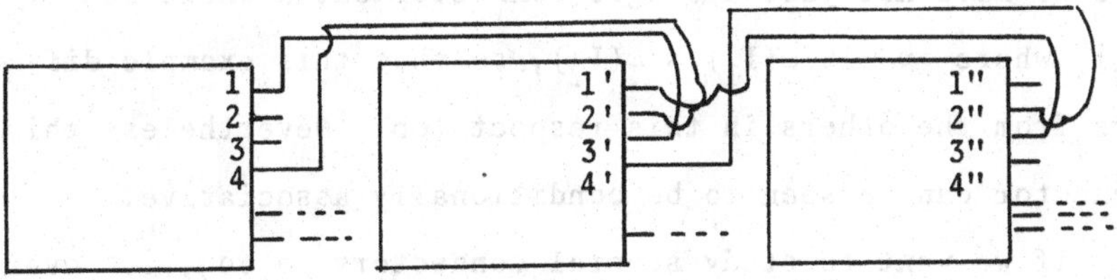

Figure 7.2

Example 5. Let the generators consist of continuous functions on rectangles in the plane with sides parallel to the coordinate axes, and have bond values $b = (E, f_{\partial E})$. Here E denotes the rectangle and $f_{\partial E}$ the restriction of the continuous function to the boundary ∂E of E. Let $\omega(g) = +\infty$, all bond values the same, and the bond relation between two bond values b and b'

$$\rho: \begin{cases} (E \cap E') \subseteq (\partial E \cap \partial E') \\ f = f' \quad \text{on} \quad \partial E \cap \partial E' \end{cases} \tag{7.15}$$

This is Case 9.3 in Chapter 3 of Volume I if we choose the connection type Σ = FULL.

A typical picture would look as in Figure 7.3. For

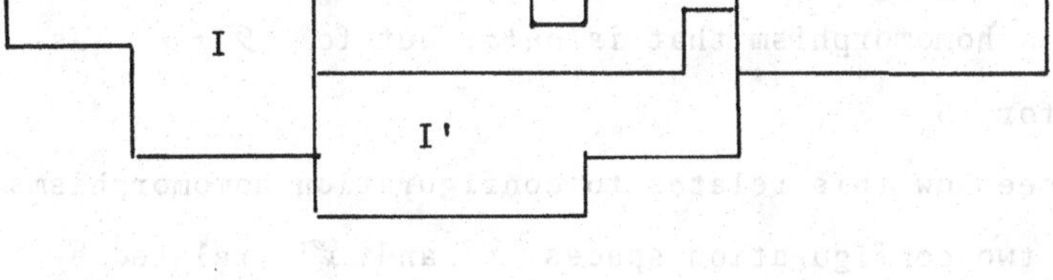

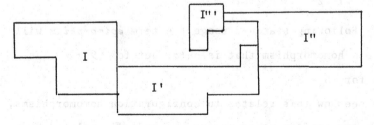

Figure 7.3

this <ρ,FULL>-regularity the conditions in Theorem 2 are not satisfied since the number of σ-active bonds is not fixed. Here we have not just a single connector but a whole family $\{\sigma_\nu\}$, where ν is $\omega(I_1) \times \omega(I_2)$, so that this example differs from the others in this respect too. Nevertheless this connector can be seen to be conditionally associative.

If we want to study several connectors $\sigma_1, \sigma_2, \ldots$ over $\mathcal{T}$ each one will have to be examined as above. The presence of several connectors gives rise to a new problem when discussing homomorphisms of image algebra; compare with the discussion in Section 4. See Notes C.

Consider two image algebras, one $\mathcal{T}$ via a set of connectors $\{\sigma\}$ and similarities S, the other one via $\{\sigma'\}$ and similarities S'.

<u>Definition 1</u>. *A mapping* $h: (\mathcal{T} \cup S \cup \{\sigma\}) \to \mathcal{T}' \cup S' \cup \{\sigma'\}$ *taking values in* $\mathcal{T}'$ *on* $\mathcal{T}$, *in* S' *on* S, *and in* $\{\sigma'\}$ *on* $\{\sigma\}$ *is called a homomorphism if* h *maps* $S \to S'$ *homomorphically and*

$$\begin{cases} \text{(i)} & h(sI) = h(s)h(I) \quad \textit{for any} \quad s \quad \textit{and} \quad I \\ \text{(ii)} & h[\sigma(I_1,I_2)] = h(\sigma)[h(I_1),h(I_2)] \quad \textit{for any} \\ & I_1, I_2, I = \sigma(I_1,I_2) \quad \textit{in} \quad \mathcal{T}. \end{cases} \tag{7.16}$$

<u>Remark</u>. Following standard usage the term *epimorphism* will indicate a homomorphism that is *onto*, but for $\mathcal{T} \to \mathcal{T}'$ as well as for $S \to S'$.

To see how this relates to configuration homomorphisms, consider two configuration spaces $\mathcal{C}$ and $\mathcal{C}'$ related by a homomorphism h. When we add two identification rules R and R' to the configuration spaces respectively, making them into image algebras $\mathcal{T} = \langle \mathcal{C}, P\rangle$, $\mathcal{T}' = \langle \mathcal{C}', R'\rangle$, when can we assert that h induces a homomorphism $H: \mathcal{T} \to \mathcal{T}'$? Here we only present a partial answer.

Theorem 3. *If the configuration homomorphism* h *satisfies* $c_1 R c_2 \Rightarrow (hc_1) R' (hc_2)$ *for all* $c_1, c_2 \in \mathcal{C}$ *the induced map* H *is an image homomorphism.*

Proof: The h map induces a map H from $\mathcal{T}$ to $\mathcal{T}'$ as the R'-equivalence class containing hc, $HI \to [hc]_{R'}$ if $I = [c]_R$. This definition is of course unique since for two $\mathcal{C}$ configurations equivalent (modulo R) h takes them into the same R' equivalence class.

To prove that H is a homomorphism we have to show that the two conditions in Definition 1 are satisfied. To see that (i) holds let $I = [c]_R$ so that

$$\begin{aligned} HsI = [hsc]_{R'} = [h(s)h(c)]_{R'} &= h(s)[h(c)]_{R'} \\ &= h(s)HI \end{aligned} \tag{7.17}$$

using the S'-covariance of the identification rule R'.

To show (ii) consider three $\mathcal{T}$ images $I_1, I_2, I = \sigma(I_1, I_2)$, and represent them by corresponding configurations c_1, c_2, $c = \sigma(c_1, c_2)$ with

$$I_1 = [c_1]_R, \quad I_2 = [c_2]_R, \quad I = [c]_R \tag{7.18}$$

where we have used the combinatory property of an equivalence relation. We then have

$$\begin{aligned} HI = [hc]_{R'} = [h\sigma(c_1, c_2)]_{R'} &= [(h\sigma)(hc_1, hc_2)]_{R'} \\ = (h\sigma)([hc_1]_{R'}, [hc_2]_{R'}) &= (h\sigma)(HI_1, HI_2)) \end{aligned} \tag{7.19}$$

as required. Q.E.D.

Remark. It is trivial that the identity mapping $id_{\mathcal{T}}: \mathcal{T} \to \mathcal{T}$ is a homomorphism. Also if $h: \mathcal{T} \to \mathcal{T}'$ and $h': \mathcal{T}' \to \mathcal{T}''$ are homomorphism, then the composition $h'' = h'h: \mathcal{T} \to \mathcal{T}''$ is also homomorphic. Indeed, h" satisfies (i) in the definition; this is just the usual fact about ordinary group homomorphisms. Also, if $I_1, I_2 \in \mathcal{T}$ such that $I = \sigma_\lambda(I_1, I_2) \in \mathcal{T}$ then $hI = \sigma'_{\lambda'}(I'_1, I'_2) \in \mathcal{T}'$ with $I'_\nu = hI_\nu$, see condition (ii). From this follows again that $h''I = h'hI = \sigma''_{\lambda''}(I''_1, I''_2)$ with $\lambda'' = h'\lambda' = h'h\lambda = h''\lambda$ and $I''_\nu = h'I_\nu = h''I_\nu$, so that (ii) holds.

Hence *image homomorphisms form a category*, similarly but in greater generality, than with the earlier definition.

Consider now an image algebra $\mathcal{T} = \mathcal{T}[S;\sigma]$ over the globally free connector σ. In general its composition table will have some undefined entries, and to express this formally we introduce

Definition 2. *The left (right) definition set for a given image* I *is given by*

$$\begin{cases} L(I) = \{J \mid \sigma(J,I) \in \mathcal{T}\} \subseteq \mathcal{T} \\ R(I) = \{J \mid \sigma(I,J) \in \mathcal{T}\} \subseteq \mathcal{T} \end{cases} \tag{7.20}$$

Hence an image J can be combined on the left to I iff $J \in L(I)$, and similarly on the right side. The definition sets have properties that can be expressed in terms of ideals, where this term is used as in other algebraic structures but with the modifications needed due to the special properties of image algebras.

Definition 3. *A set* $A \subseteq \mathcal{T}$ *is called a left ideal via* σ *if for any* $I \in A, I' \in \mathcal{T}$ *such that* $I'' = \sigma(I,I')$ *is defined we have* $I'' \in A$.

Theorem 4. *A left definition set* $L(I)$ *is covariant* $s^{-1}L(sI) = L(I), \forall s$. *If* σ *is characterized by active, individually recognizable bonds, then* $L(I)$, $(R(I))$, *is a left (right) ideal.*

Proof: We have from (7.20) and using the conditional distributive property (7.6)

$$\begin{aligned} s^{-1}L(sI) &= s^{-1}\{J|\sigma(J,sI) \in \mathscr{T}\} \\ &= s^{-1}\{J|s\sigma(s^{-1}J,I) \in \mathscr{T}\} \end{aligned} \tag{7.21}$$

and using the fact that $\mathscr{T}$ is closed under similarities

$$\begin{aligned} &= s^{-1}\{J|\sigma(s^{-1}J,I) \in \mathscr{T}\} \\ &= s^{-1}\{sK|\sigma(K,I) \in \mathscr{T}\} \\ &= \{K|\sigma(K,I) \in \mathscr{T}\} = L(I), \end{aligned} \tag{7.22}$$

so that $L(I)$ is covariant.

To prove that $L(I_0)$ is a left ideal choose arbitrary images $I \in L(I_0)$ and $I' \in \mathscr{T}$ such that $I'' = \sigma(I',I) \in \mathscr{T}$. We should show that $I'' \in L(I_0)$. Recalling that σ is globally free it is enough to show that ρ is satisfied for the bonds established by σ between I'' and I_0. But σ is conditionally associative according to Theorem 1 so that the total connector is the same for $\sigma(I'',I_0)$ as for $\sigma(I', (I,I_0))$. This implies that no bond is established between I' and I_0, only between I' and I, and between I and I_0. These two sets of bonds satisfy ρ so that

$$I'' = \sigma(I',I) \in L(I_0) \tag{7.23}$$

and $L(I_0)$ is indeed a left ideal. Q.E.D.

If the homomorphism h maps $\mathcal{I}$ into $\mathcal{I}'$ this implies relations between ideals in $\mathcal{I}$ and $\mathcal{I}'$, namely as follows.

<u>Theorem 5</u>. *Consider a homomorphism* h: $\mathcal{I}[S;\sigma] \to \mathcal{I}'[S';\sigma']$ *and a left (right) ideal* A' *in* $\mathcal{I}$, *then* $A = h^{-1}A'$ *is a left (right) ideal.*

<u>Proof</u>: To show that A is a left ideal consider two arbitrary images $I \in A$, $I' \in \mathcal{I}$ such that $I'' = \sigma(I,I') \in \mathcal{I}$. Then $hI'' \in \mathcal{I}'$ and

$$hI'' = h\sigma(I,I') = \sigma'(hI,hI'). \tag{7.24}$$

But $hI \in hA = hh^{-1}A' = A'$ so that $\sigma'(hI,hI') \in A'$, or $hI'' \in A'$ and then $I'' \in h^{-1}A' = A$ so that A is a left ideal; see Definition 3. Hence we have shown that the pre-image of A' is a left ideal just as A' itself. Of course, h^{-1} need not be one-valued. Q.E.D.

One would also like to know how the definition sets behave under homomorphisms. If h: $\mathcal{I}[S;\sigma] \to \mathcal{I}[S',\sigma']$ and $J \in L(I)$, so that $\sigma(J,I) \in \mathcal{I}$, then $h\sigma(J,I) = \sigma'(hJ,hI) \in \mathcal{I}'$ so that $hJ \in L'(hI)$. This means that

$$hL(I) \subseteq L'(hI). \tag{7.25}$$

We believe that (7.29) cannot always be strengthened to an equality.

The role of conditional right (left) zeros is similar to that of conditional units: $o \in \mathcal{I}$ is a conditional right zero if

$$\sigma(I,o) \in \mathcal{I} \Rightarrow \sigma(I,o) = o \tag{7.26}$$

The set O of conditional right zeros is S-invariant since

if $o \in O$

$$\sigma(I,so) \in \mathscr{T} \Rightarrow \sigma(s^{-1}I,o) \in \mathscr{T} \Rightarrow \sigma(s^{-1}I,o) = o \Rightarrow \Rightarrow \sigma(I,so) = so \Rightarrow so \in O. \tag{7.27}$$

Also O is a conditional right ideal, since if $o \in O$ and $I' = \sigma(I,o) \in \mathscr{T}$, then

$$\sigma(I'',I') \in \mathscr{T} \Rightarrow \sigma(I'',I') = o \Rightarrow \sigma(I'',I') \in O. \tag{7.28}$$

3.8. Homomorphisms for given global regularity

In this section we shall investigate the homomorphisms further under the added condition that the configuration spaces involved are given over a fixed family Γ of generator skeletons and a fixed connection type Σ; see Notes A. We shall then put $h\sigma = \sigma$ identically.

In this case it is of special interest to consider homomorphisms of configuration spaces that *respect external bonds* in the sense that two configurations with different external bonds will not be mapped into the same element.

Theorem 1. *Consider a homomorphism* $h: \mathscr{C} \to \mathscr{C}'$ *respecting external bonds, so that* $hc_1 = hc_2$ *implies* $B_e(c_1) = B_e(c_2)$, *and define a relation* R *on* $\mathscr{C} \times \mathscr{C}$ *by*

$$c_1 R c_2 \quad iff \quad hc_1 = hc_2. \tag{8.1}$$

Then R *is an identification rule and hence defines an image algebra* $\mathscr{T}$ *over* $\mathscr{C}$ *and* h *induces an image homomorphism* H *from* $\mathscr{T}$ *to* $\mathscr{T}' = <\mathscr{C}',\text{EQUAL}>$.

Proof: For the first part of the statement we must verify the four conditions of Definition 1.1 in Volume I, Chapter 3. It is obvious that R is an equivalence. If cRc' then c

and c' will have the same external bonds since $\underline{h}$ respects external bonds. If cRc' we have $hc = hc'$, so that for an arbitrary similarity $\underline{s}$ we have $h(sc) = h(s)h(c) = h(s)h(c') = h(sc')$ since $\underline{h}$ is homomorphic. Thus $(sc)R(sc')$. Finally, if $c = \sigma(c_1,c_2)$ and $c' = \sigma(c_1',c_2')$ are regular and c_1Rc_1', c_2Rc_2', this means that $hc_1 = hc_1'$, $hc_2 = hc_2'$. But since h is homomorphic, see (ii) in Definition 4.1,

$$\begin{aligned} hc &= h\sigma(c_1,c_2) = \sigma[hc_1,hc_2] \\ &= \sigma[hc_1',hc_2'] = h\sigma(c_1',c_2') = hc' \end{aligned} \tag{8.2}$$

so that cRc', and all four conditions hold in order that R be a legitimate identification rule.

Then $\mathcal{T} = [\mathcal{L}]_R$ is well defined and is related to $\mathcal{T}'$ by the map H that takes an $I = [c]_R$ into $[hc]_{EQUAL} \in \mathcal{T}'$. Note that the identification rule EQUAL in $\mathcal{T}$ makes images consist of single configurations. Also note that H is uniquely defined due to condition (8.1). Of course we put $Hs = hs$. Now for any $I \in \mathcal{T}$, $I = [c]_R$

$$\begin{aligned} H(sI) &= [h(sc)]_{EQUAL} = [h(s)h(c)]_{EQUAL} = [H(s)h(c)]_{EQUAL} \\ &= h(s)[h(c)]_{EQUAL} = H(s)H(I) \end{aligned} \tag{8.3}$$

so that (i) holds in Definition 7.1. If I_1,I_2, and $I = \sigma(I_1,I_2)$ are all in $\mathcal{T}$ it follows that

$$H[\sigma(I_1,I_2)] = [h\sigma(c_1,c_2)]_{EQUAL} \tag{8.4}$$

where the $\mathcal{L}$-regular configurations c_1 and c_2 have been selected so that $I_1 = [c_1]_R$, $I_2 = [c_2]_R$. Then (8.4) yields immediately

$$H[\sigma(I_1,I_2)] = [\sigma(hc_1,hc_2)]_{EQUAL}$$
$$= \sigma\{[hc_1]_{EQUAL},[hc_2]_{EQUAL}\} = \sigma(HI_1,HI_2) \tag{8.5}$$

so that (ii) holds in the definition. Q.E.D.

This theorem enables us to construct image homomorphisms directly from the examples of configuration homomorphisms encountered earlier. It can also be generalized in the following manner which is close to Theorem 1.2 in Volume I, Chapter 3, with the modifications motivated by the change to a more general definition of homomorphism. We still operate with the same global regularity for $\mathscr{C}$ and $\mathscr{C}'$.

Consider an epimorphism respecting external bonds $h: \mathscr{T} \to \mathscr{T}'$, with $\mathscr{T} = \langle \mathscr{C},R\rangle$, $\mathscr{T}' = \langle \mathscr{C}',R'\rangle$, and introduce the factor group $F = S/N$, where N is the normal subgroup

$$N = \{s|h(s) = e'\}. \tag{8.6}$$

Introduce the relation R_o on $\mathscr{C}\times\mathscr{C}$ by

$$c_1R_oc_2 \quad \text{iff} \quad h[I(c_1)] = h[I(c_2)]. \tag{8.7}$$

It is an equivalence cruder than R and hence defines a partition whose equivalence classes are unions of $\mathscr{T}$-images. On these classes we let the element of F operate in the natural manner: any $f \in F$ corresponds to a coset of N and we apply one of the elements of this coset to an $\mathscr{T}$-image in the equivalence class. This gives a unique result since if the two S-similarities s_1 and s_2 belong to the same coset we have $h(s_1) = h(s_2)$ so that if I_1 and I_2 are the two images mentioned, so that $h(I_1) = h(I_2)$, this implies

$$h(s_1I_1) = h(s_1)h(I_1) = h(s_2)h(I_2) = h(s_2I_2). \tag{8.8}$$

But relation (8.8) says that $(s_1I_1)R_o(s_2I_2)$, see (8.7), so that the result of our definition is unique and the writing fJ makes sense if J is one of the equivalence classes.

With F as the similarity group we can show that R_o is an identification rule as we did for Theorem 1. Let $\mathcal{I}_o$ be the resulting image algebra.

When can we guarantee that $\mathcal{I}_o$ is isomorphic to $\mathcal{I}'$? It is clear that the relation that $\underline{h}$ induces between $\mathcal{I}_o$ and $\mathcal{I}_1$ is bijective (remember that $\underline{h}$ is assumed to be surjective) and the same holds for the relation between F and S'. That it is homomorphic is seen as in the last proof. Consider now $h^{-1} = k$; see Notes B.

Obviously $k:S \to F$ is a group homomorphism. How about $k: \mathcal{I}' \to \mathcal{I}_o$? Given a $I' \in \mathcal{I}'$ and $s' \in S'$ there exist $I \in \mathcal{I}$, $s \in S$ with $h(I) = I'$, $h(s) = s'$. Let J be the $\mathcal{I}_o$-image containing I and let f be the coset of N containing s. Then

$$h(fJ) = h(f)h(J) = s'I' \tag{8.8a}$$

so that

$$k(s'I') = fJ = k(s')k(I') \tag{8.9}$$

and the first relation in Definition 7.1 holds.

Now let $I_1', I_2', I' = \sigma(I_1', I_2')$ all belong to $\mathcal{I}'$. Using the surjective property again we can find $\mathcal{I}$-images I_1, I_2 which are h-mapped into the primed images. Let J_1, J_2 be the $\mathcal{I}_o$-images containing the unprimed $\mathcal{I}$-images. If we knew that $I = \sigma(I_1, I_2) \in \mathcal{I}$, then $hI = \sigma(hI_1, hI_2) = \sigma(I_1', I_2') = I'$ so that

$$kI' = k\sigma(I_1', I_2') = \sigma(kI_1', I_2') \tag{8.10}$$

as needed to guarantee that k is homomorphic. In general we do not know this; we need an additional condition.

We shall assume that $\mathcal{T}$ is globally free over $\{\sigma\}$ and that h *respects bond conditions*. By the latter we shall mean that bonds of hI images in $\mathcal{T}'$ satisfy ρ' only if the homologous (remember that external bond structure is respected) bonds of I satisfy ρ. It then follows local regularity holds for $\sigma(I_1,I_2)$ so that k is homomorphic and h: $\mathcal{T}_0 \cong \mathcal{T}'$ is an isomorphism. Summing up we have

Theorem 2. *Consider an epimorphism* h: $\mathcal{T} \to \mathcal{T}'$ *respecting external bonds and bond relations and where* $\mathcal{T}$ *is globally free. Define the relation* R_0 *on* $\mathcal{C} \times \mathcal{C}$ *by*

$$c_1 R_0 c \quad iff \quad h[I(c_1)] = h[I(c_2)]. \tag{8.11}$$

Then R_0 *defines an identification rule cruder than* R *and the resulting image algebra* $\mathcal{T}_0$ *is isomorphic to* $\mathcal{T}'$.

Conversely let us start with an identification rule R_0 cruder than R and consider the natural map h: $\mathcal{T} \to [\mathcal{T}]_{R_0}$. Put $hs \equiv s$, $h\sigma \equiv \sigma$. Then we have

Theorem 3. *The natural map* h: $\mathcal{T} = \langle \mathcal{C}, R\rangle \to [\mathcal{T}]_{R_0} = \langle \mathcal{C}, R_0\rangle$ *is an epimorphism respecting external bonds and bond relations.*

Proof: The natural map is automatically surjective. It is clear that, for arbitrary $I \in \mathcal{T}$,

$$h(sI) = [sI]_{R_0} = s[I]_{R_0} = shI = h(s)h(I) \tag{8.12}$$

since R_0 is an identification rule. Second, if $I_1, I_2, I = \sigma(I_1,I_2) \in \mathcal{T}$ it follows that

$$hI = [\sigma(I_1,I_2)]_{R_0} = \sigma([I_1]_{R_0},[I_2]_{R_0}) = \sigma[hI_1,hI_2]. \quad (8.13)$$

Finally, since $[I]_{R_0}$ has the same external bonds as I it is clear that the natural map respects external bonds. But here $\rho = \rho'$ so that the bond relations are respected. Q.E.D.

We shall now construct an image algebra of use for the *weakest regularities*. Let Λ be a set of arbitrary cardinality and consider all functions $f:\Lambda \to \mathbb{N}$ with $f(\lambda) = 0$ for all but a finite number of λ-values. The set of such functions forms a commutative semi-group F under addition pointwise $f+g$: $(f+g)(\lambda) \to f(\lambda) + g(\lambda)$, $\lambda \in \Lambda$.

Given a group S of transformations $s:\Lambda \to \Lambda$, consider an arbitrary S-invariant congruence relation $\underline{r}$ over F. In other words, $frg \Rightarrow f(s\cdot)rg(s\cdot)$, $\forall s \in S$, and $f_1rg_1, f_2rg_2 \Rightarrow (f_1+f_2)r(g_1+g_2)$.

Introduce $\mathcal{T} = \langle\Lambda, S, \text{DISCRETE}, R\rangle$ with $c = (\lambda_1,\lambda_2,\ldots)Rc' = (\lambda_1',\lambda_2',\ldots)$ if $f_c r f_{c'}$. Here

$$f_c(\lambda) = \#(\lambda|\lambda \in c) \quad (8.14)$$

so that, since we only consider finite configurations, $f_c \in F$, so that (8.14) makes sense. This R is a legitimate identification rule according to Definition 1.1 in Chapter 3, Volume I. Indeed, it is obviously an equivalence, and, since $\mathcal{R}$= FREE generators can be treated as nullary, condition (ii) in the definition holds trivially. Condition (iii) follows from the fact that r is S-invariant. The last condition (iv), is satisfied since for regularity DISCRETE a combination of two configurations c and c' means simply that the two contents are joined by a disjoint union. This implies that

$$f_{\sigma(c,c')} = f_c + f_{c'}, \tag{8.15}$$

corresponding to addition in our semi-group F. Of course σ establishes no bond connections in the present case. But (8.15) implies, since r is a congruence over the semi-group, that if c_1Rc_1', c_2Rc_2' we have

$$\begin{cases} f_{c_1} r f_{c_1'}, \quad f_{c_2} r f_{c_2'} \\ \Rightarrow f_{\sigma(c_1,c_2)} = (f_{c_1}+f_{c_2}) = (f_{c_1'}+f_{c_2'}) = f_{\sigma(c_1',c_2')} \\ \Rightarrow \sigma(c_1,c_2) R \sigma(c_1',c_2'). \end{cases} \tag{8.16}$$

Hence R is an identification rule and the above $\mathscr{T}$ is well-defined. Such an $\mathscr{T}$ is called a *discrete image algebra,* the name being motivated by the following

Theorem 3. *Consider an arbitrary image algebra* $\mathscr{T}_0 = \langle \mathscr{C}_0, \mathscr{R}_0 \rangle$ *with regularity* $\mathscr{R}$ = DISCRETE. *Then it is isomorphic to some discrete image algebra* $\mathscr{T}$, $\mathscr{T} \cong \mathscr{T}_0$.

Proof: Starting from $\mathscr{T}_0 = \langle G,S,\text{DISCRETE},R_0 \rangle$ label all the g's in G by some arbitrary label λ so that $G \leftrightarrow \Lambda$. If G is finite Λ can be taken as $\{1,2,\ldots n\}$, if it is denumerable $\Lambda = \mathbb{N}$, and so on for higher cardinality. We disregard all information contained in the generators, and only insist that the λ-labelling separate elements in G.

All generators in G can be changed to nullary ones without changing the regularity as will often be convenient to assume done when dealing with discrete regularity. Since combination in $\mathscr{C}_0 = \mathscr{C}(\langle G,S,\text{DISCRETE}\rangle)$ means adding frequencies we can describe any $c \in \mathscr{C}_0$ by the function $f_c \in F$, constructed as above, and satisfying the relation (8.15).

Defining an equivalence r in the semi-group F by

$$f_1 r f_2,\ f_1 = f_{c_1} \text{ and } f_2 = f_{c_2},\ \text{if } c_1 R_0 c_2 \tag{8.17}$$

it follows that r is S-invariant, since R_0 is, and that r is a congruence over the semi-group, because of property (iv) of Definition 1.1, Volume I, Chapter 1. Hence r has the properties required and leads to a discrete image algebra $\mathcal{T}$.

Consider now the map $h: \mathcal{T}_0 \leftrightarrow \mathcal{T}$ induced by $\mathcal{C}_0 \leftrightarrow F \leftrightarrow \mathcal{C} = \langle F, S, \text{DISCRETE}\rangle$. Note that this map is S-invariant and that $h(\sigma(c_1,c_2) = \sigma(h(c_1),h(c_2))$ so that h is a homomorphism. But h is bijective so that $h^{-1} = k$ is uniquely defined. But k is also a homomorphism so that it is also an isomorphism and $\mathcal{T} \cong \mathcal{T}_0$. Q.E.D.

Remark. We mention in passing that one can also exploit the generator index $\alpha \in A$. If $G^\alpha = Sg_\alpha$, so that each generator class is generated by a single prototype g_α, for some $\mathcal{T}_0$ with DISCRETE regularity, we can proceed as follows. Let the functions f be defined on A and take as values finite sets (allowing repetition) $(s_1, s_2, \ldots)$, $s_i \in S$, with an arbitrary numbers of elements. Define the "sum" as

$$(f_1+f_2)(\alpha) = f_1(\alpha) \cup f_2(\alpha) \tag{8.18}$$

consider the map $f{:}f(c) \to (s_1, s_2, \ldots)$, where the $g_i = s_i g_\alpha$ are the generators in c, and g_α is the prototype in G^α. Then addition as in (8.18) corresponds to combination of the corresponding configurations. We can then go ahead as above, except that the semi-group F is different and now has the binary operation in (8.18). This possibility will not be pursued further.

Instead we shall now study certain polynomials on image algebras and some set operations. We shall start with the simplest case when the regularity is as weak as possible, $\mathscr{R}$ = DISCRETE. Theorem 3 tells us that we can assume $\mathscr{I}$ to be a free image algebra over some set Λ as described above without loss of generality. But then $\mathscr{I}$ forms a commutative monoid distributive relative to the unary operations s. We have just these two base operations from which polynomials can be formed.

As an example, let G consist of half planes, and $S = EG(2)$ and R identifies intersections of half planes. At this time we do not ask for simple descriptions, which would require $\mathscr{R}$ = <UNEQUAL,FULL>; instead we choose just $\mathscr{R}$ = DISCRETE. Images then mean convex sets and we have two base operations: Euclidean motions (similarities) and intersection of sets.

Another simple example is when G consists of functions $X \to \mathbb{R}$ or $\mathbb{C}$, $\mathscr{R}$ = DISCRETE, s means multiplication by real or complex non-zero scalars and R identifies sums of functions. Then $\mathscr{I}$ becomes just a vector space.

For DISCRETE image algebras all polynomials are entire, but this will not be true when we go ahead to more stringent regularities.

With configurations as in Example 3.1, where Σ = LINEAR, and with R identifying functions on their domains we have only two binary base operations in addition to the unary one $s: I \to sI$ that is always present. For example, we cannot use the connector in-bond of I_1, to out-bond of I_2, out-bond of I_1 to in-bond of I_2, since this would offend against the global regularity of Σ. We have $\sigma_{\text{left}}(I_1, I_2)$ and

$\sigma_{right}(I_1,I_2)$, the first of which concatenates I_1 to the left of I_2 (whenever possible) and the second one concatenates I_1 to the right of I_2 (whenever possible). These functions are partial since σ_{left} is defined only when the right endpoint and function value of I_1 agrees with the left endpoint and function value of I_2. Hence σ_{left} and σ_{right} are both partial base operations and so will be the polynomials formed from them. We have $\sigma_{left}(I_1,I_2) = \sigma_{right}(I_2,I_1)$.

Now let configurations be as in Example 3, where Σ = TREE and with R identifying the ordered sequence of lowest nodes. Then many binary base operations are possible, for example $\sigma_i(I_1,I_2)$ which connects the in-bond of I_1 to the ith out-bond of I_2 when possible. This requires that I_2 has at least i out-bonds and that the ith out-bond value of I_2 equals the in-bond value of I_1. Specializing further to Σ = BINARY TREE we get four basic operations and of course the unary s-operation.

More generally, for a fixed Σ and two images $I_1,I_2 \in \mathcal{I}$, a binary operation f first computes subsets of $B_s^e(I_1)$ and $B_s^e(I_2)$ in terms of the coordinates that label the external bonds. These two sets of bonds are connected in a way prescribed by f, not offending against $\mathcal{R}$, whenever possible into $\sigma(I_1,I_2)$. From these base operations and s we build the polynomials we need.

Consider now two image algebras $\mathcal{I}_1$ and $\mathcal{I}_2$ of equal global regularity and connected by the homomorphism $h: \mathcal{I}_1 \to \mathcal{I}_2$. A polynomial $p = p(I_1,I_2,I_3,\ldots)$ on $\mathcal{I}_1$ is defined in terms of base operations that have meaning on $\mathcal{I}_2$

as well for $hI_1, hI_2, hI_3, \ldots$.

The connector σ can then connect hI_1 and hI_2 within $\mathcal{T}_2$, and the unary s-operator of course also preserves external bond structure, so that $p(hI_1, hI_2, hI_3, \ldots)$ is defined as soon as $p(I_1, I_2, I_3, \ldots)$ is defined. In addition it is clear that

$$p(hI_1, hI_2, hI_3, \ldots) = hp(I_1, I_2, I_3, \ldots). \tag{8.19}$$

Note that a polynomial is defined as an element in $\mathcal{L}(\mathcal{R})$ only if all the successive operations (in the order indicated) needed to calculate it are well defined.

Now recall the meaning of *combine* and *span*, see Section 6.

<u>Theorem 4</u>. *The two set operations* combine *and* span *are constant on images if* R *is a strong* (see Notes C) *identification rule:*

$$\begin{cases} \underline{\text{combine}}\ (c_1, c_2, \ldots) \equiv \underline{\text{combine}}\ (c_1', c_2', \ldots)(\text{mod } R) \\ \underline{\text{span}}\ (c_1, c_2, \ldots) \equiv \underline{\text{span}}\ (c_1', c_2', \ldots)(\text{mod } R) \end{cases} \tag{8.20}$$

if $c_i \equiv c_i'$ (mod R), $i = 1, 2, \ldots$.

<u>Proof</u>: Consider $A = \underline{\text{combine}}(c_i)$, $A' = \underline{\text{combine}}(c_i')$. The general element a in A can then be written as in (6.1) with some connector σ and similarities $s_1, s_2, \ldots$. Form the same combination a' of the c_i' configurations. It then follows that a' is also regular since R is a strong identification rule, and using conditions (iii), (iv) of Definition 1.1 in Volume I, Chapter 3, and that $a \equiv a'(\text{mod } R)$. We treat the general element from A' in the same way and the first statement in (8.20) is established.

To prove the second relation we only have to repeat the first one since span is the iteration closure of combine. Hence both set operations are constant on images modulo R. Q.E.D.

For globally free regularity any identification rule is a strong identification rule. Indeed, if $c_i R c_i'$; $i = 1,2,$, and if $\sigma(c_1,c_2) = c$ is regular then $\sigma(c_1',c_2')$ must also be regular since only local regularity is needed and $B_e(c_i) = B_e(c_i')$; they have the same external bonds.

It then makes sense (still working with a strong R) to keep the same symbols for the set operations $2^{\mathcal{T}} \to 2^{\mathcal{T}}$ for images as we used for configurations so that we shall write

$$\text{\underline{combine} } A,\ \text{\underline{span} } A,\ A \in 2^{\mathcal{T}} \tag{8.21}$$

and the functions in (8.21) take values in $2^{\mathcal{T}}$.

This set operation span, defined for image sets, is covariant with respect to S: span SA = S span A, which follows directly from (6.1) and the fact that span is the iteration closure of combine. It also behaves as in (i) and (ii) of Theorem 6.1, since the operations on configurations were just seen to be constant on images.

In particular, for A and $B \in 2^{\mathcal{T}}$, introduce the binary operation "⊕" by

$$A \oplus B = \text{\underline{span} } (A \cup B). \tag{8.22}$$

Theorem 5. *For globally free regularity over* $\{\sigma\}$ *the set* $\mathcal{J} = 2^{\mathcal{T}}$ *is a commutative monoid over "⊕" with "⊕" distributive relative to the similarities.*

Proof: Consider three subsets A, B, and C of $\mathcal{J}$. Then $(A \oplus B) \oplus C$ consists of all images that contain configurations of the form (6.9) where the c's have been selected from A, B, and C. But this is then the same as $A \oplus (B \oplus C)$, so that $\oplus$ is associative. The empty image $e \equiv \{\phi\} (\text{mod } R)$ plays the role of unit element, and commutativity is obvious. Q.E.D.

Example 1. Let the generators be half-planes in R^2, $\mathcal{R}$ = DISCRETE, S = EG(2), and R identification by intersection as before. Then images represent convex sets. If A and B both consist of a single image I_1 and I_2 respectively, $A \oplus B$ will consist of the empty image, all sets congruent to A or to B, and to intersections of an arbitrary number of sets congruent to A or B.

Example 2. Let the generators be real-valued functions on [0,1], let s perform multiplication by non-zero real numbers, and let R identify sums of the functions appearing in the configurations. Then span means linear closure and $\oplus$ computes the linear closure of two sets of functions.

Example 3. Let generators be rewriting rules in a finite state language, $\mathcal{R}$ = LINEAR, and R identify grammatical strings with the same external bonds. Then images means grammatical phrases. If A and B each consists of a single phrase, $A = \{I_1\}$, $B = \{I_2\}$, $A \oplus B$ consists of the empty string and all grammatical phrases that can be obtained by concatenating a number of I_1-copies and I_2-copies.

With the same assumptions as in Theorem 5, fix A and form the function

$$f_A : \mathcal{J} \to \mathcal{J}, \quad f_A(B) \to A \oplus B. \tag{8.23}$$

Since "$\oplus$" is associative we have the composition rule

$$f_{A_2} f_{A_1} = f_{A_2 \oplus A_1} \tag{8.24}$$

so that f_A is a representation of the semigroup $\mathscr{I}$ by functions $\mathscr{I} \to \mathscr{I}$. While configurations can be said to be formulas, the *images and sets of images can be viewed as functions represented by those formulas.*

Returning to the notion of configuration projections in the last section, we will study what happens when the identification rule identifies generator indices in the following sense.

<u>Definition 1</u>. *If* Σ *is monotonic we shall say that the identification rule identifies generator indices if for* $c_1, c_2 \in \mathscr{L}(\mathscr{R})$ *the statement* $c_1 R c_2$ *implies*

$$(P_\alpha c_1) R (P_\alpha c_2);\ \forall \alpha \tag{8.25}$$

and

$$\sigma^A(c_1) = \sigma^A(c_2). \tag{8.26}$$

<u>Remark</u>. The statement that the two connectors in (8.26) are the same should be interpreted as follows. Let us write the two configurations

$$\begin{cases} c_1 = \sigma^A(c_1)(c_{11}^{\alpha_1}, c_{12}^{\alpha_2}, \ldots) \\ c_2 = \sigma^A(c_2)(c_{21}^{\alpha_1}, c_{22}^{\alpha_2}, \ldots) \end{cases} \tag{8.27}$$

as connections of P_α projections for various α-values, so that

$$c_{1i}^{\alpha_i} R\, c_{2i}^{\alpha_i},\ c_{1i}^{\alpha_i} = P_{\alpha_i} c_1,\ c_{2i}^{\alpha_i} = P_{\alpha_i} c_2. \tag{8.28}$$

Then $\sigma^A(c_1)$ is the connector combining the subconfigurations of c_1 together, and similarly $\sigma^A(c_2)$ for the second configuration. The external bonds of c_{1i}^{i} and c_{2i}^{i} are the same; see Volume I, 3.1, so that the statement in (8.25) makes sense.

Remark 2. In the special case $\mathcal{R}$ = DISCRETE when no connections are established the condition (8.26) plays no role and we need only consider (8.25).

Then it is possible to extend the projections P_B to the image algebra; see Section 5.

Theorem 6. *If* Σ *is monotonic and* R *identifies generator indices, for any* $I \in \mathcal{T}$ *the set*

$$P_B I = \{P_B c | c \in I\} \tag{8.29}$$

belongs to a single image in $\mathcal{T}$, *so that* P_B *maps* $\mathcal{T}$ *into itself.*

Proof: Consider two configurations $c_1 = \sigma_1(g_{11}, g_{12}, \ldots)$, $c_2 = \sigma_2(g_{21}, g_{22}, \ldots)$ in I so that $c_1 R c_2$. Then for any $\alpha \in A$ we have (8.25). Forming $P_B c_1$ and $P_B c_2$ means that in (8.27) we drop all subconfigurations whose indices are not in B and, at the same time, leave out their bonds. This means that $P_B c_1$ and $P_B c_2$ can be written

$$\begin{cases} P_B c_1 = \sigma(c_k', \ k = 1,2,\ldots) \\ P_B c_2 = \sigma(c_k'', \ k = 1,2,\ldots) \end{cases} \tag{8.30}$$

with the same σ in both cases, with $c_k' = P_{\alpha_k} c_1, c_k'' = P_{\alpha_k} c_2$, and with $c_k' R c_k''$; $k = 1,2,\ldots$. But then, according to Definition 1.1 in Volume I, Chapter 3, the rule R identifies $P_B c_1$ with $P_B c_2$ so that $P_B I$ in (8.29) is uniquely determined. Q.E.D.

Corollary. *Under the conditions of Theorem 6 images have a canonical representation*

$$I = \sigma^A(I)(I_{\alpha_1}, I_{\alpha_2}, \ldots) \tag{8.31}$$

with $I_\alpha \in \mathcal{T}^\alpha$.

This follows directly from the above proof with I_{α_k} as the image containing the regular configuration c_k'. The values of α appearing in (8.31) are those for which

$$P_\alpha I \neq \phi \tag{8.32}$$

so that the set $(\alpha_1, \alpha_2, \ldots)$ can be written

$$\alpha(I) = \{\alpha | P_\alpha I \neq \phi\}. \tag{8.33}$$

The representation in (8.31) subdivides the image into subconfigurations with constant generator index: *the regular behavior is analyzed into homogeneous elements.*

An indirect characterization of projection operators in image algebras will be attempted as follows.

Theorem 7. *Let* Σ *be monotonic and assume that for each* $\alpha \in A$ *there is an epimorphism* $P_\alpha\colon \mathcal{T} \to \mathcal{T}_\alpha$

$$P_{\alpha_1} P_{\alpha_2} I = \begin{cases} e & \text{if } \alpha_1 \neq \alpha_2 \\ P_\alpha I & \text{if } \alpha_1 = \alpha_2 = \alpha \end{cases} \tag{8.34}$$

Then

$$\begin{cases} \text{(i)} \quad P_\alpha = \mathrm{id}_{\mathcal{T}^\alpha} & \text{when restricted to } \mathcal{T}^\alpha \\ \text{(ii)} \quad \mathcal{T}^{\alpha_1} \cap \mathcal{T}^{\alpha_2} = \{e\} & \text{if } \alpha_1 \neq \alpha_2 \\ \text{(iii)} \quad I = \sigma_\alpha I^\alpha & \text{with } I^\alpha = P_\alpha I. \end{cases} \tag{8.35}$$

Proof: Recall that $\mathcal{T}^{\alpha}$ is the subimage algebra <G^{α},S,Σ,ρ> and consider an image $I \in \mathcal{T}^{\alpha}$. Then, since P_{α} is surjective (on $\mathcal{T}_2$) there exists (at least) one I_o such that $I = P_{\alpha}I_o$. But then

$$P_{\alpha}I = P_{\alpha}P_{\alpha}I_o = P_{\alpha}I_o = I \tag{8.36}$$

which proofs (i).

To see that (ii) holds say that I is in both $\mathcal{T}^{\alpha_1}$ and $\mathcal{T}^{\alpha_2}$. We can then write $I = P_{\alpha_2}I_o$, $I_o \in \mathcal{T}^{\alpha_2}$ so that, since $I \in \mathcal{T}^{\alpha_1}$ and (8.34) holds,

$$I = P_{\alpha_1}I = P_{\alpha_1}P_{\alpha_2}I_o = e. \tag{8.37}$$

Now let $I = [c]_R = [\sigma(g_1,g_2,\ldots)]_R$, and decompose c after α-classes of generators,

$$c = \underset{\alpha}{\sigma}\, c^{\alpha} \tag{8.38}$$

where c^{α} are all regular since Σ is monotonic. Then

$$I = \underset{\alpha}{\sigma}(I^{\alpha}), \quad I^{\alpha} = [c^{\alpha}]_R, \tag{8.39}$$

using (iv) of Definition 1.1 in Volume I, Chapter 3. But P_{β} was assumed to be homomorphic so that

$$P_{\beta}I = \underset{\alpha}{\sigma}(P^{\beta}I^{\alpha}) \tag{8.40}$$

where $P^{\beta}I^{\alpha} = e$ for all α except $\alpha = \beta$. Hence

$$P_{\beta}I = \underset{\alpha}{\sigma}(e,e,\ldots I^{\beta},e,\ldots) = I^{\beta} \tag{8.41}$$

so that (iii) holds. Q.E.D.

3.9. Representations by image isomorphisms

One of the tasks in abstract pattern theory is to *relate general classes of patterns to concrete patterns.* Just as the representation theory of groups studies homomorphic mappings into spaces of linear operators, representation theory for regular structures deals with homomorphisms into certain concrete regular structures. For the latter we fix $\mathcal{R} = \langle\rho,\Sigma\rangle$, both local and global regularity. In this section we shall examine three types of regularity and begin with the case when the bond relation ρ is EQUAL and the connection type Σ is LINEAR.

Is it possible to give conditions for an image algebra, in terms of its similarity and composition tables, to be isomorphic to one with prescribed regularity $\mathcal{R}$? In the present case $\mathcal{R}$ would make the new images have $\omega_{in} = \omega_{out} = 1$, the external bonds should fit each other by equality, and the images should be linked together from left to right. We shall now show that this is possible.

Recall the functions L and R that map $\mathcal{T} \to 2^{\mathcal{T}}$ by $L: I \to L(I)$ and $R: I \to R(I)$ respectively, see Definition 7.2.

Theorem 1. *In order that* $\mathcal{T}[S,\sigma]$ *be isomorphic to an image algebra* $\mathcal{T}'[S,\sigma']$ *with in-arity and out-arity one, and* σ' *meaning "concatenate to the right" and with* <EQUAL,LINEAR>-regularity, *it is necessary and sufficient that*

$$R \equiv RLR. \tag{9.1}$$

Proof: To prove the indirect part of the theorem assume that h maps $\mathcal{I}$ isomorphically onto an image algebra $\mathcal{T}'$ with <EQUAL,LINEAR>-regularity. Then I'-images have

$\omega_{in}(I') = \omega_{out}(I') = 1$ with corresponding bond values $\beta_{in}(I')$ and $\beta_{out}(I')$. Because of the isomorphism in order that $\sigma(I_1, I_2) \in \mathcal{T}$ it is necessary and sufficient that $\sigma'(hI_1, hI_2) \in \mathcal{T}'$. The latter occurs iff $\beta_{out}(hI_1) = \beta_{in}(hI_2)$ so that

$$R(I) = \{A \in \mathcal{T} | \beta_{in}(hA) = \beta_{out}(hI)\}. \tag{9.2}$$

Then, applying the function L to both sides of (9.2),

$$LR(I) = \{B \in \mathcal{T} | \beta_{out}(hB) = \beta_{out}(hI)\} \tag{9.3}$$

and, now applying R to both sides of (9.3),

$$RLR(I) = \{C \in \mathcal{T} | \beta_{in}(hC) = \beta_{out}(hI)\} = R(I). \tag{9.4}$$

But (9.4) is the condition (9.1) that we wanted to establish so that the necessary part of the theorem has been proved.

The direct part of the proof is harder. It requires us to exhibit an image algebra $\mathcal{T}'$ of <EQUAL,LINEAR>-regularity and an isomorphism $h: \mathcal{T} \leftrightarrow \mathcal{T}'$ if (9.1) holds. Let us define $\mathcal{T}'$ as consisting of all the images of $\mathcal{T}$ but with bonds changed as indicated below, if necessary with additional markers making the resulting I'-images distinct after bonds have been changed. Let h be the natural map between $\mathcal{T}$ and $\mathcal{T}'$. Let $B_s(I') = \{in, out\}$ *with bond values for* $I' = hI$ *given as subsets in* $\mathcal{T}$

$$\begin{cases} \beta_{in}(I') = RL(I) \subset \mathcal{T} \\ \beta_{out}(I') = R(I) \subset \mathcal{T} \end{cases} \tag{9.5}$$

if $RL(I) \neq \phi$ and $R(I) \neq \phi$. In the opposite cases let

$$\begin{cases} \beta_{in}(I') = n_{in} & \text{if } RL(I) = \phi \\ \beta_{out}(I') = n_{out} & \text{if } R(I) = \phi \end{cases} \tag{9.6}$$

where n_{in} and n_{out} are abstract symbols to be understood as different from each other and from all subsets of $\mathcal{T}$. Then the same S remains a similarity group of $\mathcal{T}'$, since $\beta_{out}(I_1') = \beta_{in}(I_2')$ means $R(I_1) = RL(I_2) \neq \phi$ if $I_1' = hI_1$, $I_2' = hI_2$. But definition sets are covariant with respect to similarities (see Section 7) so that

$$\begin{aligned}\beta_{out}(sI_1') &= R(sI_1) = sR(I_1) = sRL(I_2) = rsL(I_2) \\ &= RL(sI_2) = \beta_{in}(sI_2').\end{aligned} \tag{9.7}$$

Give $\mathcal{T}'$ the connection type LINEAR via concatenation to the right.

To see that $\mathcal{T}$ and $\mathcal{T}'$ are isomorphically related by h assume that $\sigma(I_1,I_2) \in \mathcal{T}$ so that

$$\begin{cases} I_1 \in L(I_2), \text{ hence } L(I_2) \neq \phi \\ I_2 \in R(I_1), \text{ hence } R(I_1) \neq \phi \end{cases} \tag{9.8}$$

Applying R to the first relation and then using the second we get

$$R(I_1) \subseteq RL(I_2) \subseteq RLR(I_1). \tag{9.9}$$

But $R = RLR$ so that

$$R(I_1) = RL(I_2) \rightarrow \beta_{out}(hI_1) = \beta_{in}(hI_2). \tag{9.10}$$

But this implies $\sigma'(I_1',I_2') \in \mathcal{G}'$ as requested.

On the other hand if $\sigma'(I_1',I_2') \in \mathcal{T}'$ we have $R(I_1) = RL(I_2) \neq \phi$. But this implies that $L(I_2) \neq \phi$ so that there exists an image J in $\mathcal{T}$ such that $\sigma(J,I_2) \in \mathcal{T}$ and then of course $I_2 \in RL(I_2)$. Therefore

$$I_2 \in RL(I_2) = R(I_1) \tag{9.11}$$

so that $\sigma(I_1,I_2) \in \mathscr{T}$ as requested. Q.E.D.

As an example let us consider an $\mathscr{T}$ with generators with $\omega(g) = 3$, symmetric regularity, and the connection type indicated in Figure 9.1, with an arbitrary number of vertical pairs of generators. The images have arity four.

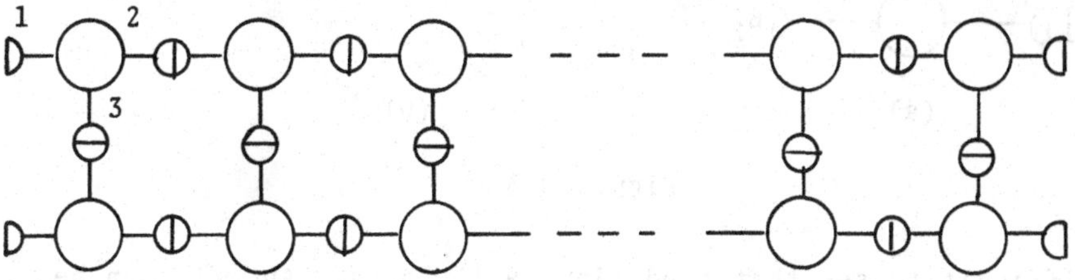

Figure 9.1

Let generators denote linear functions f on intervals [a,b] of Z with nondegenerate support (containing more than a single point) and let the 1-bond (see Figure 9.1) be (a,f(a)), the 2-bond be (b,f(b)) and the 3-bond be [a,b]. With ρ as EQUAL, let R identify pairs of functions on the same interval, schematically as in Figure 9.2.

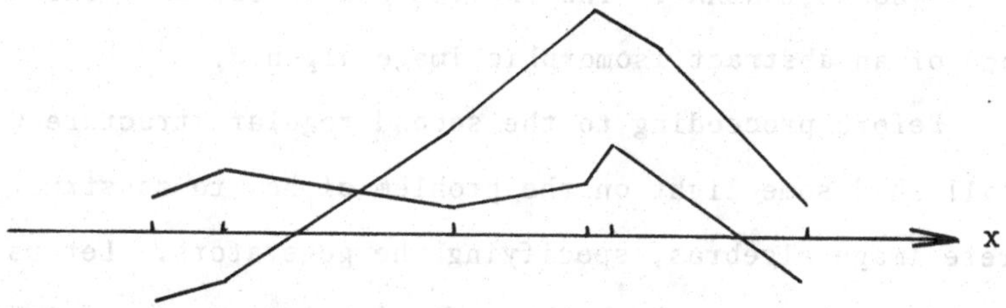

Figure 9.2

The resulting $\mathscr{T}[S;\sigma]$ with S = translation group on the real line and σ as continuous concatenation to the right of functions taking as values 2-vectors, it is not

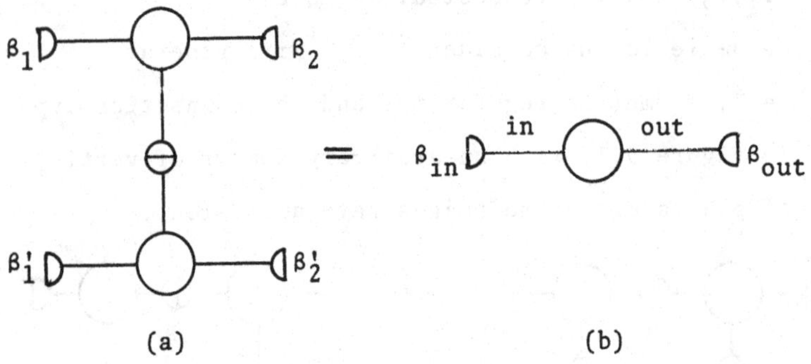

Figure 9.3

difficult to see that condition (9.1) holds. An $\mathcal{T}'$ can be constructed by taking as G' the macrogenerators the $\mathcal{C}$-configurations in (a) of Figure 9.3, of arity four, and for which the 3-bond satisfies ρ. Make them into g' generators as in (b), with arity two, with directed regularity and with bond the vector $\beta_{in} = (\beta_1,\beta_1')$ and with out-bond $\beta_{out} = (\beta_2,\beta_2')$. R' identifies vector valued functions. The new $\mathcal{T}'$ is isomorphic to the old $\mathcal{T}$.

In this case we could implement the isomorphism by giving a concrete image algebra $\mathcal{T}' = \langle G',S,\mathcal{R}',R'\rangle$ with $\mathcal{R}' = \langle$EQUAL,LINEAR$\rangle$. The theorem guarantees only the existence of an abstract isomorphic image algebra.

Before proceeding to the second regular structure we shall shed some light on the problem of how to construct concrete image algebras, specifying the generators. Let us introduce the natural concept of prime images; see Notes B.

<u>Definition 1</u>. *Given an abstract image algebra* $\mathcal{T} = \mathcal{T}[S;\sigma]$, *an image* $I \in \mathcal{T}$ *is said to be prime if it cannot be written in a non-trivial way as* $I = \sigma(I_1,I_2) \in \mathcal{T}$; I_1 *and* $I_2 \neq e$.

We need characterization of the numerical complexity of an image; see Volume I, p. 24, for the numerical complexity of a configuration. The latter is simply $n(c)$ = number of generators in the configuration c. For a fixed I the value of $n(c)$, $I = [c]_R$, is in general variable, and we shall let the numerical complexity of the image be

$$\max(I) = \sup n(c), \quad I = [c]_R \tag{9.12}$$

allowing also the value $+\infty$ for $\max(I)$. We then have an obvious

Lemma 1. *For a strong identification rule the numerical complexity of images is a conditionally super additive function via* σ

$$\max[\sigma(I_1,I_2)] \geq \max(I_1) + \max(I_2), \text{ if } \sigma(I_1,I_2) \in \mathscr{T}. \tag{9.13}$$

Proof: Assume $I = \sigma(I_1,I_2) \in \mathscr{T}$ and that both I_1 and I_2 have finite bounded complexity. Select regular configurations c_1 and c_2 such that $I_1 = [c_1]_R$, $I_2 = [c_2]$, $I = [\sigma(c_1,c_2)]$. Find c_1' and c_2' from the images I_1 and I_2 respectively so that

$$\begin{cases} \max(I_1) = n(c_1') \\ \max(I_2) = n(c_2') \end{cases} \tag{9.14}$$

Since $c_1 R_1' c_1', c_2 R c_2'$ and since $c = \sigma(c_1,c_2)$ is regular and equivalent to c, it follows since R is a strong identification rule that $\sigma(c_1',c_2') \in \mathscr{L}$ and $n[\sigma(c_1',c_2')] = n(c_1') + n(c_2')$. Recalling the definition (9.12) this shows that

$$\begin{aligned}[\sigma(c_1',c_2')]_R = I \Rightarrow \max(I) &\geq n(c_1') + n(c_2') \\ &= \max(I_1) + \max(I_2)\end{aligned} \tag{9.15}$$

so that (9.13) holds. If one or both images have infinite numerical complexity, pick the prototypes c_1 and c_2 such that their numerical complexity is greater than some arbitrarily chosen large number M. Repeating the argument it follows that $\max(I) = +\infty$. Q.E.D.

Let us consider the case when $\mathcal{I} = \mathcal{I}[S;\sigma]$ and introduce the set

$$\mathcal{I}_{\text{prime}} = \{I \in \mathcal{I} \mid I \text{ is a prime image}\}. \tag{9.16}$$

It is clear that $\mathcal{I}_{\text{prime}}$ is an S-invariant set, $S\mathcal{I}_{\text{prime}} = \mathcal{I}_{\text{prime}}$, so that the prime images form a pattern. See Notes C. Can one represent arbitrary images as compositions via σ of prime images? A partial answer is given by

Theorem 2. *If $\mathcal{I}$ has only images of finite numerical complexity relative to a strong identification rule, and if empty configurations are not regular, than any of its non-prime images I can be decomposed via σ as finite combinations of prime images.*

Proof: If $I \notin \mathcal{I}_{\text{prime}}$ there exist $I_1, I_2 \in \mathcal{I}$ such that $I = \sigma(I_1, I_2)$. If both $I_1, I_2 \in \mathcal{I}_{\text{prime}}$ we are done. Otherwise at least one of the I_1, I_2 is non-prime and can be decomposed. For example if I_1 is non-prime there exist I_3, I_4 such that $I_1 = \sigma(I_3, I_4)$. However since $\max(I_2) \geq 1$ we get from Lemma 1 $\max(I_1) < \max(I) < \infty$, so that the numerical complexities of the images involved decrease. Hence the chain of successive decompositions must end after a finite number of steps, and we have arrived at a decomposition of the type stated. Q.E.D.

Of course nothing guarantees in general that the decomposition is unique. Also, it should be noted that finite numerical complexity is a very strong requirement. For example, if the image algebra has conditional units, say $u \in \mathcal{T}$ with $\sigma(I,u) = I$, then $\max(I) = +\infty$, so that Theorem 2 does not apply.

When Theorem 2 applies we can use the images in $\mathcal{T}_{prime}$ as generators for $\mathcal{T}$ and get in this way a concrete specification of the generators.

Continuing the study of representations of abstract image algebras we shall now examine the case when the local regularity is given by the bond relation ρ = INCLUSION. We let the connection type be Σ = LINEAR as before.

The main result for this case is stated in the following theorem.

<u>Theorem 3</u>. *In order that* $\mathcal{T}[S;\sigma]$ *be isomorphic to an image algebra* $\mathcal{T}'[S;\sigma']$, *with* σ' = *'concatenate to the right', and with* <INCLUSION,LINEAR>*-regularity, it is necessary and sufficient that*

$$\{R(I_1) \supseteq \bigcap_{M \in L(I_2)} R(M) \text{ and } L(I_2) \neq \phi\} \Longleftrightarrow I_1 \in L(I_2) \tag{9.17}$$

<u>Proof of necessity</u>: If the stated isomorphism $\mathcal{T} \cong \mathcal{T}'$ holds consider two images I_1 and I_2 in $\mathcal{T}$ with the corresponding $\mathcal{T}'$ images $I_1' = hI_1$, $I_2' = hI_2$. If $R(I_1) \supseteq R(K)$ for any $K \in \mathcal{T}$ with $K \in L(I_2)$ then the corresponding relations hold in the $\mathcal{T}'$ image algebra: $R(I_1') \supseteq R(K')$ for any $K' \in \mathcal{T}'$ with $K' \in L(I_2')$ etc. We have

$$L(I_2') = \{M' | \beta_{out}(M') \subseteq \beta_{in}(I_2')\} \tag{9.18}$$

so that the intersection set in (9.17) can be written as the intersection of all R(M') with M' as in (9.18). Hence

$$\bigcap_{M' \in L(I_2')} R(M') = \{K' \mid \beta_{out}(K') \supseteq \bigcup_{n'} \beta_{out}(M')\} \qquad (9.19)$$

with M' as in (9.18). Therefore

$$\bigcap_{M' \in L(I_2')} R(M') \ni I_2' \qquad (9.20)$$

and, using (9.17) and the isomorphism and assuming the left inclusion in (9.17) to hold,

$$I_2 \in R(I_1) \qquad (9.21)$$

which is equivalent to

$$I_1 \in L(I_2). \qquad (9.22)$$

On the other hand if $I_1 \in L(I_2)$ so that $I_1' \in L(I_2')$, then we get automatically

$$\bigcap_{M' \in L(I_2')} R(M') \subseteq R(I_1') \qquad (9.23)$$

and the isomorphism gives the left side relation of (9.17).

<u>Proof of sufficiency</u>: Now assume that the condition holds and let us construct an image algebra $\mathcal{T}'$ and an isomorphism h between $\mathcal{T}$ and $\mathcal{T}'$.

Following the construction in the proof of Theorem 1, but modified to take into account the fact that local regularity is now governed by the bond relation ρ = INCLUSION, we let $\mathcal{T}'$ consist of the images in $\mathcal{T}$, if necessary with markers to separate images that have become equal to each other after bonds have been changed. The bonds of an I' corresponding to I will be given by

$$\beta_{out}(I') = [R(I)]^c \text{ if } R(I) \neq \phi, = n_{in} \text{ else}$$
$$\beta_{in}(I') = [\bigcap_{M\in L(I)} R(M)]^c \tag{9.24}$$

using the construction from Theorem 1 with the set n_{in} from a space distinct from $\mathcal{T}$.

The map $h: \mathcal{T} \leftrightarrow \mathcal{T}'$ is bijective and S-covariant; the latter is shown as in the case of <EQUAL,LINEAR>-regularity.

To show that h is homomorphic we also have to show that $\sigma(I_1,I_2) \in \mathcal{T} \Rightarrow \sigma'(I_1',I_2') \in \mathcal{T}'$ with σ' meaning the connector "concatenate to the right". But

$$\sigma(I_1,I_2) \in \mathcal{T} \Rightarrow I_1 \in L(I_2) \text{ and } I_2 \in R(I_1) \tag{9.25}$$

so that

$$\beta_{out}(I_1') = [R(I_1)]^c \tag{9.26}$$

since

$$I_2 \in R(I_1) \text{ so that } R(I_1) \neq \phi. \tag{9.27}$$

Also, because of the condition (9.17), we have

$$R(I_1) \supseteq \bigcap_{M\in L(I_2)} R(M) = [\beta_{in}(I_2')]^c \tag{9.28}$$

and we have shown

$$\beta_{out}(I_1') \subseteq \beta_{in}(I_2') \Leftrightarrow \beta_{out}(I_1')\rho\beta_{in}(I_2') \tag{9.29}$$

so that $\sigma'(I_1',I_2') \in \mathcal{T}'$.

Now let us show that h^{-1} is homomorphic, so that

$$\sigma'(I_1',I_2') \in \mathcal{T}' \Rightarrow \sigma(I_1,I_2) \in \mathcal{T}. \tag{9.30}$$

But the left side of (9.30) means that $\beta_{out}(I_1') = [R(I_1)]^c$ since $n_i \nsubseteq$ any set in $\mathcal{T}$, and $\beta_{out}(I_1') \subseteq \beta_{in}(I_2')$, so that

$$R(I_1) \supseteq \bigcap_{M \in L(I_2)} R(M) \neq \phi \tag{9.31}$$

and $L(I_2)$ cannot be empty. Using condition (9.17) we see that $I_1 \in L(I_2)$, or equivalently, that $\sigma(I_1,I_2) \in \mathcal{T}$. Q.E.D.

In this theorem we have used the operation $R \cap L\colon \mathcal{T} \to 2^{\mathcal{T}}$ defined by

$$(R \cap L)(I) = \bigcap_{M \in L(M)} R(M) \tag{9.32}$$

In this notation condition (9.17) *takes the attractive form*

$$\sigma(I_1,I_2) \in \mathcal{T} \Longleftrightarrow \{R(I_1) \supseteq (R \cap L)(I_2)\}. \tag{9.33}$$

When we try to apply the method used for Theorem 1 and 3 to deal with regularity of <EQUAL,TREE>-type we encounter a new difficulty. In the two previous cases we have the connection type LINEAR, so that $\omega_{in} = \omega_{out} = 1$: the arities were fixed. For the connection type TREE the in-arity varies from image to image, so that when we construct the new image algebra $\mathcal{T}'$ (see below) the generators would have variable in-arity.

To deal with this we shall let the images in $\mathcal{T}'$ have infinite in-arity with denumerable cardinality. Whether all the in-bonds can actually be connected or not is irrelevant but can be decided by inspecting the given composition table.

We start by proving a simple result that is valid generally for any connector and which will be needed for the proof of Theorem 4.

<u>Lemma 3</u>. *In order that* $I \in LR(I)$ *(or* $I \in RL(I)$*) it is necessary and sufficient that* $R(I) \neq \phi$ *(or* $L(I) \neq \phi$*).*

Proof: It is clear that if $R(I)$ is empty, so is $LR(I)$ and cannot contain I. On the other hand, if $R(I) \neq \phi$ it contains some image, say J, so that $\sigma(I,J) \in \mathscr{T}$. But then $I \in LR(I)$ so that the statement in the lemma is true. Q.E.D.

We can now state

Theorem 4. *In order that a given image algebra* $\mathscr{T}[S;\sigma_1,\sigma_2,\ldots]$ *be isomorphic to one* $\mathscr{T}[S;\sigma_1',\sigma_2',\ldots]$ *with* <EQUAL,TREE>*-regularity where* $\sigma_k'(I',J')$ *concatenates the out-bond of* I' *to the* kth *in-bond of* J'*, it is necessary and sufficient that the definition sets in the composition table satisfy*

$$R_\ell(I) = R_\ell L_k R_k(I), \quad \forall I \in \mathscr{T}; \ \ell,k = 1,2,\ldots \tag{9.34}$$

Proof of necessity: Assume that $\mathscr{S}[S;\sigma_1,\sigma_2,\ldots] \cong \mathscr{T}'[S;\sigma_1',\sigma_2',\ldots]$ with the isomorphism h mapping $\sigma_i \leftrightarrow \sigma_i'$, and with $\mathscr{T}'[S;\sigma_1',\sigma_2'\ldots]$ of <EQUAL,TREE>-regularity, and where $\sigma_k'[I',J']$ means the connector connect the out-bond of I' to the kth in-bond of J' (if there is one).

Before starting the main part of the proof let us make some introductory remarks. Consider three images $I_1',I_2',I_3' \in \mathscr{T}'$ and the combination indicated in Figure 9.4 and where it is assumed that the bond values fit for the closed bonds between I_3' and I_2' and between I_1' and I_2'. The resulting image I' then also belongs to $\mathscr{T}$.

In the figure I_1' has been joined to the second in-bond of I_2', say in general to the kth in-bond. Hence the result can be written $\sigma_k(I_1',I_2')$. The ℓth (third in the figure) in-bond of I_3' is joined to the out-bond of I_2', so that

$$I' = \sigma_\ell[\sigma_k(I_1',I_2'),I_3']. \tag{9.35}$$

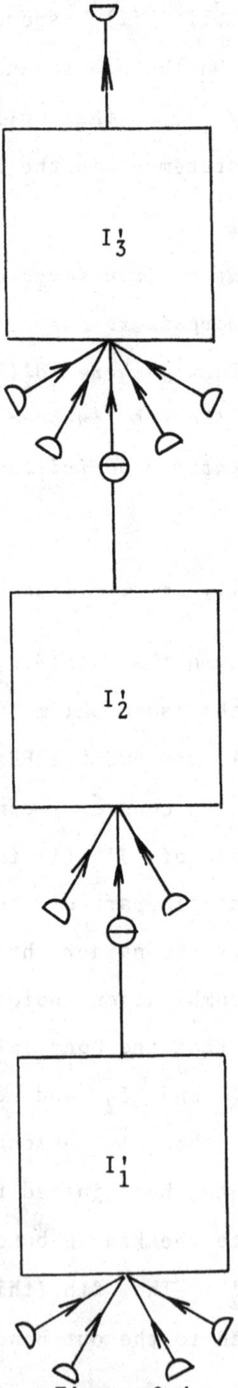

Figure 9.4

On the other hand, if we first connect the out-bond of I_1' to the ℓth in-bond of I_2' we get $\sigma_\ell(I_2',I_3')$. Now connect the out-bond of I_1' to the image we just got, and according to the diagram. But now the in-bond of $\sigma_\ell(I_2',I_3')$ that was the kth of I_2' has got a different label since it is preceded to the left by $\ell-1$ of the in-bonds of I_3'. Hence its new label is $k+\ell-1$ and we get

$$I' = \sigma_{k+\ell-1}[I_1',\sigma_\ell(I_2,I_3)]. \tag{9.36}$$

The two expressions in (9.35) and (9.36) are equal which proves an *association relation*

$$\sigma_\ell[\sigma_k(I_1',I_2'),I_3'] = \sigma_{k+\ell-1}[I_1',\sigma_\ell(I_2',I_3')]. \tag{9.37}$$

The association relation shows that if we ask that

$$\sigma_\alpha[\sigma_\beta(I_1',I_2'),I_3'] = \sigma_{\alpha'}[I_1',\sigma_{\beta'}(I_2',I_3')] \tag{9.38}$$

then we get $\beta' = \alpha$, $\alpha' = \alpha+\beta-1$. The relation

$$(\alpha,\beta) \rightarrow (\alpha+\beta-1,\alpha) \tag{9.39}$$

defines the pair $(\sigma_{\alpha'},\sigma_{\beta'})$ associated to the given pair $(\sigma_\alpha,\sigma_\beta)$.

In particular $(\sigma_\alpha,\sigma_\beta)$ is associated with itself if and only if $\alpha = \beta = 1$. This case, when conditional associativity holds, has been encountered previously in this chapter.

To prove the relation (9.34) we note that

$$\begin{cases} R_k(I) = \{J \mid \beta_{in,k}(J) = \beta_{out}(I)\} \\ L_k(L) = \{M \mid \beta_{out}(M) = \beta_{in,k}(L)\}. \end{cases} \tag{9.40}$$

This gives us, just as before, if $R_\ell(I) \neq \phi$,

$$\begin{cases} L_k R_k(I) = \{L \mid \beta_{\text{out}}(L) = \beta_{\text{out}}(I)\} \\ R_\ell L_k R_k(I) = \{K \mid \beta_{\text{in},}(K) = \beta_{\text{out}}(I)\} \end{cases} \tag{9.41}$$

Hence $R_\ell(I) = R_\ell L_k R_k(I)$ if $R_\ell(I) \neq \phi$. If $R_\ell(I) = \phi$ then both sides of the relation to be established reduce to ϕ so that the assertion holds.

Proof of sufficiency for Theorem 4: We now assume that the condition stated in the theorem holds and proceed to construct a new image algebra $\mathcal{T}'$ isomorphic to $\mathcal{T}$ and of <EQUAL, TREE>-regularity. It will have the same similarity group as $\mathcal{T}$.

Let us take as images of $\mathcal{T}'$ just the images of $\mathcal{T}$ but with bonds as will be described. If needed we mark the new images with labels to make the map $\mathcal{T} \to \mathcal{T}'$ bijective just as in the proofs of the two earlier isomorphism theorems.

Let any $I' \in \mathcal{T}'$ have out-arity one with the out-bond value

$$\beta_{\text{out}}(I') = [R_1(I), R_2(I), R_3(I), \ldots], \quad hI = I', \tag{9.42}$$

where we use the definition sets

$$\begin{cases} R_k(I) = \{J \mid \sigma_k(I,J) \in \mathcal{T}\} \\ L_k(I) = \{J \mid \sigma_k(J,I) \in \mathcal{T}\} \end{cases} \tag{9.43}$$

with $k = 1,2,3,\ldots$. Note that the bond value in (9.42) is a vector whose components are subsets of $\mathcal{T}$. If some of the R-sets are empty we use distinct labels as in the two earlier theorems.

We shall let any $I' \in \mathcal{T}'$ have infinite in-arity with the cardinality of the natural numbers. Their bond values shall be

$$\beta_{in,k}(I') = [R_1L_k(I), R_2L_k(I), R_3L_k(I), \ldots]. \qquad (9.44)$$

The definitions in (9.42) and (9.43) are unique since to any I' there corresponds exactly one I.

Since definition sets are covariant with respect to similarities it can be shown, as we did earlier, that EQUAL is a legitimate bond relation for the bonds introduced for the $\mathcal{T}'$ images.

Let us now prove that the corresponding map $h: \mathcal{T} \to \mathcal{T}'$ is homomorphic. Covariance with respect to S is clear. Assume $\sigma_k(I_1, I_2) \in \mathcal{T}$ so that $I_2 \in R_k(I_1)$ and $I_1 \in L_k(I_2)$. We have to show that I_1' can be combined with I_2' via the connector σ_k'. But this means equality of the out-bond value of I_1'

$$\beta_{out}(I_1') = [R_1(I_1), R_2(I_1), R_3(I_1), \ldots] \qquad (9.45)$$

with the kth in-bond value of I_2'

$$\beta_{in,k}(I_2') = [R_1L_k(I_2), R_2L_k(I_2), R_3L_k(I_2), \ldots]. \qquad (9.46)$$

Hence, we have to show

$$R_\ell L_k(I_2) = R_\ell(I_1);\; k,\ell = 1,2,\ldots\ . \qquad (9.47)$$

Since $I_1 \in L_k(I_2)$, we get, applying the operator R_ℓ to each side, $R\ (I_1) \subseteq R_\ell L_k(I_2)$. But applying the operator L_k to the relation $I_2 \in R_k(I_1)$ we get

$$R_\ell(I_1) \subseteq R_kL_kR_k(I_2). \qquad (9.48)$$

Combined with (9.34) this gives us the relation that was to be proved. Hence h is homomorphism.

It remains to show that h^{-1} is also homomorphic. The similarity covariance is again obvious. Let us now assume that $\sigma_k'(I_1',I_2') \in \mathcal{T}'$, with $I_1' = hI_1$, $I_2' = hI_2$. This means that $\beta_{out}(I_1') = \beta_{in,k}(I_2')$ or

$$R_\ell(I_1) = R_\ell L_k(I_2), \quad k,\ell = 1,2,\ldots \tag{9.49}$$

with $R_\ell(I_2)$ not empty. But then according to Lemma 2

$$I_2 \in R_\ell L_k(I_2) = R_\ell(I_1) \tag{9.50}$$

so that $\sigma(I_1,I_2) \in \mathcal{T}$. Q.E.D.

CHAPTER 4
SOME TOPOLOGY OF IMAGE ALGEBRAS

4.1. A topology for configurations

Combinatory regularity is algebraic in character and can be studied from the perspective of partial universal algebra. At the same time it supports other mathematical structures, for example, measures, and, as we shall see below, topologies. By this we mean at the moment notions of neighborhood, convergence, and continuity, not the topologies that characterize global regularity in terms of the connection type.

We shall introduce a topology on the set of all finite regular configurations and the associated images. Of course there is nothing unique about this topology and we may prefer to use others depending upon context. This one is the finest one that we are likely to use, however, and hence deserves special attention.

Let G and S be second countable Hausdorff spaces with the similarities forming a topological group and *such that* sg *is continuous with respect to the product topology on* S × G. This will induce a topology on

$$\mathscr{C}_\infty = \bigcup_{n=1}^{\infty} \mathscr{C}_n \qquad (1.1)$$

by introducing neighborhoods on each $\mathcal{C}_n$ obtained from the product topology on $G \times G \times \ldots G$ (n times), *with fixed but arbitrary connector* σ,

$$N(c_o) = \{c = \sigma(g_1, g_2, \ldots g_n) | g_i \in N_i(g_i^o)\} \tag{1.2}$$

where $c_o = \sigma(g_1^o, g_2^o, \ldots g_n^o) \in \mathcal{C}_n$, and the N_i are arbitrary neighborhoods of g_i^o in G. All the neighborhoods of the form (1.2) and for $n = 1,2,\ldots$ determine our topology on $\mathcal{C}_\infty$. In other words we form $\mathcal{C}_\infty$ as the topological sum of the $\mathcal{C}_n$.

<u>Lemma 1</u>. *The function* sc *is continuous jointly in* s *and* c, $s \in S$, $c \in \mathcal{C}_\infty$.

<u>Proof</u>: Writing $c = \sigma(g_1, g_2, \ldots g_n)$ we have $sc = \sigma(s\,g_1, s\,g_2, \ldots s\,g_n) = \sigma(g_1^o, g_2^o, \ldots g_n^o) = c_o$ where we have denoted $s\,g_i$ by g_i^o. For any neighborhood $N(c_o)$ as in (1.2) we can choose a neighborhood $N_S(s)$ of s in S and another $N(c)$ of c, such that

$$s' \in N_S(s) \;\&\; c' \in N(c) \Rightarrow s'c' \in N(c_o). \tag{1.3}$$

To see that this is possible we use the fact that sg is jointly continuous in s and g which means that we can find neighborhoods $N_S(s)$ and $N_i(g_i)$ such that for given $N_i(g_i^o)$

$$s' \in N_S(s) \;\&\; g_i' \in N_i(g_i) \Rightarrow s'g_i' \in N_i(g_i^o). \tag{1.4}$$

Combining these relations we see that (1.3) holds for

$$N(c) = \{\sigma(g_1, g_2, \ldots g_n) | g_i \in N_i(g_i),\ i=1,2,\ldots n\} \tag{1.5}$$

Q.E.D.

Lemma 2. *For a fixed number* r *and a given connector* σ, *the function* $c = \sigma(c_1, c_2, \ldots c_r)$ *is continuous jointly in* $c_1, c_2, \ldots c_r$.

Proof: Of course we consider only those c_i that are regular together with the resulting c. Writing

$$c_i = \sigma_i(g_{i1}, g_{i2}, \ldots) \in \mathscr{C}_{n_i}\ ;\ i = 1,2,\ldots r, \tag{1.6}$$

any given neighborhood $N(c)$ as in (1.2) is given in terms of neighborhoods $N_{ij}(g_{ij})$ of the generators involved. Introducing the neighborhoods

$$N_i(c_i) = \{\sigma_i(g'_{i1}, g'_{i2}, \ldots) | g'_{i1} \in N_{i1}(g_{i1}), \\ g'_{i2} \in N_{i2}(g_{i2}), \ldots\} \tag{1.7}$$

it follows that $c'_i \in N_i(c_i), \forall i \Rightarrow c \in N(c)$. Q.E.D.

4.2. A topology for images

Adding an identification rule R to the configuration space $\mathscr{C}_\infty$, we choose as a topology in the resulting image algebra $\mathscr{T}_\infty = \langle \mathscr{C}_\infty, R \rangle$ the identification topology making the map $\mathscr{C}_\infty \to \mathscr{T}_\infty$ continuous; see Schubert (1968), p. 34.

Theorem 1. *The image algebra with the above topology is a topological image algebra if* R *is an open equivalence, in the sense that*

(i) sI *is jointly continuous in* s *and* I.

(ii) *for fixed connector* σ *the function* $I = \sigma(I', I'')$ *is jointly continuous in* I' *and* I".

(iii) *any polynomial on* $\mathscr{T}_\infty$ *is continuous jointly in all its arguments.*

Proof: Consider the maps

$$\begin{cases} f: S \times \mathscr{C}_\infty \to S \times \mathscr{T}_\infty, & f(s,c) \to (s,[c]) \\ g: S \times \mathscr{T}_\infty \to \mathscr{T}_\infty & g(s,I) \to sI \end{cases} \tag{2.1}$$

where f is continuous because of the way the topology was introduced on $\mathscr{T}$. But the composition of g with f is

$$gf(s,c) = g(s,[c]) = s[c] = [sc]. \tag{2.2}$$

But sc is continuous on $S \times \mathscr{C}_\infty$ (see Lemma 1.1) and the projection function $[\cdot]$ is also continuous. Hence $g = sI$ is continuous as stated in (i); see Schubert, ibid.

To prove (ii) we proceed in a similar manner, but now with the maps

$$\begin{cases} f: \mathscr{C}_\infty \times \mathscr{C}_\infty \to \mathscr{T}_\infty \times \mathscr{T}_\infty, & f(c_1,c_2) \to ([c_1],[c_2]) \\ g: \mathscr{T}_\infty \times \mathscr{T}_\infty \to \mathscr{T}_\infty & , g(I_1,I_2) \to \sigma(I_1,I_2). \end{cases} \tag{2.3}$$

Here $[\cdot,\cdot]_{R\times R}$ means equivalence classes on $\mathscr{C}_\infty \times \mathscr{C}_\infty$ modulo $R \times R$. Note that g need only be a partial function, in which case the second relation in (2.3) should be restricted appropriately. Also note that the correspondence

$$(\mathscr{C}_\infty \times \mathscr{C}_\infty)/(R \times R) \leftrightarrow (\mathscr{C}_\infty/R) \times (\mathscr{C}_\infty/R) \tag{2.4}$$

is not only bijective but also topological; see Schubert (1968), p. 43. The map

$$(c_1,c_2) \to [c_1,c_2]_{R\times R} \tag{2.5}$$

is continuous which makes f continuous. But the composition of g with f can be written

$$gf(c_1,c_2) = g([c_1],[c_2]) = [\sigma(c_1,c_2)]. \tag{2.6}$$

Since $\sigma(c_1,c_2)$ is continuous according to Lemma 1.2 and again using the continuity of $[\cdot]$ it follows that gf is continuous and hence g which proves (ii).

If the diagonal in $\mathscr{C}_\infty \times \mathscr{C}_\infty$ given by $\{(c_1,c_2)|c_1Rc_2\}$ is closed the topological space $\mathscr{T}_\infty$ is Hausdorff.

The statement in (iii) follows by repeated application of (i) and (ii). Of course the polynomial will usually be partial so that the statement holds only when restriction is made to the appropriate domain. Q.E.D.

Images can be formed by a *prototype selector* $p: \mathscr{C}_\infty \to \mathscr{C}_\infty$ which assigns to each configuration c its prototype $p(c)$. Note that here we are talking about prototypes for configurations, not for images as in Volume I, p. 104. The images then result as the element of the quotient map. In general a prototype selector can behave very drastically unless care is taken to make it smooth.

If we have access to a continuous prototype selector p this induces a continuous map $\phi: \mathscr{T} \to \mathscr{C}$.

Indeed, the map R which identifies configuration by their equivalence modulo R is continuous, and we have just assumed that $p = \phi R$ is continuous. It then follows that ϕ must be continuous; see Schubert (ibid).

Consider a sequence I_ν in $\mathscr{T}_\infty$ converging to some image I in the same image algebra. Then, introducing the prototypes $c_\nu = \phi(I_\nu)$, $c = \phi(I)$ we have $c_\nu \to c$, or in the terminology of Siwiec (1971) the map $R: \mathscr{C}_\infty \to \mathscr{T}_\infty$ is sequence-covering. Now we can use this property to prove the statements (i), (ii), (iii) in the theorem, so that the conclusions hold as long as we can find a continuous prototype selector.

To be able to find continuous prototype selectors we may need topologies coarser than the one described.

In passing we mention that the proof of the theorem could also have been based on the result (see Siewiec (ibid)) that any open map of a first countable space is sequence-covering. For completeness we reproduce his proof of this statement.

Let $R: \mathscr{C}_\infty \to \mathscr{T}_\infty$ be open and $I_n \to I$ in $\mathscr{T}_\infty$. Without loss of generality we can assume that the I_n are distinct. There is a configuration $c \in R^{-1}I$ such that RU is a neighborhood of I for every neighborhood U of c. Denote $C_n = R^{-1}y_n$ and consider a decreasing open base $\{U_n\}$ for c. For every i there is an m_i such that C_n intersects U_n for all $n \geq m_i$. We now let $c_j \in C_j$ for $j \leq m_1$ and $c_j \in C_j \cap U_i$ for all j such that $m_i \leq j < m_{i+1}$. If N is a neighborhood of c there is an i_o such that $c \in U_{i_o} \subset N$ so that $c_j \in U_{i_o} \subset N$ for all $j \geq m_{i_o}$: c_i tends to c, so that R is sequence-covering as stated.

4.3. Some examples

Let us consider some examples, starting with discrete image algebras, $\mathscr{R}$= DISCRETE in the first two examples.

Example 1. Let generators be half-planes, $S = EG(2)$, and let R identify intersections of the half-planes in the configuration. Then images represent convex sets, actually polygons, and the whole plane $u = R^2$ plays the role of a unit. In G we choose the natural topology: $g_\nu \to g$ if the half-planes g_ν tend to the half-plane g considered as sets.

To show that the equivalence R defines an open map it is enough to show that $R^{-1}RC$ for any open set C in $\mathscr{C}_\infty$ is open; see Schubert (1968), p. 37. But $\mathscr{C}_n \cap R^{-1}RC$ can be seen directly to be open for any n and hence the statement follows.

On the other hand, if $I_\nu \to I$ in this topology, then there is a natural number N such that from some ν on the I_ν can be represented with N generators. This is so because the various $\mathscr{C}_n$ are not topologically connected with each other by the definition in (2). But then, with some enumeration of the generators, these generators converge to certain fixed generators and the statement follows.

The induced topology in $\mathscr{T}_\infty$ means that $I_\nu \to I$ iff the convex polygons I_ν have a number of sides tending to the number of sides of I and $I_\nu \to I$ considered as sets. Indeed, if this is so, then we can choose $c_\nu \in I_\nu$, $c \in I$ with eventually $c_\nu = (g_{\nu 1}, g_{\nu 2}, \ldots g_{\nu N})$ and $c = (g_1, g_2, \ldots g_N)$. If $I_\nu \to I$ as sets we can choose the g's such that $g_{\nu_j} \to g_j$; $j = 1,2,\ldots N$, and convergence holds in the topology of $\mathscr{T}_\infty$.

Example 2. Let the generators be functions ax^p with real non-zero a, and $p \in \mathbb{N}$, with similarities as multiplication by non-zero reals, and R identifying sums of the generators seen as functions on $[0,1]$. Then images represent polynomials in x, $x \in [0,1]$, and the polynomial identically zero plays the role of a unit in this $\mathscr{T}_\infty$.

The topology defined on this image algebra means that $I_\nu \to I$ if the degrees of I_ν tend to the degree of I and that $I_\nu(x) \to I(x)$ pointwise, all $x \in [0,1]$. To see that this is so let $I_\nu(\cdot)$, have degrees tending to that of I and $I_\nu(\cdot) \to I(\cdot)$ pointwise. Then we can find configurations as in the previous example with converging generators. It follows that $I_\nu \to I$ in the topology of $\mathcal{T}_\infty$.

Conversely if $I_\nu \to I$ in this topology we can immediately conclude that I_ν contains configurations with the numbers of generators tending to that of some $c \in I$. Since these generators, for an appropriate choice, converge as $\nu \to \infty$ it follows that $I_\nu(\cdot) \to I(\cdot)$ pointwise.

Example 3. Let generators be all linear real-valued functions on closed bounded intervals on the real line, S = the translation group on $\mathbb{R}$, ρ asking that boundary points and boundary values be equal, Σ = LINEAR. Then images represent continuous piecewise linear functions on closed bounded intervals. Functions defined on a single point play the role of conditional units.

Convergence in $\mathcal{T}_\infty$ means that $I_\nu \to I \in \mathcal{T}_\infty$ if and only if the "degrees" of I_ν tend to that of I, the supports of I_ν tend to the support of I, and the functions $I_\nu(\cdot)$ tend pointwise to $I(\cdot)$. Proof as above.

Once the topology has been introduced on $\mathcal{T}_\infty$ we can speak of open sets, Borel sets, and so on, which can be used to construct σ-algebras needed to support measures on the image algebra. This enables us to treat rigorously such constructs as the set $\mathcal{P}(\mathcal{T})$ of all Borel measures on $\mathcal{T}$, the set of measures $\{P_s;\ s \in S\}$ generated from a $P \in \mathcal{P}(\mathcal{T})$ and translating P by s, and properties of the "convolutions" that P generates as the conditional probability measure of $I = \sigma(I_1, I_2)$ (given that it is defined) where I_1, I_2 are independent and have probability measures P_1 and P_2 respectively, and where the conditioning subset has positive measure.

CHAPTER 5
METRIC PATTERN THEORY

5.1. Regularity controlled probabilities

Given a set of laws $\mathcal{R}$ for the regular structure, they induce natural probability measures over the configuration space $\mathcal{C}(\mathcal{R})$ and associated image algebras. This topic - metric pattern theory - was introduced in Section 2.10 of Volume I and we shall pursue it further in this chapter, extend the results to great generality and deepen some of them. When doing this we shall concentrate our attention on the configurations and neglect the corresponding questions for images; see Notes A. A reader can therefore in the present chapter think of the identification rule R as EQUAL, treating images as identical to configuration. Important advances have been made in metric pattern theory after the appearance of Volume I, some of which are contained in two reports by Hwang and Thrift, see Bibliography; much of this chapter is devoted to presenting their results.

When configurations are generated according to $\mathcal{R} = \langle\Sigma,\rho\rangle$ the process can be analyzed in terms of successive choices: *structural choices*. Among these we mentioned three that will appear often.

1. The choice of the number of generators $\#(c)$.
2. The choice of content(c) when $\#(c)$ has been chosen.
3. The choice of connector σ for given content.

Let us first mention three heuristic principles. They will be given in precise form in what follows. These principles constitute our model of *regularity controlled probabilities*, (see Notes B) a term that will be used, with some abuse of terminology, also when we deal with softened regularity (see below). The construction of a probability measure over the regular structures is based on the first principle that *all choices are made conditionally independent*. What event we use for the conditioning will depend upon the problem at hand; several examples will be given below.

The second principle says that *the choices are made according to probability measures that are conditionally identical*. Again the conditioning event can vary from case to case.

The third principle says that the choice of closing or leaving open a bond couple (that has not yet been dealt with) should have *a probability depending only upon the two bond values involved.*

To exemplify these principles let us consider finite configurations from a finite generator space G. Let the probability measure Q be defined on G and the non-negative *acceptance function* A on $B \times B$ for couples of bond values β_1 and β_2.

Conditioned by the size $\#(c) = n$ we are then led to

$$P[c|\#(c)] = Z^{-1} \prod_{i=1}^{n} Q(g_i) \prod_{k,\ell} A(\beta_k,\beta_\ell). \tag{1.1}$$

In (1.1) the subscript $\underline{i}$ enumerates the $n = \#(c)$ generators belonging to the fixed set content(c). The subscripts k and ℓ enumerate all the bonds of the generators involved. Z is a normalizing constant whose value should be selected so that P has total measure one; see Notes B.

It is important to realize that this measure *can associate positive probabilities to non-regular configurations*. Indeed, a $\underline{c}$ for which (1.1) is positive can offend against local regularity, if $A(\beta_k,\beta_\ell) > 0$ when $\beta_k \rho \beta_\ell \neq \text{TRUE}$. It can also offend against global regularity if the resulting connector $\sigma \notin \Sigma$. When we want to emphasize this possibility we shall speak of P as describing *softened regularity*.

A second example, when we insist on strict regularity, we mention the measure given as

$$P[c|\#(c),\mathscr{R}] = P[c|\mathscr{L}_n(\mathscr{R})] = \begin{cases} P[c|\#(c)], & \text{if } c \in \mathscr{L}(\mathscr{R}) \\ 0 & \text{else} \end{cases} \tag{1.2}$$

Of course the normalizing constant Z must then be readjusted to give total measure one.

As a third example consider

$$P[c|\#(c),\text{content}(c)] = Z^{-1} \prod_{k,\ell} A(\beta_k,\beta_\ell) \tag{1.3}$$

for softened regularity and the analog of (1.2) for strict regularity.

As a fourth example we give the measure over $\mathscr{L}(\mathscr{R})$ and do not restrict the value of n

$$P[c|\mathscr{L}(\mathscr{R})] = \begin{cases} Z^{-1}\pi_n \prod_{i=1}^{n} Q(g_i) \prod_{k,\ell} A(\beta_k,\beta_\ell) & \text{if } c \in \mathscr{L}(\mathscr{R}) \\ 0 \quad \text{else} \end{cases} \tag{1.4}$$

It is clear that (1.4) leads to strict regularity but can be modified to softened regularity as before.

In a last example we assume $\#(c)$ and the connector σ to be fixed

$$P[c|\#(c),\sigma] = \begin{cases} Z^{-1} \prod_{i=1}^{n} Q(g_i) & \text{if } \sigma(g_1,\ldots g_n) \in \Sigma_n \\ 0 & \text{else} \end{cases} \tag{1.5}$$

Note that here we get the value zero if the g_i's selected do not have bond structures that fit locally and globally via the fixed σ. For softened regularity we get of course the modified form

$$P[c|\#(c),\sigma] = Z^{-1} \prod_{i=1}^{n} Q(g_i) \prod_{(k,\ell)\in\sigma} A(\beta_k,\beta_\ell) . \tag{1.6}$$

Many other conditionings will arise (see Notes D) but will not be treated here.

All of these examples were for finite generator space G (and hence $\text{card}(B) < \infty$). In the opposite case the measures will be defined by Radon-Nikodym derivatives and with densities that will be denoted by small letters. For example (1.3) becomes

$$\begin{aligned} p[c|\#(c),\text{content}(c)] &= \frac{dP[c|\#(c),\text{content}(c)]}{dm(c)} \\ &= Z^{-1} \prod_{k,\ell} a(\beta_k,\beta_\ell) \end{aligned} \tag{1.7}$$

where m is some given σ-finite measure, often a Lebesgue measure in special cases, or at least simply related to Lebesgue measure.

In the dynamic study of pattern formation over time we shall only deal with *Markov type dynamics*. Say that again we look at the finite case with fixed configuration size n and a time parameter $t \in \mathbb{N}$. Then the configuration c_t at time t should form a Markov chain over the finite state space $\mathscr{C}_n(\mathscr{R})$ for strict regularity. A special case that will receive a good deal of attention is when the bond choices opening and closing, are of birth-and-death type, controlled by intensities $\lambda(\beta_k, \beta_\ell)$ for closing an open bond, and $\mu(\beta_k, \beta_\ell)$ for opening a closed bond. This can also be generalized to allow for introduction (birth) of a new generator, and for deletion (death) of a generator that is already in content(c_t).

Once the regularity controlled probabilities have been introduced, we shall study a variety of limiting problems. The first *limit problem* deals with the case when the softening of the regularity is controlled by a parameter θ, reminiscent of kT in statistical mechanics. In that context k would be Boltzmann's constant and T absolute temperature, but we shall speak of θ as an abstract "temperature" irrespective of its possible interpretations. In particular we shall see what happens with the probability measures when θ drops to zero: what are the *frozen patterns* and how are they approached by *cold patterns*.

The second *limit problem* is concerned with what can be said about P when $\#(c)$ becomes large. One would hope that the preliminary results to be given in Section 15 could be extended and established generally. This represents a major area of research in metric pattern theory and is also connected with the attempt to find "laws of large numbers" and "central limit theorems" for regular structures. Some surprising results have been found recently and will be presented in the later sections.

5.2. Conditioning by regularity

Questions of measurability and related topics have played a subordinated role in matric pattern theory so far - the real difficulties lie elsewhere. An exception is the rigorous definition of regularity controlled probabilities when the bond relation ρ is such that the set in $B \times B$ (where $\beta_1 \rho \beta_2$ holds) has Q^2-measure zero. This happens for example when $B = \mathbb{R}$, ρ = EQUAL, and Q is continuous. The problem in this case has become known as *conditioning on the diagonal*. It was pointed out in Volume I, Section 2.10, that the usual definition of conditional probability, based on the Radon-Nikodym derivative, is not adequate for the present purpose. Instead, it was argued, one should introduce the probabilities conditioned by ρ via a limit process reminiscent of the older way of defining conditional probability before Radon-Nikodym derivatives were used for this purpose. In the cited reference it was shown that the limit existed for the case stated above, but only with strong assumptions. We shall extend these results now.

Say that we are on the real line and that our measure Q is absolutely continuous w.r.t. a fixed measure m

$$f(x) = \frac{Q(dx)}{m(dx)} \tag{2.1}$$

and consider the measure P_ε with $\varepsilon > 0$

$$\frac{P_\varepsilon(dx)}{m(dx)} = \frac{f_\varepsilon(x) f(x)}{\int f_\varepsilon(x) f(x) m(dx)} \tag{2.2}$$

where

$$f_\varepsilon(x) = \frac{1}{2\varepsilon} \int_{x-\varepsilon}^{x+\varepsilon} f(x) m(dx). \tag{2.3}$$

This is just the procedure suggested in the cited reference. We shall treat two cases, first when $X = \mathbb{R}$ (or any $\mathbb{R}^n$) and then when X is a metric space.

Assume $X = \mathbb{R}$, m = Lebesgue measure, and $f \in L_2[(-\infty,\infty)]$. With the notation

$$\phi_\varepsilon(x) = \frac{1}{\varepsilon}\,\phi\left(\frac{x}{\varepsilon}\right); \quad \varepsilon > 0; \tag{2.4}$$

for any frequency function ϕ, define

$$I_\varepsilon(x) = \frac{1}{2\varepsilon}\, I_{[-1,1]}\left(\frac{x}{\varepsilon}\right) \tag{2.5}$$

for the frequency function I corresponding to $R(-1,1)$. Then we can write f_ε as the convolution

$$f_\varepsilon(x) = (I_\varepsilon * f)(x). \tag{2.6}$$

We shall define, *generalizing* (2.2) *in a natural way*,

$$\frac{P_\varepsilon(dx)}{m(dx)} = \frac{(\phi_\varepsilon * f)(x) f(x)}{\int(\phi_\varepsilon * f)(x) f(x) m(dx)} \tag{2.7}$$

Clearly (2.7) is well-defined. In fact, $\phi_\varepsilon \in L^1(m)$ and $f \in L^2(m)$, hence $\phi_\varepsilon * f \in L^2(m)$ and $(\phi_\varepsilon * f) f \in L^1(m)$. By the fact that $\|f*\phi_\varepsilon - f\|_2 \to 0$ as $\varepsilon \to 0$, see e.g. Stein (1970), for each fixed Borel set A

$$\left|\int_A (\phi_\varepsilon * f) f - \int_A f^2\right|^2 \le \left(\int_A |\phi_\varepsilon * f - f|^2\right) \int_A f^2 \to 0 \tag{2.8}$$

as $\varepsilon \to 0$.

Therefore we have the following result; see Notes A.

<u>Theorem 1</u>. *Assume* $f \in L_2$ *and let* $\{P_\varepsilon\}$ *be defined by* (2.7) *and* P_0 *by*

$$\frac{dP_0}{dm}(x) = \frac{f^2(x)}{\int f^2}. \tag{2.9}$$

Then, $P_\varepsilon \to P_0$ *weakly as* $\varepsilon \to 0$ *and* P_0 *is independent of the choice of* $\{\phi_\varepsilon\}$.

Remark. There is another approach of the problem. Let us write

$$f_\varepsilon(x) = \frac{1}{m(B(x,\varepsilon))} \int_{B(x,\varepsilon)} f, \tag{2.10}$$

where $B(x,\varepsilon)$ denotes the ball with center x and radius ε. The *maximal function* of f is defined by

$$M(f)(x) = \sup_{\varepsilon>0} \frac{1}{m(B(x,\varepsilon))} \int_{B(x,\varepsilon)} f. \tag{2.11}$$

By the fact that $M(f) \in L^2(m)$, see Stein (1970), and $f_\varepsilon \to f$ a.e., we have

$$P_\varepsilon \to P_0 \quad \text{weakly.} \tag{2.12}$$

Q.E.D.

We now turn to the second case and assume that X is a complete, separable metric space with a regular Borel measure m. Also we have to assume, a bit artificially, the following

Condition A: There exists $K > 0$ such that for any measurable set E and for any covering $\mathscr{L}$, consisting of open balls with $\sup_i m(B_i) < \infty$, of E, there exists a countable disjoint subcollection $\{B_n\}$ of $\mathscr{L}$ such that $\Sigma m(B_n) \geq Km(E)$. Also assume $m(B) < \infty$ for any ball with finite radius, then $M(f) \in L^2(m)$ and $\|Mf\| \leq K_0\|f\|$ (K_0 depends only on K). The proof is the same as in Segal-Kunge (1978).

Consider the linear functional $F_\varepsilon: L^2(G) \to R$ defined by $f_\varepsilon(g) = \int g_\varepsilon f$, where g_ε is obtained from g as f_ε was from f as f_ε is. Then we have

$$|F_\varepsilon(g)| \leq \|g_\varepsilon\| \, \|f\| \leq \|M(g)\| \, \|f\| \leq (K_0\|f\|)\|g\|, \tag{2.13}$$

so that $\|F_\varepsilon\| \leq K_0\|f\|$. Let $F(g) = \int gf$. We try to prove $F_\varepsilon(f) \to F(f)$ as $\varepsilon \downarrow 0$. Notice that if g is continuous then $g_\varepsilon \to g$ pointwise which leads to $F_\varepsilon(g) \to F(g)$. If f can be approximated by continuous functions in L^2, then

$$|F_\varepsilon(f)-F(f)| \leq |F_\varepsilon(t)-F_\varepsilon(g)| + |F_\varepsilon(g)-F(g)| + |F(g)-F(f)|$$
$$\leq 2K_0\|f-g\| + |F_\varepsilon(g)-f(g)| \to 0. \tag{2.14}$$

Since f is a density, $f \in L^1(m) \cap L^2(m)$. For any $\varepsilon > 0$ choose M large enough such that $\|f_2\| < \varepsilon$, where $f_2 = f^1{}_{[|f|>M]}$ and $f_1 = f-f_2$. Clearly $f_1, f_2 \in L^1 \cap L^2$. By an application of Lusin's theorem, there exists a bounded continuous function $g \in L^1 \cap L^2$ such that

$$\int |g-f_1| < \frac{\varepsilon^2}{2M} \quad \text{and} \quad |g| \leq M. \tag{2.15}$$

Hence

$$\int |g-f| \leq \|g-f_1\| + \varepsilon \tag{2.16}$$

and

$$\|g-f_1\| = (\int|g-f_1|^2)^{\frac{1}{2}} \leq (2M\int|g-f_1|)^{\frac{1}{2}} \leq (2M\cdot \frac{\varepsilon^2}{2M}) = \varepsilon. \tag{2.17}$$

Hence f can be approximated by continuous functions in $L^2(m)$. To sum up, we have

Theorem 2. *Under Condition A,* P_ε *defined by*

$$\frac{dP_\varepsilon}{dm}(x) = \frac{\frac{f(x)}{m(B(x,\varepsilon))}\int_{B(x,\varepsilon)} f(y)m(dy)}{\frac{f(x)}{m(B(x,\varepsilon))}\int_{B(x,\varepsilon)} f(y)m(dy))m(dx)} \tag{2.18}$$

converges as $\varepsilon \downarrow 0$ *to* P_0 *with density* $f^2\|f\|^{-2}$.

Remark. The covering assumption is satisfies for Lebesgue measure in $\mathbb{R}^n$. Still, it sounds very artificial.

What will happen if $f \in L^1$ but not L^2? Let's go back to the one dimensional case with P_ε defined by (2.2).

If $f \in L^2[-a,a]$ for any finite interval $[-a,a]$ but $f \notin L^2$, then by Fatou's Lemma it is easily seen that $P_\varepsilon([-a,a]) \to 0$. Therefore $\{P_\varepsilon\}$ is not tight. If $f \in L^2([a,\infty)\quad(-\infty,b])$, then $\{P_\varepsilon\}$ is tight. Moreover if P_o exists then $P_o[b,a] = 1$.

If $f \notin L^2$, then $\frac{dP_\varepsilon}{dm}(x) \to 0$ a.e. Is it possible that P_o exists and $P_o << m$? The following example is informative.

Let ϕ be the uniform density on $[0,1)$ which is regarded as a circle. The sequence $\{r_n\}$ of all dyadics is ordered by: Put $r_o = 0$. If the nth level, i.e. $\{\frac{m}{2^n} | 0 \leq m < 2^n\}$, has been ordered in its natural order and to the index k, then the $(n+1)^{th}$ level has indices from $k+1$ to $k+2^{n+1}$ in its natural order. Now define

$$f(x) = \sum_0^\infty C_n b_n^{-1}\ \phi\left(\frac{x+r_n}{b_n}\right), \tag{2.19}$$

where $C_n > 0$, $b_n > 0$, $\Sigma C_n = 1$ and $\Sigma C_n^2 b_n^{-1} = \infty$. Then,

$$\int f = \Sigma C_n \int b_n^{-1}\ \phi\left(\frac{x+r_n}{b_n}\right) = 1 \tag{2.20}$$

and

$$\begin{aligned}\int f^2 &\geq \Sigma C_n^2 \int b_2^{-2}\ \phi^2\left(\frac{x+r_n}{b_n}\right) = \Sigma C_n^2 b_n^{-1} \int b_n^{-1}\ \phi\left(\frac{x+r_n}{b_n}\right) \\ &= \Sigma C_n^2 b_n^{-1} = \infty.\end{aligned} \tag{2.21}$$

Now, define

$$f_n(x) = \sum_0^{(n-1)^{th}\text{ level}} C_m b_m^{-1}\ \phi\left(\frac{x+r_m}{b_m}\right) \tag{2.22}$$

with corresponding "distribution" function F_n. Let $g_n = f - f_n$, $G_n = F - F_n$. Notice that G_n and g_n are invariant under translation by 2^{-n}. Let us divide the integral

$$(2\varepsilon)^{-1} \int (F(x+\varepsilon)-F(x-\varepsilon))f(x)dx \tag{2.23}$$

into four parts:

$$\begin{cases} A_1 = (2\varepsilon)^{-1} \int (G_n(x+\varepsilon)-G_n(x-\varepsilon))g_n(x)dx \\ A_2 = (2\varepsilon)^{-1} \int (F_n(x+\varepsilon)-F_n(x-\varepsilon))f_n(x)dx \\ A_3 = (2\varepsilon)^{-1} \int (G_n(x+\varepsilon)-G_n(x-\varepsilon))f_n(x)dx \\ A_4 = (2\varepsilon)^{-1} \int (F_n(x+\varepsilon)-F_n(x-\varepsilon))g_n(x)dx. \end{cases} \tag{2.24}$$

By Fatou's lemma, $A_1 \to \infty$. Also $A_2 \to \int f_n^2(x)dx$. But

$$\begin{aligned} A_3 &\le \sup_x |f_n(x)|(2\varepsilon)^{-1} \int (G_n(x+\varepsilon)-G_n(x-\varepsilon))dx \\ &\to \sup_x |f_n(x)| \int g_n < \infty, \end{aligned} \tag{2.25}$$

see Stein (1970) and

$$A_4 \le \sup_x \left|\frac{F_n(x+\varepsilon)-F_n(x-\varepsilon)}{2\varepsilon}\right| \int g_n < \infty, \tag{2.26}$$

since for fixed n, $\sup_x |f_n(x)|$ and $\sup_x \left|\frac{F_n(x+\varepsilon)-F_n(x-\varepsilon)}{2\varepsilon}\right|$ are finite. Now consider the characteristic function of P_ε, and choose any convergent subsequence of P_ε. Then

$$\begin{aligned} &\lim_{\varepsilon\to 0} \int e^{itx}\, dP(x) \\ &= \lim_{\varepsilon\to 0} \frac{(2\varepsilon)^{-1} \int e^{itx}[G_n(x+\varepsilon)-G_n(x-\varepsilon)]g_n(x)dx}{(2\varepsilon)^{-1} \int [G_n(x+\varepsilon)-G_n(x-\varepsilon)]g_n(x)dx} = \psi(t). \end{aligned} \tag{2.27}$$

For each fixed n, $[G_n(x+\varepsilon)-G_n(x-\varepsilon)]g_n(x)$ is invariant under translation of 2^{-n}. Hence $\psi(t)$ is invariant under translation of 2^{-n} for any n. This implies that $\psi(t)$ is the characteristic function of the uniform distribution. Therefore we can conclude in this example that $P_\varepsilon \to P_0$ as $\varepsilon \to 0$, where P_0 is the uniform distribution.

This is not likely to be the last word about conditioning on the diagonal but it will have to be enough for the time being. Instead we shall return to the regularity controlled probability measures discussed in the last section and derive some simple but basic properties for them.

We shall begin by some introductory remarks following Thrift (1977). Consider a directed graph with a set S of vertices enumerated by a subscript i; $i = 1,2,\ldots n$; and edges (or arcs) forming a set A

$$A \subset S \times S - \{(i,i) \mid i \in S\}. \tag{2.28}$$

We shall need the set, for fixed i, that are reached directly from the vertex i:

$$s(i) = \{j \mid (i,j) \in A\}. \tag{2.29}$$

In pattern theoretic terms the vertices will often be generators and the arcs describe the connector of some configuration, but variations of this interpretation will occur sometimes.

We assume that associated with each $1,\ldots,n$ there is a stochastic variable X_k and that the joint frequency function of $X_1,\ldots,X_n$ is given by

$$p(x_1,\ldots,x_n) = \prod_{i\in S} \prod_{j\in s(i)} f_{ij}(x_i,x_j). \tag{2.30}$$

We assume p is positive on $\mathbb{R}^n$. For a given $i \in S$, the *neighborhood* of i is given by

$$N(i) = \{j \in S | j \in s(i) \text{ or } i \in s(j)\}. \tag{2.31}$$

For a subset $c \subset S$ the *interior* of c is given by

$$in(c) = \{k \in c | N(k) \subset c\}. \tag{2.32}$$

The *boundary* of c is given by

$$bd(c) = c - in(c). \tag{2.33}$$

We now fix i,c where $i \notin c$. Let $p(x_i | x_j, j \in c)$ be the conditional frequency function of X_i given X_j, $j \in c$, which is given by

$$p(x_i | x_j : j \in c) = \frac{\left(\prod_{k \in S-c-\{i\}} \int_{x_k} dx_k\right) \prod_{k \in S} \prod_{j \in s(k)} g_{kj}(x_k, x_j)}{\left(\prod_{k \in S-c} \int_{x_k} dx_k\right) \prod_{k \in S} \prod_{j \in s(k)} g_{kj}(x_k, x_j)}. \tag{2.34}$$

Lemma 1. *Let* $c \in S$ *and* $i \in S-c$ *be fixed. Then*

$$p(x_i | x_j : j \in c) = p(x_i | x_j : j \in bd(c)).$$

Proof: Write:

$$\prod_{k \in S} \prod_{j \in s(k)} g_{kj}(x_k, x_j)$$
$$= \left(\prod_{k \in in(c)} \prod_{j \in s(k)} g_{kj}(x_k, x_j)\right)\left(\prod_{k \in S-in(c)} \prod_{j \in s(k)} g_{kj}(x_k, x_j)\right). \tag{2.35}$$

Note that $s(k) \subset c$ if $k \in in(c)$. Also

$$\prod_{k \in S-in(c)} \prod_{j \in s(k)} g_{kj}(x_k, x_j)$$
$$= \left(\prod_{k \in S-c} \prod_{j \in s(k)} g_{kj}(x_k, x_j)\right)\left(\prod_{k \in bd(c)} \prod_{j \in s(k)} g_{kj}(x_k, x_j)\right). \tag{2.36}$$

Note that $s(k) \subset S-in(c)$. if $k \in S-c$. Further

$$\prod_{k\in bd(c)}\prod_{j\in s(k)} g_{kj}(x_k,x_j)$$

$$= \Big(\prod_{k\in bd(c)}\ \prod_{j\in s(k)-c} g_{kj}(x_k,x_j)\Big)\Big(\prod_{k\in bd(c)}\ \prod_{j\in s(k)\cap c} g_{kj}(x_k,x_j)\Big). \tag{2.37}$$

Combining (2.35), (2.36) and (2.37) we get

$$p(x_i|x_j:j\in c) = \frac{\Big(\prod_{k\in S-c-\{i\}}\int_{x_k} dx_k\Big)\prod_{k\in S}\ \prod_{j\in s(k)} g_{kj}(x_k,x_j)}{\Big(\prod_{k\in S-c}\int_{x_k} dx_k\Big)\prod_{k\in S}\ \prod_{j\in s(k)} g_{kj}(x_k,x_j)} \tag{2.38}$$

which can be written as

$$\frac{\Big(\prod_{k\in S-c-i}\int_{x_k} dx_k\Big)\Big(\prod_{k\in S-c}\ \prod_{j\in s(k)} g_{kj}(x_k,x_j)\Big)}{\Big(\prod_{k\in S-c}\int_{x_k} dx_k\Big)\Big(\prod_{k\in S-c}\ \prod_{j\in s(k)} g_{kj}(x_k,x_j)\Big)} \tag{2.39}$$

$$\frac{\prod_{k\in bd(c)}\ \prod_{j\in s(k)-c} g_{kj}(x_k,x_j)}{\prod_{k\in bd(c)}\ \prod_{j\in s(k)-c} g_{kj}(x_k,x_j)}$$

where the factors not involved in the integrations have been cancelled.

Recall the remark that $s(k) \subset S - in(c)$ if $k \in S-c$ and note that in the above expression there are no x_k, k in(c), variables remaining, Hence we have proved

$$p(x_i|x_j:j \in c) = p(x_i|x_j:j \in bd(c)). \tag{2.40}$$

Q.E.D.

This lemma means that the distribution of X_i given X_j $j \in c$ is the same as the distribution of X_i given X_j, $j \in bd(c)$. Intuitively, we expect a $j \in bd(c)$ will have a stochastic influence on i if and only if there is a "chain" from j to i outside c. This notion is made concrete by the following definition: we say there is a *chain* from j to i if there is a sequence $j, a_1, \ldots, a_m, i$ from S with

$$\begin{cases} a_1 \in N(j) \\ a_{k+1} \in N(a_k) \quad k = 1,\ldots,m-1 \\ i \in N(a_m). \end{cases} \tag{2.41}$$

If $j \in c$, $i \notin c$ we say $j, a_1, \ldots, a_{m,i}$ is a *chain outside* c if $a_k \in S-c$, $k = 1,\ldots,m$.

With c,i fixed, define

$$K(c,i) = \{K \subset S-c-\{i\} \mid K \subset \text{in}(c\ K)\}. \tag{2.42}$$

Lemma 2. *If* $K_1 \subset \text{in}(c \cup K_1)$ *and* $K_2 \subset \text{in}(c \cup K_2)$ *then* $K_1 \cup K_2 \subset \text{in}(c \cup K_1 \cup K_2)$.

Proof: Follows directly from the relation $\text{in}(A \cup B) \supset \text{in}(A) \cup \text{in}(B)$. Q.E.D.

From this lemma it follows that there exists a maximal element in $K(c,i)$, namely

$$K^*(c,i) = \bigcup_{K \in K(c,i)} K. \tag{2.43}$$

Lemma 3. *Let* $c \in S$, $i \in S-c$ *be fixed. Then*

$$p(x_i | x_j : j \in c) = p(x_i | x_j : j \in \text{bd}(c \cup K)) \forall K \in K(c,i). \tag{2.44}$$

Proof: Fix $K \in K(c,i)$. Then $\text{bd}(c \cup K) \subset c \subset c \cup K$. Let $p(x_j : j \in K)$ denote the marginal frequency function of $(X_j : j \in K)$. From Lemma 1 we get

$$p(x_i | x_j : j \in \text{bd}(c \cup K)) = p(x_i | x_j : j \in c \cup K). \tag{2.45}$$

Also,

$$p(x_i|x_j:j\in c) = \Big(\prod_{k\in K} \int_k dx_k\Big) p(x_i|x_j:j\in c\cup K)p(x_j:j\in K)$$

$$= \Big(\prod_{k\in K} \int_k dx_k\Big) p(x_i|x_j:j\in bd(c\cup K))p(x_j:j\in K)$$

$$= p(x_i|x_j:j\in bd(c\cup K)), \tag{2.46}$$

since $K \subset in(c\cup K)$. Q.E.D.

In particular, we get (see Thrift (1977), (1979) and Notes B)

Theorem 3. *We have*

$$p(x_i|x_j:j \in c) = p(x_i|x_j:j \in bd(c \cup K^*(c,i))). \tag{2.47}$$

The notion of a chain from j to i outside c mentioned earlier is now seen to be the defining property for $bd(c\ K^*(c,i))$, as seen in the following.

Lemma 4. *Given* c, $i \in S-c$, $j \in bd(c \cup K^*(c,i))$ *if and only if there exists a chain* $j,a_1,\ldots,a_m,i$ *outside* c.

The proof was given in Thrift (1977) where further references can be found.

An important special case that we shall return to repeatedly is when the functions $g_{ij}(x)$ in (2.30) are of the form

$$\text{constant} \times \exp[-\tfrac{1}{2}x^THx] \tag{2.48}$$

where H is some non-negative definite quadratic form. In other words we deal with Gaussian configurations. Since Gaussian distributions have linear regression one can express the conditional probability relations above in terms of linear relations. For example, if F is a subset of S and

$k \notin F$ then it can be shown that

$$E[X_k | X_j ; j \in F] = E[X_k | X_j ; j \in \partial F(k)]. \tag{2.49}$$

where we use the boundary of F w.r.t. k

$$\partial F(k) = \{f \in F | \exists \text{ a path } f, j_1, j_2, \ldots k \tag{2.50}$$

such that $j_1, j_2, \ldots k \subset S-F\}$. See Thrift (1979). A definition completely analogous to (2.50) will be used in later sections for the boundary of w.r.t. a set of vertices. Similarly for sets of edges.

5.3. Frozen patterns: finite G and n

The "temperature" θ will be introduced for regular structures in analogy with the way it enters the Gibbs ensembles in physics. When G and the configurations are finite, the generators have finite arity, and when $\#(c)$ and σ are fixed, we shall assume that the total "energy" $H(c)$ is the sum of all interaction energies

$$H(c) = \sum_{i=1}^{n} k(g_i) + \sum_{(k,\ell)\in\sigma} h(\beta_k, \beta_\ell). \tag{3.1}$$

In (3.1) the first sum represents the self interactions.

To relate the energies to Q and A from Section 1 we put with normalizing constants K_1 and K_2

$$\begin{cases} Q(g) = K_1 \exp[-\frac{k(g)}{\theta}] \\ A(\beta', \beta'') = K_2 \exp[-\frac{h(\beta', \beta'')}{\theta}] \end{cases} \tag{3.2}$$

so that (1.6) shows that the probability $P_\theta(c)$ associated with c is proportional to

$$\exp[-\frac{H(c)}{\theta}] . \tag{3.3}$$

It is easy to see how the probability measure P_θ behaves when the temperature drops to zero. Writing out the normalization constant explicitly in (3.3) we have

$$P_\theta(c) = \frac{\exp[-\frac{H(c)}{\theta}]}{\sum_{c'} \exp[-\frac{H(c')}{\theta}]} \tag{3.4}$$

where the sum in the denominator is extended over all configurations. Defining the *minimum energy set*

$$M = \{c' | H(c') = \min_c H(c)\} \tag{3.5}$$

we can express (3.4) as

$$P_\theta(c) = \frac{\exp[-\frac{H(c)-m}{\theta}]}{N + \sum_{c' \notin M} \exp[\frac{H(c')-m}{\theta}]} \tag{3.6}$$

with

$$m = \min_c H(c'), \quad N = \#M. \tag{3.7}$$

Note that in (3.6) the ratios in numerator and denominator are non-negative. In the numerator they are zero iff $c \in M$; in the sum in the denominator they are all strictly positive. Hence we can announce a simple but illuminating result as

Theorem 1. *For finite* G *and* c's, *and when* n *and* σ *are fixed, the limiting probability measure is uniform on the minimum energy set* M

$$\lim_{\varepsilon \downarrow 0} P_\theta(c) = \begin{cases} \frac{1}{N} & \text{if } c \in M \\ 0 & \text{else} \end{cases} \tag{3.8}$$

Example. Let G = {1,2,3} with arity two, both bond values equal to g itself, g = 1,2,3. Further let

$$k(g) = 0, \quad g = 1,2,3 \tag{3.9}$$

and with the interaction matrix

$$h(\beta',\beta");\beta',\beta" = 1,2,3 = \begin{bmatrix} 7 & 3 & 2 \\ 3 & 7 & 5 \\ 2 & 5 & 7 \end{bmatrix} \tag{3.10}$$

with Σ = CYCLIC, σ with all bond couples closed in a cycle, and $n = 4$ we get the minimum energy configuration

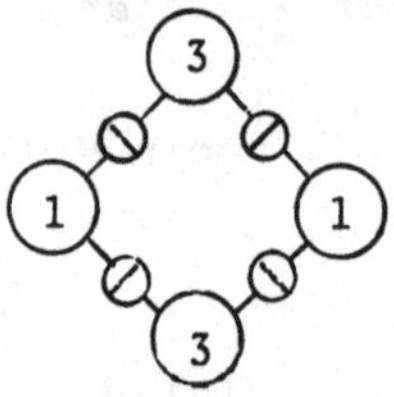

Figure 3.1

and its one-step cyclic permutation. Clearly these are the only two frozen patterns, each of which receives 50% probability. Notice that these frozen patterns correspond to a local regularity governed by the bond relation

$$\beta\rho\beta' = \text{TRUE} \Longleftrightarrow (\beta=1\wedge\beta'=3) \vee (\beta=3\wedge\beta'=1) \tag{3.11}$$

<u>Remark 1</u>. Theorem 1 indicates the importance of finding the minimum energy states. Some information of how this can be done can be found in Section 3.8 of Volume I.

<u>Remark 2</u>. When the limiting measure P exists we shall call the configurations in its support the *frozen patterns*

$$\text{set of frozen patterns} = \text{support}(P). \tag{3.12}$$

5.4. Frozen patterns; infinite G and finite n

The completely finite case, treated in the previous section is quite straightforward. The *half-finite case, when the generator space is infinite but the configurations are still finite,* presents greater difficulties. We shall describe some progress, mainly due to Hwang (1978).

Say that $\#(c)$ is fixed to n, σ is also fixed, and that G has been parametrized as the real line. This means that the total energy can be written as $H(x)$, $x \in \mathbb{R}^n$, where $x = (x_1, x_2, \ldots, x_n)$ represents the configuration with generators $x_1, x_2, \ldots x_n$. The function H will be restricted by three assumptions, the first of which is

$$H \text{ is continuous and } m\{H(x) < a\} > 0 \text{ if } \inf_x H(x) > a. \quad (4.1)$$

We get the P_θ measures in terms of the Radon-Nikodym derivatives with respect to some measure m, here assumed to be a probability measure,

$$\frac{P_\theta(dx)}{m(dx)} = Z^{-1} \exp\left[-\frac{H(x)}{\theta}\right] = f_\theta(x) \quad (4.2)$$

where

$$Z = \int_{\mathbb{R}^n} \exp\left[-\frac{H(x)}{\theta}\right] m(dx). \quad (4.3)$$

To be able to discuss limiting measures we must first make sure that the family $\{P_\theta\}$ is tight. To shed some light on this prove a negative result.

Theorem 1. *If* H *does not have a minimum the family* $\{P_\theta\}$ *cannot be tight.*

Proof: We shall do it indirectly, assuming that there exists a sequence of θ-values such that $P_\theta \to P$ weakly when θ

runs through this sequence. Select a decreasing sequence $\{a_k\}$ such that

$$\lim_{k\to\infty} a_k = m = \inf_x H(x) \tag{4.4}$$

and such that all a_k are continuity points for the stochastic variable $H(x)$ when x is distributed according to P.

Then we can write

$$P_\theta[H(x) \geq a_k] = \frac{\int_{H(x)\geq a_k} \exp\left[-\frac{H(x)}{\theta}\right] m(dx)}{\int_{\mathbb{R}^n} \exp\left[-\frac{H(x)}{\theta}\right] m(dx)} \tag{4.5}$$

which is at most equal to (recall that m is a normed measure here)

$$\frac{1}{\int_{\mathbb{R}^n} \exp\left[-\frac{H(x)-a_k}{\theta}\right] m(dx)} . \tag{4.6}$$

This in turn is at most equal to the expression

$$\frac{}{\int_{H(x)<a_k} \exp\left[-\frac{H(x)-a_k}{\theta}\right] m(dx)} . \tag{4.7}$$

What happens with (4.7) when $\theta \downarrow 0$? For each x in the region of integration the argument of exp tends to $-\infty$. Also note that the m-measure of the region is positive for h large enough. The bounded convergence theorem then implies that (4.7) tends to zero.

Hence

$$\lim_{\theta\downarrow 0} P_\theta[H(x) \geq a_k] = 0 \tag{4.8}$$

but $P_\theta \to P$ weakly so that

$$P[H(x) \geq a_k] = 0 \tag{4.9}$$

for h large enough. Recalling the definition of the $\{a_k\}$ sequence we get $P(\mathbb{R}^n) = 0$, a contradiction. Q.E.D.

This forces the following assumption upon us:

$$\min_x H(x) \quad \text{exists} \tag{4.10}$$

and we can assume w.l.g. that its value is zero.

With similar notation as in Section 3 we put

$$\begin{cases} M = \{x \mid H(x) = 0\} \\ m(M) = m \end{cases}$$

Theorem 2. *Under assumptions* (4.1), (4.10) *and* $m > 0$ *the limit* P *exists and coincides with the uniform measure over the minimum energy configuration, that is*

$$P(A) = \frac{1}{m}\, m(A \cap M). \tag{4.11}$$

Proof: We have, using (4.2),

$$f_\theta(x_o) = \frac{1}{(13)+(14)} \tag{4.12}$$

where

$$(13) = \int_{H(x) \neq H(x_o)} \exp\left[-\frac{H(x)-H(x_o)}{\theta}\right] m(dx) \tag{4.13}$$

and

$$(14) = \int_{H(x) = H(x_o)} \exp\left[-\frac{H(x)-H(x_o)}{\theta}\right] m(dx). \tag{4.14}$$

Separe cases according to whether x_o is in M or not. If x_o is not a minimum energy configuration then (4.13) is at least

$$\int_{H(x) < H(x_o)} \exp\left[-\frac{H(x)-H(x_o)}{\theta}\right] m(dx) \to \infty \text{ as } \theta \downarrow 0. \tag{4.15}$$

In the opposite case, $x_o \in M$, (4.13) is equal to

$$\int_{H(x)>0} \exp\left[-\frac{H(x)}{\theta}\right] m(dx) \to 0 \quad \text{as} \quad \theta \downarrow 0 \tag{4.16}$$

while (4.14) is just $m(M) = m$.

Hence

$$\lim_{\theta \downarrow 0} f_\theta(x_o) = f(x_o) = \begin{cases} \frac{1}{m} & \text{if } x_o \in M \\ 0 & \text{else} \end{cases} \tag{4.17}$$

According to Scheffé's theorem (see Notes A) this guarantees that $P_\theta \to P$ weakly where P means the uniform distribution w.r.t. $m(dx)$ over M. Q.E.D.

Theorem 2 is informative but it does not tell us what happens in the degenerate but interesting case $m = m(M) = 0$. One would expect P to concentrate (if it exists) on M and we shall look more carefully at *how* this happens in two cases: when M is finite and when it is the union of smooth manifolds.

But first let us note that if we assume

$$\text{we can find a positive } \varepsilon \text{ such that } \{x \mid H(x) \le \varepsilon\} \text{ is compact} \tag{4.18}$$

then $\{P_\theta\}$ is tight. This is almost immediate since

$$P_\theta[H(x) > \varepsilon] \le \frac{1}{\int_{H(x)\le\varepsilon} \exp\left[-\frac{H(x)-}{\theta}\right] m(dx)} . \tag{4.19}$$

As $\theta \downarrow 0$ this tends to zero and this implies tightness.

Now let M be the finite set with the elements $x_1, x_2, \ldots x_r$. We shall assume further that

$$H \in C^3(\mathbb{R}^n),\ \frac{m(dx)}{\mu(dx)} = f(x) \text{ is continuous; } \mu \text{ is Lebesgue measure in } \mathbb{R}^n. \tag{4.20}$$

Under this condition we can announce

Theorem 3. *For the given set* M *of minimum energy configurations let us assume that* (4.20) *holds and that* $\det H''(x_i) \neq 0$ *for all* $i = 1,2,\ldots,r$ *and that for some* k *we have* $f(x_k) > 0$. *Then the limiting measure is given by*

$$P(x_i) = \frac{f(x_i)[\det H''(x_i)]^{-1/2}}{\sum_{j=1}^{r} f(x_j)[\det H''(x_j)]^{-1/2}}. \tag{4.21}$$

Note. In (4.20) we use the notation for the Hessian

$$H''(x) = \left\{\frac{\partial^2 H(x)}{\partial x_i \partial x_j};\ i,j = 1,2,\ldots n\right\} \tag{4.22}$$

where for the moment x_i stands for the ith coordinate, not the element of M as above and later.

Proof: Let A_i be a closed neighborhood of x_i such that it contains no other element of M. Then from (4.2) again

$$P(A_i) = \frac{\int_{A_i} \exp\left[-\frac{H(x)}{\theta}\right] m(dx)}{\int_{\mathbb{R}^n} \exp\left[-\frac{H(x)}{\theta}\right] m(dx)}. \tag{4.23}$$

This expression will now be treated by a variation of Laplace's method:

Lemma 1. *Let the real valued function* h *on* $\mathbb{R}^n$ *be* C^3, $h(0) = 0$, $h(x) > 0$ *for* $x \neq 0$ *and with* $\det(h''(0)] \neq 0$, *with* h *bounded away from zero at infinity. Then for a continuous and* L_1*-function* ϕ *over* $\mathbb{R}^n$ *we have*

$$\lim_{\theta\downarrow 0}(2\pi\theta)^{-n/2}\int_{\mathbb{R}^n}\exp\left[-\frac{h(x)}{\theta}\right]\phi(x)\,dx = \phi(0)\{\det[h''(0)]\}^{-\frac{1}{2}}. \quad (4.24)$$

Proof of Lemma 1: Given any positive

$$\left|\int_{\mathbb{R}^n} - \int_{|x|\le\delta} \exp\left[-\frac{h(x)}{\theta}\right]\phi(x)\,dx\right| \le \exp\left(-\frac{\varepsilon_o}{\theta}\right)\int_{|x|>\delta}|\phi(x)|\,dx. \quad (4.25)$$

We just have to choose ε_o so small that $h(x) > \varepsilon_o$ for $|x| > \delta$. Note that (4.25) implies that

$$(4.25) = 0[(2\pi\theta)^{n/2}]. \quad (4.26)$$

Recalling that h has a unique minimum at $x = 0$ and $h \in C^3$ we see that $h''(0)$ is a symmetric and positive definite $n \times n$ matrix. Let λ be the smallest eigenvalue of $h''(0)$. If ε is chosen smaller than λ, then the matrices

$$\begin{cases} h''(0) + \varepsilon I \\ h''(0) - \varepsilon I \end{cases} \quad (4.27)$$

are also positive definite. Now pick $\delta < \varepsilon$ such that the inner products satisfy (use the Taylor expansion up to quadratic terms)

$$\begin{cases} \frac{1}{2}([h''(0)-\varepsilon I]x,x) \le h(x) \le \frac{1}{2}([h''(0)+\varepsilon I]x,x) \\ |x| \le \delta \end{cases} \quad (4.28)$$

With the notation

$$\begin{cases} m_1 = \min_{|x|\le\delta} \phi(x) \\ m_2 = \max_{|x|\le\delta} \phi(x) \end{cases} \quad (4.29)$$

we get from (4.28) the bounds

$$(2\pi\theta)^{-n/2}\int_{|x|\le\delta}\exp\left\{-\frac{([h''(0)-\varepsilon I]x,x)}{2\theta}\right\}m_1\,dx$$

$$\le (2\pi\theta)^{-n/2}\int_{|x|\le\delta}\exp\left[-\frac{h(x)}{\theta}\right]\phi(x)\,dx \tag{4.30}$$

$$\le (2\pi\theta)^{-n/2}\int_{|x|\le\delta}\exp -\frac{([h''(0)+\varepsilon I]x,x)}{2\theta}\, m_2\, dx.$$

Now let $\theta \downarrow 0$ and use the values of the usual Gaussian integrals. We get from (4.30)

$$m_1\{\det[h''(0)+\varepsilon I]\}^{-1/2}$$

$$\le \underline{\lim}_{\theta\downarrow 0}\,(2\pi\theta)^{-n/2}\int_{|x|<\delta}\exp\left[-\frac{h(x)}{\theta}\right]\phi(x)\,dx \tag{4.31}$$

$$\le \overline{\lim}_{\theta\downarrow 0}\,(2\pi\theta)^{-n/2}\int_{|x|<\delta}\exp\left[-\frac{h(x)}{\theta}\right]\phi(x)\,dx$$

$$\le m_2\{\det[h''(0)-\varepsilon I]\}^{-1/2}.$$

Combining (4.31) with (4.26) we arrive at the bounds

$$m_1\{\det[h''(0)+\varepsilon I]\}^{-1/2}$$

$$\le \underline{\lim}_{\theta\downarrow 0}\,(2\pi\theta)^{-n/2}\int_{\mathbb{R}^n}\exp\left[-\frac{h(x)}{\theta}\right]\phi(x)\,dx \tag{4.32}$$

$$\le \overline{\lim}_{\theta\downarrow 0}\,(2\pi\theta)^{-n/2}\int_{\mathbb{R}^n}\exp\left[-\frac{h(x)}{\theta}\right]\phi(x)\,dx$$

$$\le m_2\{\det[h''(0)-\varepsilon I]\}^{-1/2}.$$

When we make $\varepsilon \downarrow 0$, so that m_1 and $m_2 \to \phi(0)$, we find that the left and right side of (4.32) tend to

$$\phi(0)\{\det[h''(0)]\}^{-1/2} \tag{4.33}$$

as stated in the Lemma. Q.E.D.

The lemma can now be applied directly to (4.23) to complete the proof of Theorem 3. Q.E.D.

Example 1. Let G have generators of arity two and be parametrized by the real line, $G = \mathbb{R}$, with both bond values equal to g itself. With Σ = LINEAR let us define the function h appearing in (3.1) by

$$h(x_i, x_j) = \phi(x_i + x_j) \tag{4.34}$$

where the smooth and non-negative function ϕ has $\phi(x) = 0$ iff $x = 0$. Let k also be smooth and non-negative with $k(x) = 0$ iff $x = c$ or $-c$. What are the frozen patterns?

With $\#(c) = n$ we should solve

$$H(x) = \sum_{i=1}^{n-1} \phi(x_{i+1}+x_i) + \sum_{i=1}^{n} k(x_i) = \min. \tag{4.35}$$

Since the H-function is non-negative but can take the value zero it is clear that there are two frozen patterns, namely c_1 and c_2 as

$$\begin{cases} C_1 = (c,-c,c,-c,\ldots) \\ C_2 = (-c,c,-c,c,\ldots). \end{cases} \tag{4.36}$$

This corresponds to the bond relation for strict regularity

$$\beta'\rho\beta'' = \text{TRUE} \Leftrightarrow (\beta'=c \wedge \beta''=-c) \vee (\beta'=-c \wedge \beta''=c). \tag{4.37}$$

The limiting probabilities for c_1 and c_2 can be calculated from equation (4.21).

The relation in (4.37) is suggestive and points to an important area of research, so far scarcely touched. Say that we start with some regularity controlled probability measure, for example the one in equation (1.4). We know when and how a limiting measure is reached for lowered temperature = stricter regularity. Can one give conditions in order that *the regularity of the frozen patterns can be described (locally) by some bond relation* ρ? In spite of its manifest importance this problem will not be investigated here.

Note that if we change the set up in Example 1 by letting $k \equiv 0$, then M will consist of a smooth manifold:

$$x_1 = -x_2 = x_2 = -x_3 = \dots \tag{4.38}$$

and Theorem 3 does not apply.

We now turn to the second part of the half-finite case when the set M of minimum energy configurations is the union of a finite number of smooth manifolds. Since this is considerably more difficult to analyze we shall begin by some preliminary considerations, again following Hwang (1978) closely.

In addition to the previous assumptions, we also assume that each component of M is a smooth manifold (or C^3-manifold). These manifolds may be of different dimensions. We also assume M has finitely many components. An interesting question arises: "Will the limiting probability measure concentrate on the highest dimensional manifolds?". When θ is small enough, the major contribution is in a small neighborhood of M. Since the gradient of H at each point of M is 0, we cannot use the implicit function theorem. In a small neighborhood of M we shall therefore change the coordinate system to a polar coordinate system along M and write the limiting probability measure in terms of some intrinsic measures on the manifolds.

Let M be a k-dimensional compact smooth manifold in R^n. Then by the tubular neighborhood theorem (Milnor-Stasheff (1974), there exists a tubular neighborhood $T(\varepsilon)$ of M such that for any z in $T(\varepsilon)$, z can be written as $m + v$, where m is a point on M and $v \perp M$ at m with $|v| < \varepsilon$. The map $z \to (m,v)$ is a diffeomorphism.

Now in $T(\varepsilon)$ we are going to change $d\mu = dx_1 \ldots dx_n$ to the polar coordinates as done in Weyl (1939). Consider local coordinates of M, $m = m(u^1,\ldots,u^k)$. We can determine n-k normal vectors $\mathcal{N}(1),\ldots,\mathcal{N}(n-k)$ such that

$$\frac{\partial m}{\partial u^i} \cdot \mathcal{N}(j) = 0, \quad \mathcal{N}(i) \cdot \mathcal{N}(j) = \delta_{ij} \tag{4.39}$$

and $\mathcal{N}(i)$ depends on $(u^1,\ldots,u^k)$ smoothly. Then for any $z \in T(\varepsilon)$,

$$\begin{aligned} z &= z(u^1,\ldots,u^k,t_1,\ldots,t_{n-k}) \\ &= m(u^1,\ldots,u^k) + t_1\mathcal{N}(1)+\ldots+t_{n-k}\mathcal{N}(n-k), \end{aligned} \tag{4.40}$$

where $|(t_1,\ldots,t_{n-k})| < \varepsilon$. In terms of $u^1,\ldots,u^k$, $t_1,\ldots,t_{n-k}$

$$\begin{aligned} dx_1 \ldots dx_n = &\left|\det\left(\frac{\partial z}{\partial u^1},\ldots,\frac{\partial z}{\partial u^k},\mathcal{N}(1),\ldots,\mathcal{N}(n-k)\right)\right| \\ &dt_1 \ldots dt_{n-k}du^1 \ldots du^k. \end{aligned} \tag{4.41}$$

Since it is a matter only of notation, (4.41) is independent of the local coordinates. We obtain the basis

$$\frac{\partial m}{\partial u^1},\ldots,\frac{\partial m}{\partial u^k}, \mathcal{N}(1),\ldots,\mathcal{N}(n-k). \tag{4.42}$$

Write

$$\frac{\partial \ (i)}{\partial u^\alpha} = \sum_\beta G_\alpha^\beta(i) \frac{\partial m}{\partial u^\beta} + \overline{\mathcal{N}}, \tag{4.43}$$

where $\bar{\mathscr{N}}$ is a linear combination of $\mathscr{N}(i)$'s. Then

$$\frac{\partial z}{\partial u^\alpha} = \frac{\partial}{\partial u^\alpha}(m + t_1\mathscr{N}(1) + \ldots + t_{n-k}\mathscr{N}(n-k))$$

$$= \frac{\partial m}{\partial u^\alpha} + t_1\frac{\partial \mathscr{N}(1)}{\partial u^\alpha} + \ldots + t_{n-k}\frac{\partial \mathscr{N}(n-k)}{\partial u^\alpha} \tag{4.44}$$

$$= \sum_\beta\Big(\delta_\alpha^\beta + \sum_{i=1}^{n-k} t_i G_\alpha^\beta(i)\Big)\frac{\partial m}{\partial u^\beta} + \hat{\mathscr{N}}_\alpha,$$

where $\hat{\mathscr{N}}_\alpha$ is a linear combination of $\mathscr{N}(i)$'s. Further,

$$\Big|\det\Big(\frac{\partial z}{\partial u^1}, \ldots, \frac{\partial z}{\partial u^k}, \mathscr{N}(1), \ldots, \mathscr{N}(n-k)\Big)\Big|$$

$$= \Big|\det\Big(\sum_\beta\Big(\delta_1^\beta + \sum_{i=1}^{n-k} t_i G_1^\beta(i)\Big)\frac{\partial m}{\partial u^\beta} + \hat{\mathscr{N}}_1, \ldots, \tag{4.45}$$

$$\sum_\beta\Big(\delta_k^\beta + \sum_{i=1}^{n-k} t_i G_k^\beta(i)\Big)\frac{\partial m}{\partial u^\beta} + \hat{\mathscr{N}}_k, \mathscr{N}(1), \ldots, \mathscr{N}(n-k)\Big)\Big|$$

This can be written as

$$\Big|\det\Big(\sum_\beta\Big(\delta_1^\beta + \sum_{i=1}^{n-k} t_i G_1^\beta(i)\Big)\frac{\partial m}{\partial u^\beta}, \ldots$$

$$\sum_\beta\Big(\delta_k^\beta + \sum_{i=1}^{n-k} t_i G_k^\beta(i)\Big)\frac{\partial m}{\partial u^\beta}, \mathscr{N}(1), \ldots, \mathscr{N}(n-k)\Big)\Big| \tag{4.46}$$

$$= \Big|\det\Big[\delta_\alpha^\beta + \sum_i t_i G_\alpha^\beta(i)\Big]\Big|\,\Big|\det\Big(\frac{\partial m}{\partial u^1}, \ldots, \frac{\partial m}{\partial u^k},$$

$$\mathscr{N}(1), \ldots, \mathscr{N}(n-k)\Big)\Big|.$$

Equation (4.46) holds, since

$$\Big(\frac{\partial m}{\partial u^1}, \ldots, \frac{\partial m}{\partial u^k}, \mathscr{N}(1), \ldots, \mathscr{N}(n-k)\Big)\begin{pmatrix} \{x_\alpha^\beta\}_{k\times k} & 0 \\ 0 & I_{(n-k)\times(n-k)} \end{pmatrix}$$

$$= \Big(\sum_\beta x_1^\beta\frac{\partial m}{\partial u^\beta}, \ldots, \sum_\beta x_k^\beta\frac{\partial m}{\partial u^\beta}, \mathscr{N}(1), \ldots, \mathscr{N}(n-k)\Big) \tag{4.47}$$

where

$$x_\alpha^\beta = \delta_\alpha^\beta + \sum_{i=1}^{n-k} t_i G_\alpha^\beta(i). \tag{4.48}$$

Let $g_{\alpha\beta} = \frac{\partial m}{\partial u^\alpha} \frac{\partial m}{\partial u^\beta}$. Using the fact that $(\det(A_1,\ldots,A_n))^2 = \det\{A_i A_j\}$ and (4.40), we have

$$\left(\det\left(\frac{\partial m}{\partial u^1},\ldots,\frac{\partial m}{\partial u^k},\mathcal{N}(1),\ldots,\mathcal{N}(n-k)\right)\right)^2 = \det\{g_{\alpha\beta}\}. \tag{4.49}$$

Write $|\det\{x_{\alpha\beta}\}| = |x_{\alpha\beta}|$ in the following. Then (4.46) becomes

$$|\delta_\alpha^\beta + \sum_i t_i G_\alpha^\beta(i)|\,|g_{\alpha\beta}|^{1/2}. \tag{4.50}$$

Consider $|g_{\alpha\beta}|^{1/2} du^1 \ldots du^k$, which is determined by the local coordinate (U,Φ) where $\Phi(u^1,\ldots,u^k) = m \in U \subseteq M$. Now consider another local coordinate (V,ϕ) where $\phi(v^1,\ldots,v^k) = m \in V \subseteq M$. Since Φ,ϕ are homeomorphisms, we will use $(u^1,\ldots,u^k)$ and $\Phi(u^1,\ldots,u^k)$ interchangeably. In $U \cap V$, we have

$$\begin{aligned} h_{ij} &= \frac{\partial m}{\partial v^i}\frac{\partial m}{\partial v^j} = \left(\sum_k \frac{\partial m}{\partial u^k}\frac{\partial u^k}{\partial v^i}\right)\left(\sum_\ell \frac{\partial m}{\partial u^\ell}\frac{\partial u^\ell}{\partial v^j}\right) \\ &= \sum_{k\ell} g_{k\ell}\frac{\partial u^k}{\partial v^i}\frac{\partial u^\ell}{\partial v^j}. \end{aligned} \tag{4.51}$$

So $|h_{\alpha\beta}| = |g_{\alpha\beta}| \left|\frac{\partial u^k}{\partial v^j}\right|^2$, and

$$\begin{aligned} |h_{\alpha\beta}|^{1/2} dv^1 \ldots dv^k &= |g_{\alpha\beta}|^{1/2}\left|\frac{\partial u^k}{\partial v^j}\right| dv^1 \ldots dv^k \\ &= |g_{\alpha\beta}|^{1/2}\, du^1 \ldots du^k. \end{aligned} \tag{4.52}$$

The transition law holds in the intersection. The measure on U defined by $|g_{\alpha\beta}|^{1/2}\, du^1 \ldots du^k$ is independent of the local coordinates, i.e. if we choose another (V,ϕ), then

(V,ϕ) and (U,ϕ) define the same measure on the intersection $U \cap V$.

Let $\{(U_\alpha,\Phi_\alpha)\}$ be the atlas of the compact manifold M. For each (U_α,Φ_α) we have a local measure defined on U_α by $|g_{ij}|^{1/2}du^1 \ldots du^k$. By the Riesz representation theorem there exists a unique positive linear functional λ_α on $C_c(U_\alpha)$ such that

$$\lambda_\alpha f = \int_{U_\alpha} f|g_{ij}|^{1/2}du^1 \ldots du^k. \tag{4.53}$$

Because of the transition law, λ_α and λ_β are equal on $C_c(U_\alpha \cap U_\beta)$. Using a partition of unity, there exists a unique positive linear functional λ on $C_c(M) = C(M)$ such that the restriction of λ to each $C_c(U_\alpha)$ is equal to λ_α. Using the Riesz representation theorem again, we can find a unique measure $\mathcal{M}$ on M corresponding to λ such that $\mathcal{M}$ restricted to U_α is the same as the original local measure defined by $|g_{ij}|^{1/2}du^1 \ldots du^k$. Since M is compact, $\mathcal{M}(M) < \infty$. The measure $\mathcal{M}$ is independent of the local coordinates, so we call $\mathcal{M}$ the intrinsic measure on the manifold M.

After these preliminaries we are ready to announce the main result.

Theorem 4. Hwang (1978). *Assume that* M *has finitely many components and each component is a compact smooth manifold. The energy function* H *and probability* Q *should satisfy the conditions* (4.1), (4.10), (4.18), (4.20) *and* $m = 0$. *If the density* f *in* (4.20) *is not identically zero on the highest dimensional manifolds and* $\det \frac{\partial^2 H}{\partial t^2}(u) \neq 0$ *for* $u \in M$, *then the limiting probability measure concentrates on the highest dimensional manifolds and can be written as:*

$$\frac{dP}{d\mathscr{M}}(u) = \frac{f(u)\left(\det \frac{\partial^2 H}{\partial t^2}(u)\right)^{-1/2}}{\int_N f(u)\left(\det \frac{\partial^2 H}{\partial t^2}(u)\right)^{-1/2} d\mathscr{M}}$$

where $\mathscr{M}$ *is the sum of intrinsic measures on the highest dimensional manifolds.*

Proof: Let $\{M_\ell\}_1^q$ be the components of M and g be a bounded continuous function from $\mathbb{R}^n$ to $\mathbb{R}$. Consider

$$\int_{R^n} \exp\left(\frac{-H(z)}{\theta}\right) f(z)g(z)dz. \tag{4.54}$$

As in the proof in Lemma 4.1, the difference between

$$\sum_{\ell=1}^q \int_{T_\ell(\varepsilon)} \exp\left(\frac{-H(z)}{\theta}\right) f(z)g(z)dz \tag{4.55}$$

and (4.54) is exponentially small, where $T_\ell(\varepsilon)$ is an ε-tubular neighborhood of M_ℓ, $T_\ell(\varepsilon) \cap T_d(\varepsilon) = \phi$ if $\ell \neq d$, and $T_\ell(\varepsilon)$ is chosen closed.

Fix ℓ, and consider the integral

$$\int_{T_\ell(\varepsilon)} \exp\left(\frac{-H(z)}{\theta}\right) f(z)g(z)dz$$

$$= \int_{M_\ell}\int_{|t|\leq\varepsilon} \exp\left(\frac{-H(t,u)}{\theta}\right) f(t,u)g(t,u)\,|\delta_\alpha^\beta + \sum_i t_i G_\alpha^\beta(i)|$$

$$dt_1 \ldots dt_{n-k_\ell}\, d_\ell \tag{4.56}$$

where $_\ell$ is the intrinsic measure on M_ℓ and k_ℓ is the dimension of M_ℓ. For each fixed u

$$\frac{\int_{|t|\le\varepsilon} \exp\left(\frac{-H(t,u)}{\theta}\right) f(t,u)g(t,u)\,|\delta_\alpha^\beta + \sum_i t_i G_\alpha^\beta(i)|\,dt_1\ldots dt_{n-k}}{(2\pi\theta)^{\frac{n-k_\ell}{2}}}$$

$$\to f(0,u)g(0,u) \det \frac{\partial^2 H}{\partial t^2}(0,u)^{-1/2}, \tag{4.57}$$

with

$$H(t,u) = \frac{1}{2} < \frac{\partial^2 H}{\partial t^2}(0,u)t,t > + \frac{1}{6}\frac{\partial^3 H}{\partial t^3}(\bar{t},u)(t) \tag{4.58}$$

where $\bar{t} \in$ segment $(0,t)$ and

$$\frac{\partial^3 H}{\partial t^3}(\bar{t},u)(t) = \sum_{i=1}^{k_\ell}\sum_{j=1}^{k_\ell}\sum_{k=1}^{k_\ell} \frac{\partial^3 H}{\partial t_i \partial t_j \partial t_k}(\bar{t},u)t_k t_j t_i. \tag{4.59}$$

Let $\lambda(u)$ be the minimal eigenvalues of $\frac{\partial^2 H}{\partial t^2}(0,u)$. Since $\frac{\partial^2 H}{\partial t^2}(0,u)$ is positive definite and M_ℓ is compact, $\min_{u\in M_\ell} \lambda(u) = \lambda > 0$. Choose $0 < 2\delta_\ell < \lambda$, then $H(t,u) \ge \delta_\ell |t|^2 + \frac{1}{6}\frac{\partial^3 H}{\partial t^3}(\bar{t},u)(t)$. Let

$$B = \max_{ijk} \max_{\substack{|t|\le\varepsilon \\ u\in\bar{M}}} \left|\frac{\partial^3 H}{\partial t_i \partial t_j \partial t_k}(t,u)\right|, \tag{4.60}$$

then $B < \infty$. We can choose ε_ℓ small enough such that

$$\frac{1}{2}\delta_\ell |t|^2 - \frac{1}{6}\sum_{ijk} B|t_i t_j t_k| \ge 0 \text{ for any } |t| \le \varepsilon_\ell. \tag{4.61}$$

Then for any $|t| \le \varepsilon_\ell$ we have $H(t,u) \ge \frac{1}{2}\delta_\ell |t|^2$. Therefore we can replace ε by ε_ℓ and still have the same result in (4.57). Let

$$A(\theta,\ell) = \int_{u\in M}\int_{|t|\le\varepsilon_\ell} \exp\left(\frac{-H(t,u)}{\theta}\right) f(t,u)g(t,u) \left|\delta_\alpha^\beta + \sum_i t_i G_\alpha^\beta(i)\right| dt_1\ldots dt_{n-k_\ell}\, d\mathcal{M}_\ell, \tag{4.62}$$

and

$$\bar{A}(\theta,\ell) = \int_{u\in M_\ell}\int_{|t|\le\varepsilon_\ell} \exp\left(\frac{-H(t,u)}{\theta}\right) f(t,u) \left|\delta_\alpha^\beta + \sum_i t_i G_\alpha^\beta(i)\right| dt_1\ldots dt_{n-k_\ell}\, d\mathcal{M}_\ell. \tag{4.63}$$

Because

$$0 \le \exp\left(\frac{-H(t,u)}{\theta}\right) \le \exp\left(-\frac{\delta_\ell |t|^2}{2\theta}\right), \quad \mathcal{M}_\ell(M_\ell) < \infty, \tag{4.64}$$

by (4.57)-(4.64) and the dominated convergence theorem, we have

$$\frac{A(\theta,\ell)}{(2\pi\theta)^{\frac{n-k_\ell}{2}}} \to \int_{u\in M_\ell} f(0,u)\left(\det \frac{\partial^2 H}{\partial t^2}(0,u)\right)^{-1/2} d\mathcal{M}_\ell. \tag{4.65}$$

and

$$\frac{\bar{A}(\theta,\ell)}{(2\pi\theta)^{\frac{n-k_\ell}{2}}} \to \int_{u\in M_\ell} f(0,u)\left(\det \frac{\partial^2 H}{\partial t^2}(0,u)\right)^{-1/2} d\mathcal{M}_\ell. \tag{4.66}$$

Let $\max_{1\le\ell\le q} k_\ell = m$, and consider

$$\frac{\int_{R^n} \exp\left(\frac{-H(z)}{\theta}\right) f(z)g(z)dz}{\int_{R^n} \exp\left(\frac{-H(z)}{\theta}\right) f(z)dz} \approx \frac{\sum_\ell A(\theta,\ell)}{\sum_\ell \bar{A}(\theta,\ell)} = \frac{\sum_\ell A(\theta,\ell)(2\pi\theta)^{-\frac{(n-m)}{2}}}{\sum_\ell \bar{A}(\theta,\ell)(2\pi\theta)^{-\frac{(n-m)}{2}}} \tag{4.67}$$

$$= \frac{\sum_\ell A(\theta,\ell)(2\pi\theta)^{\frac{-(n-k_\ell)}{2}}(2\pi\theta)^{\frac{m-k_\ell}{2}}}{\sum_\ell \bar{A}(\theta,\ell)(2\pi\theta)^{\frac{-(n-k_\ell)}{2}}(2\pi\theta)^{\frac{m-k_\ell}{2}}}$$

which tends to

$$\frac{\sum_{k_\ell=m} \int_{M_\ell} f(0,u)g(0,u)\left(\det \frac{\partial^2 H}{\partial t^2}(0,u)\right)^{-1/2} d\mathscr{M}_\ell}{\sum_{k_\ell=m} \int_{M_\ell} f(0,u)\left(\det \frac{\partial^2 H}{\partial t^2}(0,u)\right)^{-1/2} d\mathscr{M}_\ell} . \tag{4.68}$$

Now use (4.65), (4.66) and that

$$(2\pi\theta)^{\frac{m-k_\ell}{2}} \to 0 \quad \text{if} \quad k_\ell \neq m. \tag{4.69}$$

Let $\mathscr{M} = \sum_{k_\ell=m} \mathscr{M}_\ell$, i.e.

$$\mathscr{M}(S) = \sum_{k_\ell=m} \mathscr{M}_\ell(S \cap M_\ell) \tag{4.70}$$

where S is a Borel set in N. Now (4.68) becomes

$$\frac{\int_N f(0,u)g(0,u)\left(\det \frac{\partial^2 H}{\partial t^2}(0,u)\right)^{-1/2} d\mathscr{M}}{\int_N f(0,u)\left(\det \frac{\partial^2 H}{\partial t^2}(0,u)\right)^{-1/2} d\mathscr{M}} . \tag{4.71}$$

We can regard $\mathscr{M}$ as a measure on $(R^n, \mathscr{B})$ by considering $\mathscr{M}(B) = \mathscr{M}(B \cap N)$. If we define

$$K(z) = \begin{cases} \det \frac{\partial^2 H}{\partial t^2}(0,u)^{-1/2} & \text{if} \quad z = u \in N \\ 0 & \text{otherwise} \end{cases} \tag{4.72}$$

then (4.68) becomes

$$\int_{R^n} g(z)\left(\frac{f(z)k(z)}{\int_{R^n} f(z)k(z)d}\right) d\mathscr{M}. \tag{4.73}$$

If P is defined by

$$\frac{dP}{d\mathscr{M}}(z) = \frac{f(z)k(z)}{\int_{R^n} f(z)k(z)d\mathscr{M}} \tag{4.74}$$

then $P_\theta \to P$ weakly. Clearly P concentrates on M, and there is no ambiguity in writing

$$\frac{dP}{d\mathscr{M}}(u) = \frac{f(u)\left(\det \frac{\partial^2 H}{\partial t^2}(u)\right)^{-1/2}}{\int_M f(u)\left(\det \frac{\partial^2 H}{\partial t^2}(u)\right)^{-1/2} d\mathscr{M}} \quad \text{for } u \in N. \tag{4.75}$$

Q.E.D.

5.5. Quadratic energy function

The previous results take an especially attractive form when the interaction energies are quadratic functions of the generators. The following results also hold when the configuration is an element in a separable Hilbert space, see Hwang (1978), but we shall keep to the earlier assumption $c = x \in \mathbb{R}^n$.

If H is a non-negative definite quadratic form it can happen that the set

$$\{x \mid H(x) \le \varepsilon\} \tag{5.1}$$

is not compact, namely when H is singular. We must therefore exercise some caution when applying last section's methods to this case.

We shall assume that

$$H(x) = \frac{1}{2} x^T F x - k^T x; \quad k \in \mathbb{R}^n \tag{5.2}$$

so that

$$\frac{P_\theta(dx)}{m(dx)} = \frac{\exp\left[-\frac{1}{2\theta}(x^T Fx + 2k^T x)\right]}{\int_{R^n} \exp\left[-\frac{1}{2\theta}(y^T Fy + 2k^T y)\right] dy} \tag{5.3}$$

and let m denote the standardized normal measure with the x-components i.i.d. N(0,1). Of course nothing will be changed if we add a constant c on the right side of (5.2); choosing c we can normalize the value of $\min_x H(x)$ when it exists.

When looking for a limiting measure P for the frozen patterns we can immediately assume that F is non-negative definite. Otherwise Theorem 4.1 applies and gives a negative answer.

Therefore F *will from now on be assumed to be non-negative definite.* But we can also say something about the vector k in (5.2). Assume first the $k \in \mathscr{R}(F)$ = the range space of F, so that there exists a vector x_0 such that $k = F(x_0)$. Then

$$H(x) = \frac{1}{2}\{(F[x-x_0],x-x_0)-(Fx_0,x_0)\} \tag{5.4}$$

which implies that H(x) assumes its minimum value at $x = x_0$. This is fine, we want the minimum to be attained.

On the other hand if H attains its minimum, say at some x_0, then the Fréchet derivative must be zero at x_0 when we move it along a vector $h \in \mathbb{R}^n$. But that means that

$$dH_{x_0}(h) = (Fx_0-k,h) = 0, \ \forall h \tag{5.5}$$

which implies that $Fx_0-k = 0$, or $k = Fx_0$, or $k \in \mathscr{R}(F)$.

We shall therefore assume from now on that k belongs to the range space $\mathscr{R}(F)$, say $k = Fm$. With these assumptions calculate the characteristic function $\psi_\theta(z)$ of P_θ. To do this note that the relevant linear-quadratic form can be expressed as

$$\frac{1}{\theta} H(x) + \frac{1}{2} \|x\|^2 = \frac{1}{2}\{([\frac{F}{\theta} + I]x,x) - 2(\frac{k}{\theta},x)\} \tag{5.6}$$

recalling that the m-measure in the denominator of the Radon-Nikodym dervative (5.3) corresponds to the quadratic form $\frac{1}{2} \|x\|^2$. But (5.6) equals

$$\frac{1}{2}([\frac{F}{\theta} + I][x-n_\theta], x-n_\theta) + c_0 \tag{5.7}$$

with some real constant c_0 and the vector

$$n_\theta = (\frac{F}{\theta} + I)^{-1} \frac{k}{\theta} \tag{5.8}$$

where the inverse of course exists since F is non-negative definite.

This leads to the characteristic function, see e.g. Cramér (1945), p.

$$\psi_\theta(z) = \exp\{-\frac{1}{2}([\frac{F}{\theta} + I]^{-1}z,z) + i(n_\theta,z)\}. \tag{5.9}$$

With the assumptions just introduced we get

$$n_\theta = (\frac{F}{\theta} + I)^{-1} \frac{1}{\theta} Fm = (\frac{F}{\theta} + I)^{-1}[(I + \frac{F}{\theta})m-m] \tag{5.10}$$

$$= m - (\frac{F}{\theta} + I)^{-1}m.$$

It remains to find the asymptotic behavior of

$$(\frac{F}{\theta} + I)^{-1} \tag{5.11}$$

and this is the crucial step in our analysis. To do this write M in its spectral decomposition, as

$$F = \sum_{k=1}^{p} \lambda_k E_k, \quad \lambda_k > 0, \tag{5.12}$$

with the projections E_k; note that the eigenvalues equal to zero (if there are any) have been left out. Since $m \in \mathscr{R}(F)$

belonging to the subspaces associated with $E_1, E_2, \ldots E_p$, we get

$$(\frac{F}{\theta} + I)^{-1} m = \sum_{k=1}^{p} \frac{1}{\frac{\lambda_k}{\theta} + 1} E_k m. \tag{5.13}$$

As $\theta \downarrow 0$ this tends to zero and (5.10) tends indeed to m.

Lemma 1. *If* F *is non-negative definite then*

$$\lim_{\theta \downarrow 0} (\frac{F}{\theta} + I)^{-1} \to \pi \tag{5.14}$$

where π *denotes the projection to the null space* $\mathcal{N}(F)$ *of* F.

Proof: For an arbitrary $x \in \mathbb{R}^n$ decompose it into $u+v$, $u \in \mathcal{R}(F)$, $v \in \mathcal{N}(F)$, and use the spectral decomposition (5.11) again. Then

$$(\frac{F}{\theta} + I)^{-1} v = v \tag{5.15}$$

and we already know, by the same reasoning as after equations (5.11) that

$$\lim_{\theta \downarrow 0} (\frac{F}{\theta} + I)^{-1} u = 0. \tag{5.16}$$

This proves (5.14). Q.E.D.

Combining what we have learnt we can state, using (5.9), that

$$\lim_{\theta \downarrow 0} \psi_\theta(z) = \exp\{-\frac{1}{2}(\pi z, z) + i(m,z)\} \tag{5.17}$$

and we have arrived at a pleasing conclusion stated as

Theorem 1. *In order that the frozen patterns have a well defined measure* P *it is necessary and sufficient that* F *is non-negative definite and* $k \in \mathcal{R}(F)$. *Then* P *is the Gaussian measure with covariance operator* π *and mean* m, $k = Fm$.

We now apply Theorem 1 to three cases, the first two of which are quite simple.

Example 1. Let the generators have infinite arity with the bond values of a given generator all equal to some $x \in \mathcal{R}$. Let Σ = FULL and

$$x^T F x = \sum_{k \neq \ell} (x_k - x_\ell)^2 \tag{5.18}$$

where the terms correspond to all the bond couples $k \leftrightarrow \ell$. The null space of F consists of the diagonal set, which of course is not compact,

$$D: x_1 = x_2 = \ldots = x_n. \tag{5.19}$$

To get a meaningful limit measure on the frozen patterns we should choose a vector $k \perp D$ in other words $k = (k_1, k_2, \ldots k_n)$ with

$$\sum_{\nu=1}^{n} k_\nu = 0. \tag{5.20}$$

Then the limit measure P has the covariance matrix with all entries equal to $n^{-1/2}$. All the correlation coefficients take the value 1.

The interaction energy terms in (5.18) are *attractive*: they tend to make the bond values x_k equal. What happens if we only change the minus sign in (5.18) to a plus sign, making the interactions *repelling*? Well, if $n = 2$ we get to the anti-diagonal

$$A: x_1 = -x_2 \tag{5.21}$$

and the covariance matrix

$$= \left\{ \begin{matrix} \frac{1}{\sqrt{2}} & -\frac{1}{\sqrt{2}} \\ -\frac{1}{\sqrt{2}} & \frac{1}{\sqrt{2}} \end{matrix} \right\} \tag{5.22}$$

so that the correlation coefficient takes the value -1. On the other hand if $n > 2$ the matrix F is non-singular so that all variances become zero: there is just a single frozen pattern.

<u>Example 2</u>. Let Σ be a finite square lattice in the plane, so that we could index the generators as $g_{\nu\mu}$; $\nu,\mu = 1,2,\ldots N$; $n = N$? All generators shall have arity four with bonds E,N,W,S, all whose bond values are equal to some real number, say $x_{\nu\mu}$. The interaction energy between two adjacent generators shall be of the form $(x^1+x^2)^2$ where x^1 and x^2 are the bond values of the two bonds connected via Σ. Then

$$x^T F x = \sum_{\nu=1}^{N-1} \sum_{\mu=1}^{N} (x_{\nu\mu}+x_{\nu+1\mu})^2 + \sum_{\nu=1}^{N} \sum_{\mu=1}^{N-1} (x_{\nu\mu}+x_{\nu\mu+1})^2 . \tag{5.23}$$

This F matrix is singular and its null space $\mathcal{N}(F)$ is one-dimensional and consists of fields (i.e. N^2-vectors) of the form

y	-y	y	-y	...
-y	y	-y	y	...
y	-y	y	-y	...
-y	y	-y	y	...

Again π is easily calculated and the correlation coefficients between $x_{\nu\mu}$ and $x_{\nu+p\mu+q}$ is 1 if $p+q$ is even and -1 else.

A variation of this pattern is when the generators have arity 8, meaning that each generator $g_{\nu\mu}$ has 8 neighbors connected to it, namely: $g_{\nu+1,\mu}, g_{\nu+1\mu+1}, g_{\nu\mu+1}, g_{\nu-1\mu+1}, g_{\nu-1\mu}, g_{\nu-1\mu-1}, g_{\nu\mu-1}, g_{\nu+1\mu-1}$. With the same local interaction terms it can now be seen that F has then full rank: there is just a single frozen pattern in the support of the limiting measure P.

The third example is more complicated but also more challenging for continued study.

Example 3. The generators shall now be parametrized by $\mathbb{R}^5$, say $g = (x_0, x_1, x_2, x_3, x_4)$ and be of arity 12; the bond coordinates are shown in Table 1. Note that the bond values are in $\mathbb{R} \cup \mathbb{R}^2$. The global regularity is governed by the way bonds can connect as indicated in the third columns. Local regularity shall be governed by the bond relation ρ = EQUAL. Think of a generator as having five "sites" with x_0 in the center and x_1 to the right, x_2 above the center, x_3 to the left, and x_4 below the center.

For example the generators a, shown as the set with circles, and b, shown with crosses, can be combined as in (i) of Figure 1 via the bond couple $(1,0) \to (-1,0)$: if

$$\begin{cases} x_4(a) = x_0(b) \\ x_0(a) = x_3(b) \end{cases} \tag{5.24}$$

Similarly the combination in (ii), with the bond couple $(1,1) \to (-1,-1)$, is regular if

$$\begin{cases} x_2(a) = x_3(b) \\ x_1(a) = x_4(b) \end{cases} \tag{5.25}$$

Table 1

bond coordinate of bond		bond value b	bond coordinate to be connected to	
-2	0	x_3	2	0
-1	-1	(x_3, x_4)	1	1
-1	0	(x_0, x_3)	1	0
-1	1	(x_2, x_3)	1	-1
0	-2	x_4	0	2
0	-1	(x_0, x_4)	0	1
0	1	(x_2, x_0)	0	-1
0	2	x_2	0	-2
1	-1	(x_1, x_4)	-1	1
1	0	(x_4, x_0)	-1	0
1	1	(x_2, x_1)	-1	-1
2	0	x_1	-2	0

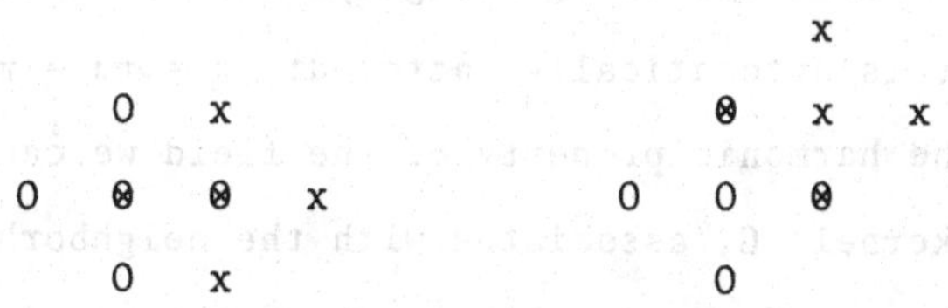

(i) (ii)

Figure 1

The idea behind this regularity is of course that the five values of generators should coincide if/when they overlap. Hence a regular configuration produces a *discrete field*, naturally indexed as $x_{st}; s,t \in \mathbb{Z}$, and we shall now consider its probability measure P_θ when the regularity controlled model in (1.5) is used with

$$Q(g) = \text{constant} \times (x_1+x_2+x_3+x_4-4x_0)^2. \tag{5.26}$$

Since G is infinite we must interpret (1.5) as a Radon-Nikodym derivative w.r.t. an m-measure which is chosen as before.

What happens when the temperature θ drops to zero? Theorem 1 tells us immediately that the limiting measure contracts to the null space of F, here given by the constraints

$$x_{s+1,t}+x_{s,t+1}+x_{s-1,t}+x_{s,t-1}-4x_{st} = 0. \tag{5.27}$$

But this is Laplace's equation in discrete form so that *the frozen patterns consist of discrete harmonic functions* on the subset of $\mathbb{Z}^2$ considered.

The theorem also tells us that the covariance operator of the limiting measure is π = projection down to the subspace described by all the equations (5.27). But a projection operator is characterized by being symmetric and idempotent so that it is automatically factored: $\pi = \pi\pi = \pi^T\pi$. However, using the harmonic property of the field we can write, with the kernel G, associated with the neighborhoods we have chosen for the Laplacian, that solves the boundary value problem for the given domain

$$x_{st} = \sum_{u,v} K(x,t;u,v)\xi_{uv} \tag{5.28}$$

summed over the boundary of the discrete square in the plane. The representation (5.28) is the general one for (discrete) *stochastic harmonic functions* but the requirement that it correspond to a projection as covariance operator restricts the covariance structure of the $\{\xi_{uv}\}$ further. We have not pursued this problem further but it clearly deserves further study.

In the last example we started with a regular structure, say $\mathcal{R}_1$, which was quite flexible. On $\mathcal{L}_n(\mathcal{R}_1)$ we introduced a measure P_θ controlled by $\mathcal{R}_1$ regularity. As $\theta \downarrow 0$ we arrived at the frozen patterns, here the harmonic ones, say $\mathcal{R}_2$-regularity, more rigid than $\mathcal{R}_1$. This is another instance of tendencies toward regularity the study of which was begun in Section 3.8 of Volume I.

5.6. Frozen patterns: infinite G and n

In the case with finite configurations the frozen patterns are reasonably well understood as shown in Sections 3-5. This is not so for infinite n, especially when the cardinality of the configurations is that of the continuum. The general theory of patterns has so far dealt almost exclusively with $\#(c) < \infty$, so that the discussion in this section is somewhat premature. Therefore we shall only argue by examples.

Suppose the image algebra represents functions, for example $I: \mathbb{R} \to \mathbb{R}$, and we want to construct it by local generators expressing constraints. With n finite this could be achieved by letting all g's be related difference operators with

$$\omega(g) = 2; \; f(\beta_{in}, \beta_{out}) = \alpha \in A. \qquad (6.1)$$

With ρ = EQUAL this leads to images consisting of solutions to a difference equation. But how would one define the regularity controlled probability measures? In the absence of better knowledge of how this should be done (see Notes A) we shall avoid the difficulty by making card(c) = denumerable with ρ = TRUE.

To fix ideas let $G = ¢$ and consider for a configuration $c = (\ldots g_{-1}, g_0, g_1, \ldots)$ the image

$$I = I(t) = \sum_{k=-\infty}^{\infty} g_k; \; g_k = s_k \frac{1}{\sqrt{2\pi}} e^{ikt}; \; s_k \in ¢; \; t \in (0, 2\pi) \tag{6.2}$$

The sum in (6.2) shall be interpreted with L_2-convergence so that we should ask that

$$\sum_{k=-\infty}^{\infty} |s_k|^2 < \infty. \tag{6.3}$$

The obvious embedding is to use complex, separable Hilbert space $\mathcal{H}$. Let the m-measure on $\mathcal{H}$ be given by the Gaussian measure with mean value vector zero and a covariance operator B of trace type; see e.g. Grenander (1963).

We must now introduce the Radon-Nikodym derivative

$$\frac{P_\theta(dx)}{m(dx)} = \frac{\exp[-\frac{1}{2\theta}(Fx,x)]}{\int_{\mathcal{H}} \exp[-\frac{1}{2\theta}(Fy,y)]m(dy)} \tag{6.4}$$

with a suitably chosen self-adjoint continuous operator F. With considerable loss of generality we shall assume that F is diagonal with respect to the system

$$\gamma_k = \frac{1}{\sqrt{2\pi}} e^{ikt} \tag{6.5}$$

and with

$$F\gamma_k = f_k\gamma_k; \quad f_k > 0. \tag{6.6}$$

What happens then for the frozen patterns? Leaving out the proof, we just state the result and refer the reader to Hwang (1978) for a more complete treatment (also see Notes B). The measure P_θ given by (6.4) will converge as the temperature drops to zero, $\theta \downarrow 0$, and (the limit P is also Gaussian, with mean zero and the covariance operator $B^{1/2}\pi B^{1/2}$, where π is the projection down to the null space $\mathcal{N}(B^{1/2}FB^{1/2})$.

This is more of less what one could expect but the method fails to tell us what happens when F is not bounded. Say that L satisfies

$$L\gamma_k = (-k^2+a^2)\gamma_k \tag{6.7}$$

so that L is formally the differential operator

$$Lx = \frac{d^2x}{dt^2} + a^2x. \tag{6.8}$$

One can then show, see the same reference, that the measures P_θ given by

$$\frac{P_\theta(dx)}{m(dx)} = \frac{\exp[-\frac{1}{2\theta}(Lx,Lx)]}{\int_{\mathscr{X}}\exp[-\frac{1}{2\theta}(Ly,Ly)]m(dy)}, \tag{6.9}$$

but where m is the Gaussian measure with $B = F$, are well defined and contract to the minimal energy images

$$M = \{x \mid Lx = 0\}. \tag{6.10}$$

With these sketchy remarks we leave the doubly infinite case with the hope that more light will be shed on this problem in the future.

5.7. Asymptotically minimum energy

The results in the previous sections have shown that the frozen patterns correspond to minimum energy and it is therefore important to find those configurations that have minimum energy. This is often possible to do at least asymptotically when the configuration becomes large, $\sigma \to \infty$, in some sense that will be made precise below.

To investigate how this can be done we return in the present section to the finite case and assume that $k \equiv 0$ in (3.1) while

$$h(\beta_1,\beta_2) = t(\alpha_1,\alpha_2) \tag{7.1}$$

where α_1 and α_2 are the respective generator indices of the generators connected via the bonds with bond values β_1 and β_2. We then have the total energy

$$H(c) = \sum_{\alpha,\beta} N_{\alpha\beta} t(\alpha,\beta) \tag{7.2}$$

where

$$N_{\alpha\beta} = \#(\text{bond couples in } \sigma \text{ with generator indices } \alpha \text{ and } \beta \text{ respectively}). \tag{7.3}$$

The *relative energy* will be

$$\sum_{\alpha,\beta} f_{\alpha\beta} t(\alpha,\beta);\ f_{\alpha\beta} = \frac{N_{\alpha\beta}}{N} \tag{7.4}$$

where N stands for the total number of bond couples in the connector σ that belongs to c.

We then only have to consider the possible values of the vector of *relative bond couple frequencies* $\{f_{\alpha\beta}(\sigma)\} = f(\sigma)$. The dimension of the vector is the square of the number of generator classes.

The connection type becomes a POSET if $\sigma_1 \leq \sigma_2$ is understood as meaning that σ_1 is a subgraph of σ_2. Remember that since G is finite and each generator has finite arity (see Section 3) it follows that arities are uniformly bounded by some constant K. Assume moreover that we have (see Kelley (1955) and Definition 1 below).

<u>Condition 1</u>. *The net associated with* Σ *goes to infinity.*

We now introduce a few definitions.

<u>Definition 1</u>. *A sequence* $\{\sigma_n\}$ *tends to infinity,* $\sigma_n \to \infty$, *if for every* $n \geq m$ *we have* $\sigma_n \geq \sigma_m$ *and* $\{\sigma_n\}$ *is cofinal* (see Notes A).

<u>Definition 2</u>. *The set* F *of feasible solutions consist of all vectors* f *that can be approached by* $\sigma_n \in \Sigma$ *in the sense that*

$$\lim_{\sigma_n \to \infty} f(\sigma_n) = f. \tag{7.5}$$

Of course F is bounded and not empty. But we can claim more as shown in Hwang (1978).

<u>Theorem 1</u>. F *is compact.*

<u>Proof</u>: Consider a sequence $\{x_k\} \in F$ with $x_k \to x$. We can find $\sigma_{nk} \in \Sigma$ and regular configurations c_{nk} associated with the connector σ_{nk} such that

$$f(\sigma_{nk}) \to x_k \quad \text{as} \quad \sigma_{nk} \to \infty. \tag{7.6}$$

We shall show that F is closed; since it is bounded this guarantees compactness.

Without loss of generality assume that $|x_k - x| < 1/k$ and choose

$$\begin{cases} \sigma_1 = \sigma_{n_1 1} \\ c_1 = c_{n_1 1} \end{cases} \tag{7.7}$$

such that $|x_1 - f(c_1)| < 1$. Assume that for $\ell \leq m$ the connectors σ_ℓ and configurations c_ℓ have been chosen that

$$\begin{cases} \sigma_\ell = \sigma_{n_\ell \ell} \\ c_\ell = c_{n_\ell \ell} \end{cases} \tag{7.8}$$

with $|x_\ell - f(c_\ell)| < 1/\ell$ with $\sigma_\ell \geq \sigma_{\ell 1}$, $\sigma_\ell \geq \sigma_{\ell-1}$. We then choose

$$\begin{cases} \sigma_{m+1} = \sigma_{n_{m+1} m+1} \\ c_{m+1} = c_{n_{m+1} m+1} \end{cases} \tag{7.9}$$

with $|x_{m+1} - x(c_{m+1})| < (m+1)^{-1}$ with $\sigma_{m+1} \geq \sigma_{m+1 1}, \sigma_{m+1} \geq \sigma_m$. This can be done since

$$\{\sigma_{nm+1};\ n = 1,2,\ldots\} \tag{7.10}$$

is cofinal. Since

$$\{\sigma_{n1};\ n = 1,2,\ldots\} \tag{7.11}$$

is cofinal so is $\{\sigma_n;\ n = 1,2,\ldots\}$. Hence $f(c_n) \to x$ as $\sigma_n \to \infty$. Q.E.D.

To get further we recall that in Volume I, Section 3.8, it was shown that for Σ = LINEAR the set F was shown to be convex. Since the relative energy in (7.4) is a linear function of f convexity is the property one would wish to establish.

To do this we shall introduce a binary operation $\circ$ from Σ to Σ such that

<u>Condition 2</u>. *The binary operation satisfies*

1: *It is commutative and associative*

2: $\sigma_1 \leq \sigma_1 \circ \sigma_2$, $\sigma_2 \leq \sigma_1 \circ \sigma_2$

3: *If* $\sigma_1 \leq \sigma_1'$ *and* $\sigma_2 \leq \sigma_2'$ *then* $\sigma_1 \circ \sigma_2 \leq \sigma_1' \circ \sigma_2'$: *it is monotomic.*

We also need

<u>Definition 3</u>. *The connection type* Σ *is called homogeneous with respect to the binary operation if*

1. *for any* $\sigma_1, \sigma_2 \in \Sigma$ *there exists a natural number* k *such that* k *copies of* σ_1 (see Notes B) *is a subgraph of* $\sigma_1 \circ \sigma_2$.

2. *for any* $\sigma_\nu' \to \infty$ $\sigma_\mu'' \to \infty$

$$\frac{N(\sigma_\nu' \circ \sigma_\mu'')}{k\, N(\sigma_\nu')} \to 1 \tag{7.12}$$

and

$$\frac{n(\sigma_\nu' \circ \sigma_\mu'')}{k\, n(\sigma_\nu')} \to 1 \tag{7.14}$$

and

$$\frac{k}{n(\sigma_\mu'')} \to 1 \tag{7.14}$$

<u>Remark</u>. In (7.12), (7.13), (7.14) we have used notation in analogy to (7.3): $N(\sigma)$ denotes the number of bonds in σ while $n(\sigma)$ denotes the number of generators.

We illustrate Definition 3 by a few special cases to which we shall return later.

We illustrate this by some special cases to which we shall return later.

<u>Example 1</u>. Let Σ = LINEAR and σ_n be the connector from LINEAR with n generators. Define $\sigma_n \circ \sigma_m = \sigma_{n\times m}$. It satis-

fies the conditions of Definition 3.

Example 2. Let Σ = SQUARE LATTICE and let σ_i be the connector of size $n_i \times m_i$; $i = 1,2$. Define $\sigma_1 \circ \sigma_2$ to be the connector of dimensions $n_1 \times n_2$ and $m_1 \times m_2$ respectively. If we choose $k = n_2 m_2$ then (7.12) becomes

$$\frac{N(\sigma_1 \circ \sigma_2)}{k\, N(\sigma_1)} = \frac{2n_1n_2m_1m_2 - (n_1n_2 + m_1m_2)}{n_2m_2[2n_1m_1 - (n_1 + m_1)]} \to 1. \tag{7.15}$$

Further (7.13) is

$$\frac{n(\sigma_1 \circ \sigma_2)}{k\, n(\sigma_1)} = \frac{n_1n_2m_1m_2}{n_2m_2n_1m_1} = 1 \tag{7.14}$$

and (7.14) finally reduces to

$$\frac{k}{n(\sigma_2)} = \frac{n_2m_2}{n_2m_2} . \tag{7.17}$$

Example 3. Assuming Σ to be homogeneous in the sense of Definition 3 let $\sigma_1 \times \sigma_2$ be the Cartesian product of graphs, see Harary (1969), and define on $\Sigma \times \Sigma$ the binary operation as

$$(\sigma_1 \times \sigma_1) \circ (\sigma_2 \times \sigma_2) = (\sigma_1 \circ \sigma_2) \times (\sigma_1 \circ \sigma_2). \tag{7.18}$$

Then $\Sigma \times \Sigma$ is homogeneous with respect to the new binary operator; the proof is left to the reader.

We are now ready for the new result due to Hwang (1976).

Theorem 2. *If* Σ *is homogeneous the set* F *is convex.*

Proof: If the vectors $x', x'' \in F$ there exist sequences σ', σ'' associated with regular configurations c', c'' such that

$$\begin{cases} f(\sigma') \to x' \\ f(\sigma'') \to x'' . \end{cases} \tag{7.19}$$

Hence

$$\frac{1}{2}x + \frac{1}{2}y = \lim\left[\frac{1}{2}\, x(\sigma') + \frac{1}{2}\, x(\sigma'')\right] = \frac{N(\sigma'')\{N_{\alpha\beta}(c')\}+N(\sigma')\{N_{\alpha\beta}(c'')\}}{2\, N(\sigma')N(\sigma'')} \tag{7.20}$$

which we rearrange slightly as

$$\frac{\frac{1}{2}\, n^2(\sigma'')\{N_{\alpha\beta}(c')\} + \frac{1}{2}\frac{N(\sigma')n(\sigma'')}{N(\sigma'')n(\sigma')}\, n(\sigma')n(\sigma'')\{N_{\alpha\beta}(c'')\}}{N(\sigma')\, n^2(\sigma'')}. \tag{7.21}$$

Apply equations (7.12) and (7.14) using the commutativity of the binary relation. This yields for the ratio in the numerator of (7.21)

$$\frac{N(\sigma')n(\sigma'')}{N(\sigma'')n(\sigma')} \to 1. \tag{7.22}$$

Consider now $(\sigma'\circ\sigma'')\circ\sigma''$ with about $1/2\, n^2(\sigma'')$ copies of c' and $1/2\, n(\sigma')n(\sigma'')$ copies of c''. Then (7.21) is approximately equal to the f-vector of the combined configuration. Hence

$$\frac{1}{2}\, x + \frac{1}{2}\, y \in F. \tag{7.23}$$

Recalling from the proof of Theorem 1 that F is closed this implies almost directly (see Notes C) that F is convex.

Q.E.D.

In the linear case it was shown that F is a polyhedron. We do not know if this statement also extends to general connection types. Note however that Example 2 and 3 can be handled by Theorem 2.

We hope that it will be possible to arrive at a more general concept of "homogeneous configurations," applicable to any Σ for which the connections are the same everywhere in the interior of configuration skeleton.

5.8. Asymptotics for large configurations

We have studied the asymptotics of regularity controlled probabilities for low temperatures, $\theta \downarrow 0$. We shall now keep the temperature constant, say with $\theta = 1$, and investigate what happens when the size $n = \#(c)$ becomes large. Our aim in this and the following sections is to show that *marginal distributions converge* (see Notes A) as the configurations are made larger.

In Volume I we began exploring this problem area and showed that for Σ = LINEAR convergence took place. The limiting measure was also obtained in closed form ibid. pp. 72-74. For a more complicated connection type Σ = SQUARE LATTICE(γ), only heuristic results were obtained. In both cases the interaction terms were quadratic, so that the regularity controlled measures are Gaussian; this assumption will be retained here, but we shall now attempt a fully rigorous analysis.

Our analytic procedure will be as follows. For n fixed we study the measure induced on $\mathscr{L}_n(\mathscr{R})$ by the given regularity. Following a time honored device, familiar to the physicists, we shall embed our regularity in cyclic ones. Of course, as n grows, one expects the influence of this embedding to have a negligible influence. Its introduction will simplify the analysis a great deal.

The reason for this is that the invariances produced by the periodicity with respect to the cyclic groups can be handled conveniently via *circulant matrices*. Since they will be used extensively, and since some of their properties are not as well known as they deserve to be, we shall present some background material in the remainder of this section. More can be found in Davis (1979) from which the following has been borrowed.

Let $F^{n\times m}$ denote the set of $n \times m$ matrices over a field F ($F = \mathbb{R}$ or $\mathbb{C}$). I_n is the identity matrix in $\mathbb{C}^{n\times n}$ and

$$\pi_n = \begin{pmatrix} 0 & 1 & 0 & \cdots & 0 \\ 0 & 0 & 1 & \cdots & 0 \\ 1 & 0 & 0 & \cdots & 0 \end{pmatrix} \in \mathbb{C}^{n\times n} \tag{8.1}$$

denotes the fundamental circulant matrix in $\mathbb{C}^{n\times n}$. A circulant matrix $C \in \mathbb{C}^{n\times n}$ is given by

$$C = \pi_n^o c_o + \ldots + \pi_n^{n-1} c_{n-1} = \sum_{t=0}^{n-1} \pi_n^t c_t \tag{8.2}$$

where $c_t \in \mathbb{C}$ $t = 0,\ldots,n-1$. The *Fourier matrix* in $\mathbb{C}^{n\times n}$ is given by

$$F_n = \frac{1}{\sqrt{n}} \begin{pmatrix} 1 & 1 & \cdots & 1 \\ 1 & \omega_n & \cdots & \omega_n^{n-1} \\ \vdots & & & \\ 1 & \omega_n^{n-1} & \cdots & \omega_n^{(n-1)(n-1)} \end{pmatrix}^* \tag{8.3}$$

where superscript * denotes complex conjugate, and $\omega_n = e^{\frac{2\pi i}{n}}$ is a principal n-th root of unity.

The notation for *Kronecker product* $\otimes$ and *sum* $\oplus$ is defined by

$$A \otimes B = \begin{pmatrix} a_{11}B & \cdots & a_{1n}B \\ \vdots & & \\ a_{m1}B & \cdots & a_{mn}B \end{pmatrix} \tag{8.4}$$

for $A \in F^{m\times n}$ and $B \in F^{p\times q}$, and

$$A \oplus B = \begin{pmatrix} A & 0 \\ 0 & B \end{pmatrix} \tag{8.5}$$

Properties of $\otimes$ and $\oplus$ are given in Marcus and Minc [1964]; for example, $(A \otimes B)(C \otimes D) = AC \otimes BD$ as long as matrix multiplication can be defined.

Let $A_0, \ldots, A_{n-1} \in ¢^{p \times p}$. Then

$$A = \sum_{t=0}^{n-1} \pi_n^t \otimes A_t \tag{8.6}$$

is a form written in analogy with (8.2), and is called a *block circulant matrix* over $¢^{p \times p}$. We now state the *inversion theorem*.

<u>Theorem 1</u>. *Let*

$$A = \sum_{t=0}^{n-1} \pi_n^t \otimes A_t, \quad A_t \in ¢^{p \times p} \tag{8.7}$$

be a nonsingular matrix in $¢^{np \times np}$. *Then*

$$\begin{cases} A^{-1} = B \\ B = \sum_{j=0}^{n-1} \pi_n^j \otimes B_j \\ B_j = \frac{1}{n} \sum_{k=0}^{n-1} \omega_n^{-jk} \left(\sum_{t=0}^{n-1} \omega_n^{tk} A_t \right)^{-1}, \; j=0,\ldots,n-1. \end{cases} \tag{8.8}$$

<u>Proof</u>: Let $\Omega_n = \text{diag}(1, \omega_n, \ldots, \omega_n^{(n-1)})$. Then Theorem 3.2.1 in Davis (1979) states that $\pi_n = F_n^* \Omega_n F_n$. Hence

$$\begin{aligned} \pi_n^t \otimes A_t &= (F_n^* \Omega_n^t F_n) \otimes A_t \\ &= (F_n^* \otimes I_p)(\Omega_n^t \otimes A_t)(F_n \otimes I_p), \end{aligned} \tag{8.9}$$

and so

$$\begin{aligned} A &= \sum_{t=0}^{n-1} \pi_n^t \otimes A_t \\ &= (F_n \otimes I_p)^* \left\{ \sum_{t=0}^{n-1} (\Omega_n^t \otimes A_t) \right\} (F_n \otimes I_p), \end{aligned} \tag{8.10}$$

and hence the inverse

$$A^{-1} = (F_n \otimes I_p)^* \left\{\sum_{t=0}^{n-1} (\Omega_n^t \otimes A_t)\right\}^{-1} (F_n \otimes I_p).$$

But

$$\left\{\sum_{t=0}^{n-1} (\Omega_n^t \otimes A_t)\right\}^{-1} = \operatorname{diag}\left(\left\{\sum_{t=0}^{n-1} \omega_n^{ts} A_t\right\}^{-1};\quad s = 0,\ldots,n-1\right). \tag{8.11}$$

Now for any sequence $C_0,\ldots,C_{n-1}$,

$$\frac{1}{n}\sum_{j=0}^{n-1} \omega_n^{js}\left(\sum_{k=0}^{n-1} \omega_n^{-jk} C_k\right) = \frac{1}{n}\sum_{j=0}^{n-1}\sum_{k=0}^{n-1} \omega_n^{j(s-k)} C_k = C_s. \tag{8.12}$$

Therefore

$$\operatorname{diag}\left(\left\{\sum_{t=0}^{n-1} \omega_n^{ts} A_t\right\}^{-1};\quad s = 0,\ldots,n-1\right)$$

$$\operatorname{diag}\left(\left\{\frac{1}{n}\sum_{j=0}^{n-1} \omega_n^{js}\left(\sum_{k=0}^{n-1} \omega_n^{-jk}\left(\sum_{t=0}^{n-1} \omega_n^{tk} A_t\right)^{-1}\right)\right\};\ s = 0,\ldots,n-1\right)$$

$$= \sum_{j=0}^{n-1} \Omega_n^j \otimes \frac{1}{n}\sum_{k=0}^{n-1} \omega_n^{-jk}\left(\sum_{t=0}^{n-1} \omega_n^{tk} A_t\right)^{-1}$$ Q.E.D.

<u>Remark</u>. The proof is really Fourier analytic, employing the discrete transform related to the characters of the cyclic group, see equation (8.3).

5.9. Spectral density matrix for Σ = LINEAR(γ).

Let the generators have arity $2p$ with $\omega_{in}(g) = \omega_{out}(g) = p$ and with real bond values. The Q measure (see Section 1) on G will be chosen as Gaussian with mean value vector zero and a covariance matrix that will also be denoted by Q and given later. The generator skeleton γ is then fixed and we shall consider the connection type Σ = LINEAR(γ) where the connector has translation invariance.

Enumerating the generators by a subscript j we shall assume that at each integer (vertex) j we have a stochastic generator g_j with

$$\begin{cases} \text{out-bonds} & X_{j,1},\dots,X_{j,p} \\ \text{in-bonds} & Y_{j,1},\dots,Y_{j,p}. \end{cases} \tag{9.1}$$

For each j, $\binom{X_j}{Y_j} \in \mathbb{R}^{2p}$ is a Gaussian vector with mean 0 and covariance matrix

$$Q = H^{-1} = \begin{pmatrix} H_{11} & H_{12} \\ H_{12}^T & H_{22} \end{pmatrix}^{-1} \tag{9.2}$$

where $H_{11}, H_{12}, H_{22} \in \mathbb{R}^{p\times p}$. These stochastic vectors are assumed to be i.i.d. The bond relation ρ shall be chosen as EQUALITY, and out-bond $X_{j,k}$ is connected to in-bond $Y_{j+a_k,k}$, $j \in \mathbb{Z}$, $k \in \{1,\dots,p\}$. By conditioning on bond relation ρ, we will obtain a stochastic process $\{X_e | e \in E(a_1,\dots,a_p)\}$. (We shall write $X_{e_{j,k}}$ for $X_{j,k}$.) The distributions of the stochastic process are given in terms of the covariance function

$$R(k) = E(X_j X_{j+k}^T) \qquad k \in \mathbb{Z},\ X_j \in \mathbb{R}^p. \tag{9.3}$$

The main task in this section is the computation of the spectral density matrix $f(e^{i\theta})$ for which

$$R(k) = \frac{1}{2\pi}\int_{-\pi}^{\pi} e^{-ik\theta} f(e^{i\theta})d\theta \qquad (9.4)$$

for the vector valued process $\{X_j | j \in \mathbb{Z}\}$ by a limiting argument: we define distributions on configurations defined cyclically with n vertices, and let $n \to \infty$. We now follow the analysis in Thrift (1979).

We first define the equivalence relation $j \sim j+nt$ on the vertices, which extends to $e_{j,k} \sim e_{j+nt,k}$ on the edges, where t ranges over $\mathbb{Z}$. A configuration is defined by taking n generators $g_0, g_1, \ldots, g_{n-1}$ where g_j has out-bonds $X_{j,1}, \ldots, X_{j,p}$ and in-bonds $Y_{j,1}, \ldots, Y_{j,p}$. The generators are positioned on the vertices of the circulant connector defined by the equivalence relation $\sim$, and bonds along the respective edges. An example is seen in Figure 9.1.

We now have a finite collection of generators, and we want to calculate the probability distribution of the configuration after conditioning on bond relation EQUAL. The bond relation EQUAL can be expressed by the compact formula

$$Y_j = E_{1,p}X_{j-a_1} + E_{2,p}X_{j-a_2} + \ldots + E_{p,p}X_{j-a_p} \quad \text{(subscripts taken mod n)}$$
$$j = 0,1,\ldots,n-1 \qquad (9.5)$$

with $E_{k,p} = \text{diag}(0,\ldots,1,\ldots 0)$, where $(0,\ldots,1,0,\ldots,0)^T$ is a p-vector with 1 in the kth place and 0 elsewhere. Equation (9.5) can be written even more compactly as

$$Y = \Big(\sum_{k=1}^{p} \pi_n^{-a_k} \otimes E_{k,p}\Big)X, \quad X = (X_0^T, \ldots, X_{n-1}^T)^T, \qquad (9.6)$$

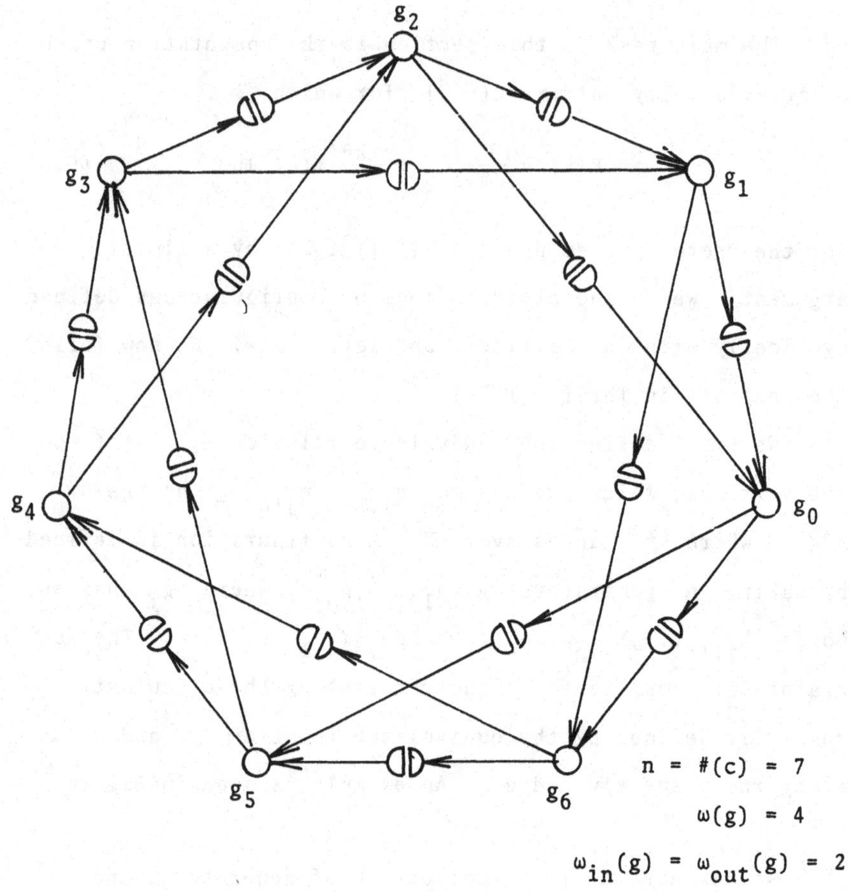

Figure 9.1

where π_n is the fundamental circulant matrix in $C^{n\times n}$. If we assume that the generators $g_0,\ldots,g_{n-1}$ are, before the conditioning via ρ, i.i.d. with covariance matrix

$$Q = H^{-1} = \begin{pmatrix} H_{11} & H_{12} \\ H_{12}^T & H_{22} \end{pmatrix}^{-1}, \tag{9.7}$$

then the joint frequency function of $\binom{X}{Y}$ is given by

$$f\binom{X}{Y} = K_0 \exp\{-\tfrac{1}{2} H\binom{X}{Y}\} \tag{9.8}$$

where K_0 is the appropriate normalizing constant, and $H\binom{X}{Y}$ is the quadratic form

$$X^T(I_n \otimes H_{11})X + X^T(I_n \otimes H_{12})Y + Y^T(I_n \otimes H_{12}^T)X + Y^T(I_n \otimes H_{22})Y. \tag{9.9}$$

To get the joint frequency function of $x \in \mathbb{R}^n$ after conditioning on bond relation EQUAL we simply substitute the expression (9.6) into (9.9) to get a quadratic form in X:

$$\begin{aligned} X^T\Big\{\pi_n^o \otimes H_{11} &+ (\pi_n^o \otimes H_{12})\Big(\sum_{k=1}^{p} \pi_n^{-a_k} \otimes E_{k,p}\Big) \\ &+ \Big(\sum_{k=1}^{p} \pi_n^{a_k} \otimes E_{k,p}\Big)(\pi_n^o \otimes H_{12}^T) \\ &+ \Big(\sum_{j=1}^{p} \pi_n^{a_k} \otimes E_{j,p}\Big)(\pi_n^o \otimes H_{22})\Big(\sum_{k=1}^{p} \pi_n^{a_k} \otimes E_{k,p}\Big)\Big\}X. \end{aligned} \tag{9.10}$$

The matrix inside the curly brackets in (9.10) is equal to

$$\begin{aligned} \pi_n^o \otimes H_{11} &+ \sum_{k=1}^{p} \pi_n^{-a_k} \otimes H_{12}E_{k,p} \\ &+ \sum_{k=1}^{p} \pi_n^{a_k} \otimes E_{k,p}H_{12}^T \\ &+ \Big(\sum_{j=1}^{p} \pi_n^{a_j} \otimes E_{j,p}H_{22}\Big)\Big(\sum_{k=1}^{p} \pi_n^{a_k} \otimes E_{k,p}\Big) \\ &\triangleq \sum_{k=-q}^{q} \pi_n^k \otimes C_k, \end{aligned} \tag{9.11}$$

where the $C_k \in \mathbb{R}^{p\times p}$ are found upon calculation of the above expression. (They depend only on H_{11}, H_{12}, H_{22}, and the parameters $(a_1,\ldots,a_p = q)$.) Let

$$C = \sum_{k=-q}^{q} \pi_n^k \otimes C_k. \tag{9.12}$$

Then

$$f(X) = K_1 \exp\{-\tfrac{1}{2} X^T CX\}, \tag{9.13}$$

with K_1 the appropriate normalizing constant, gives frequency function of X after conditioning on bond relation EQUAL.

We can remark immediately that C is a block circulant matrix with the property that $C^T = C$, which follows from $C_{-k} = C_k^T$, $k = 0,\ldots,q$. Also C is positive definite, so C is a positive definite symmetric matrix. Since C is block circulant we can invoke the Theorem 8.1. This immediately gives

$$\begin{aligned} R_n(j) &= E(X_0 X_j^T) \\ &= \frac{1}{n}\sum_{k=0}^{n-1} \omega_n^{-jk}\Big(\sum_{t=-q}^{q} C_t \omega_n^{tk}\Big)^{-1}, \quad \omega_n = e^{\frac{2\pi i}{n}}. \end{aligned} \tag{9.14}$$

The final step is to relate $R_n(j)$ to the process $\{X_j \mid j \in \mathbb{Z}\}$. We consider $R_n(j)$ as a Riemann sum which approximates

$$R(j) = \frac{1}{2\pi}\int_0^{2\pi} e^{-ij\theta}\Big(\sum_{t=-q}^{q} C_t e^{it\theta}\Big)^{-1} d\theta \tag{9.15}$$

(let $\theta_k = \frac{2\pi}{n} k$, $k = 0,\ldots,n-1$ be a partition of $[0,2\pi)$). We then define the covariance function of the $\{X_j \mid j \in \mathbb{Z}\}$ process to be given in (9.15) and the spectral density matrix

$$f(e^{i\theta}) = [\Phi(e^{-i\theta})]^{-1} = \Big(\sum_{t=-q}^{q} C_t\, e^{it\theta}\Big)^{-1}. \tag{9.16}$$

and we can state the result due to Thrift (1979).

<u>Theorem 1</u>. *The limiting covariances have a spectral density* f *given by* (9.16), *where* Φ *is a Hermitian positive definite*

matrix for every $\theta \in [-\pi,\pi)$.

Proof: The fact that Φ is Hermitian follows from $C_t^* = C_{-t}$. In (9.14) we see that $\sum_{t=-q}^{q} C_t \omega_n^{tk}$ is positive definite for every $k = 0,\ldots,n-1$ and every n. By approximation $\frac{2\pi k_i}{n_i} \to \theta$ for some sequence k_i, n_i we see that $\sum_{t=-q}^{q} C_t e^{it\theta}$ is positive semidefinite for each fixed $\theta \in [-\pi,\pi)$. To get "positive definite" suppose 0 is an eigenvalue of $\sum_{t=-q}^{q} C_t e^{it\theta}$. Now

$$H(\varepsilon) = \begin{pmatrix} H_{11}-\varepsilon I_p & H_{12} \\ H_{12}^T & H_{22} \end{pmatrix} \text{ is positive definite} \tag{9.17}$$

for $\varepsilon > 0$ sufficiently small. We also have $-\varepsilon$ is an eigenvalue of $\sum_{t=-q}^{q} C_t e^{it\theta} - \varepsilon I_p$. But if we had started with $H(\varepsilon)$ instead of H we would have concluded $\sum_{t=-q}^{q} C_t e^{it\theta} - \varepsilon I_p$ is positive semidefinite, from (2.5). This statement implies that Φ is positive definite. Q.E.D.

5.10. Factorization of the spectral density matrix

Theorem 9.1 characterized the limiting measure on $\mathscr{C}(\mathscr{R})$ induced by the regularity; the result is in terms of a stationary stochastic process taking vectors as values. In order to better understand the way the regularity has given rise to a stochastic structure over $\mathscr{C}(\mathscr{R})$ we shall factor the spectral density matrix. This factorization will lead to an important representation of the random configurations in terms of simpler randomness as will be established in Section 11.

Consider the spectral density matrix $f = \Phi^{-1}$ of the X_j, $j \in \mathbb{Z}$; and let $z = e^{-i\theta}$ so that

$$\Phi(z) = \sum_{t=-q}^{q} C_t^T z^t, \quad C_{-t} = C_t^T \quad t = 1,\dots,q. \tag{10.1}$$

We have shown that Φ is Hermitian positive definite on $|z| = 1$. Suppose that we can find matrices $\Gamma_0, \Gamma_1, \dots, \Gamma_q \in \mathbb{C}^{p\times p}$ such that

$$\Gamma(z) = \Gamma_0 + \Gamma_1 z + \dots + \Gamma_q z^q \tag{10.2}$$

has the properties that

$$\begin{cases} \text{(a)} & \det \Gamma(z) \text{ has no roots inside or on the unit circle } |z| = 1. \\ \text{(b)} & \Phi(z) = \Gamma^*(z)\Gamma(z) \quad \text{on} \quad |z| = 1. \end{cases} \tag{10.3}$$

Then it can be proved that Z_n defined by

$$Z_n = \Gamma_0 X_n + \Gamma_1 X_{n-1} + \dots + \Gamma_q X_{n-q} \qquad n \in \mathbb{Z} \tag{10.4}$$

has the properties that

$$\begin{cases} \text{(c)} & E(Z_m Z_n^*) = \delta_{n,m} I_p \qquad n,m \in \mathbb{Z} \\ \text{(d)} & E(X_m Z_n^*) = 0 \quad \text{if} \quad m < n \\ \text{(e)} & X_n = \sum_{j=0}^{\infty} A(j) Z_{n-j}, \text{ where } \sum_{j=0}^{\infty} A(j) z^j \end{cases} \tag{10.5}$$

is the power series on expanding $\Gamma(z)^{-1}$. (Since it is assumed that $\det \Gamma(z)$ has no roots inside or on the unit circle, $\Gamma(z)^{-1}$ can be computed using partial fractions.)

The above well known results can be found in Whittle [1963], pages 98-103.

The process $\{Z_n | n \in \mathbb{Z}\}$ is called the process of *innovations*. Property (c) states that the innovations are uncorrelated (stochastically independent in our case);

(d) states that the innovation Z_n is uncorrelated with the "past" $\{X_m | m < n\}$. A process defined by (10.4) is the well-known *autoregression*. The ability to *find an autoregression which explains a certain physical model* (in our case a particular bond interaction model) has proven useful in many statistical investigations. The remainder of this section will be devoted to finding a factorization (10.3b), which is equivalent to finding an autoregression.

The theory of *polynomial matrices* will be used to obtain a factorization of Φ. The main source for this subject is Dennis-Traub (1976), with Robinson (1967) discussing the problem of autoregression. If we look at Φ, we see that it is a certain type of polynomial form, namely a *quasi-polynomial matrix*. The term polynomial matrix is restricted to those of the form

$$G(z) = G_0+G_1z+\ldots+G_qz^q, \quad G_i \in ¢^{r\times s}; \tag{10.6}$$

a quasi-polynomial matrix allows negative exponents of z. In $G(z)$, q is called the degree of G ($q = \deg G$).

The following notation is from Robinson (1967). For a given quasi-polynomial matrix

$$A(z) = A_{-n}z^{-n} + \ldots + A_0 + \ldots + A_mz^m \tag{10.7}$$

where $A_j \in ¢^{r\times s}$; $j = -n,\ldots,m$; then

$$A_*(z) \triangleq [A(1/\bar{z})]^* \tag{10.8}$$

is another quasi-polynomial matrix (QPM). The subscript "*" notation has the properties that

$$\begin{cases} A_{**}(z) = A(z) \\ [A(z)B(x)]_* = B_*(z)A_*(z), \text{ as long as} \end{cases} \tag{10.9}$$

matrix multiplication is compatible. For example, if

$$\begin{cases} A(z) = \sum_{s=-m}^{n} A_s z^s, \text{ then} \\ A_*(z) = \sum_{s=-m}^{n} A_s^* z^{-s}, \text{ and} \end{cases} \tag{10.10}$$

for

$$\Phi(z) = \sum_{t=-q}^{q} C_t z^t, \quad C_{-t} = C_t^T \tag{10.11}$$

we have $\Phi_*(z) = \Phi(z)$. A QPM with square coefficients with this property is called *extended hermitian*, "extended" because for such a QPM, $\Phi(z)^* = \Phi(z)$ on $|z| = 1$; that is it is Hermitian in the usual sense on the unit circle.

For a given Φ define

$$G_\Phi(z) = z^q \Phi(z) \tag{10.12}$$

so G_Φ is an (ordinary) polynomial matrix. The subscript "R" (denoting *reversal*) notation is defined for polynomial matrices:

$$\begin{cases} G(z) = G_0 + G_1 z + \ldots + G_n z^n \\ G_R(z) = G_n^* + G_{n-1}^* z + \ldots + G_0^* z^n. \end{cases} \tag{10.13}$$

Lemma 1. *If a QPM* Φ *is extended hermitian, then* $G_\Phi = (G_\Phi)_R$, *and for every root* z_0 *of* $\det \Phi(z) = 0$ *in* $1 < |z| < \infty$ *there is a root* $1/\bar{z}_0$ *in* $0 < |z| < 1$, *and vice versa.*

Proof: By definition,

$$G_{\Phi,R}(z) = [G_\Phi(z)]_* z^{2q} = [G_\Phi(z)z^{-q}]_* z^q$$
$$= \Phi_*(z)z^q = \Phi(z)z^q = G_\Phi(z). \tag{10.14}$$

Consequently, if $\det \Phi(z_0) = 0$ with $1 < |z_0| < \infty$, then $\det G_\Phi(z_0) = 0$, which implies $\det G_{\Phi,R}(z_0) = 0$, so $\det G_\Phi(\frac{1}{\bar{z}_0}) = 0$. Q.E.D.

To fix terminology we define the following terms

determinental equation of Φ	$\det G_\Phi(z) = 0$
determinental roots of Φ	roots of $\det G_\Phi(z) = 0$
left-latent vectors (row)	$r_0^* G_\Phi(z_0) = 0$
right-latent vectors (column)	$G_\Phi(z_0)c_0 = 0$.

Define $S_1 = \{z | 1 < |z| < \infty\}$, $S_0 = \{z | 0 < |z| < 1\}$. We know that because Φ is positive definite on $|z| = 1$, there are no determinental roots on that set. We shall make a temporary assumption, which will be relaxed later: namely, Φ has $2pq$ *distinct* determinental roots

$$\begin{cases} \{z_1, \ldots, z_{pq}\} & \text{in } S_1 \\ \left\{\frac{1}{\bar{z}_1}, \ldots, \frac{1}{\bar{z}_{pq}}\right\} & \text{in } S_0. \end{cases} \tag{10.15}$$

Let $A(z)$ be a polynomial matrix with coefficients in $\mathbb{C}^{p\times p}$. For a matrix C in $\mathbb{C}^{p\times p}$ define

right functional value: $A_{(r)}(C) = A_0 + A_1C + \ldots + A_nC^n$ and
left functional value: $A_{(\ell)}(C) = A_0 + CA_1 + \ldots + C^nA_n$.

Let $A(z) = A_0 + Z_1z + \ldots + A_nz^n$, $B(z) = B_0 + B_1z + \ldots + B_mz^m$, with all coefficients in $C^{p\times p}$. Suppose B_m is nonsingular. The *divisor theorem for polynomial matrices* (Lancaster (1966), pages 47-49) states that there exists unique PMs Q_1,R_1,Q_2,R_2, where R_1 is either identically 0 or of degree less than B (and the same for R_2), such that

$$\begin{cases} A(z) = Q_1(z)B(z) + R_1(z) \\ A(z) = B(z)Q_2(z) + R_2(z). \end{cases} \tag{10.16}$$

If $R_1 = 0$ (respectively $R_2 = 0$) then $B(z)$ is called a *right divisor* of $A(z)$ (respectively a *left divisor* of $A(z)$). Let $B(z) = I-Bz$ where B is nonsingular. Then the *remainder theorem* states that the remainder R_1 is given by $A_{(r)}(B^{-1})$ and the remainder R_2 is given by $A_{(\ell)}(B^{-1})$.

The following key theorem due to Thrift will provide a factorization of Φ, an extended hermitian QPM that is positive definite on $|z| = 1$ and has $2pq$ distinct determinental roots in $S_1 \cup S_0$.

<u>Theorem 1</u>. *Under the above assumptions about* Φ, *there exists a matrix* $U_1 \in ¢^{p\times p}$ *such that*

$$\Phi(z) = (Iz^{-1}+U_1^*)\Phi_2(z)(Iz+U_1) \tag{10.17}$$

where the roots of $\det(Iz+U_1) = 0$ *are all in* S_1 *and* Φ_2 *is an extended hermitian* QPM *of degree less than* Φ.

<u>Proof</u>: We denote the determinantal roots by $z_1,\ldots,z_{pq}$, $\frac{1}{\bar{z}_1},\ldots,\frac{1}{\bar{z}_{pq}}$ where $\{z_1,\ldots,z_{pq}\} \subset S_1$ and $\left\{\frac{1}{\bar{z}_1},\ldots,\frac{1}{\bar{z}_{pq}}\right\} \subset S_0$, and $\Phi(z) = \sum_{t=-q}^{q} C_t z^t$. Note that the assumptions prescribed to Φ imply that C_q is nonsingular. We write C_t for C_t^T to simplify notation. The proof involves several steps based on three lemmas from Dennis-Traub (1976).

Consider $C_q^{-1}G_\Phi(z)$. The *block companion matrix* associated with $C_q^{-1}G_\Phi(z)$ is

$$B_\Phi = \begin{pmatrix} 0 & I_p & & & \\ & & \cdot & & \\ & & & \cdot & \\ & & & \cdot & I_p \\ -C_q^{-1}C_q^T & -C_q^{-1}C_{q-1}^T & \cdots & & -C_q^{-1}C_{q-1} \end{pmatrix} \tag{10.18}$$

We now state three propositions from the fundamental paper Dennis-Traub (1976) without proof.

Lemma 2. $\det(B_\Phi - zI_{2pq}) = \det(C_q^{-1}G_\Phi(z))$.

Lemma 3. *If* z_0 *is a root of* $\det(C_q^{-1}G_\Phi(z)) = 0$ *and* c_0 *is a right-latent vector*

$$C_q^{-1}G_\Phi(z_0)c_0 = 0, \tag{10.19}$$

then z_0 *is an eigenvalue of* B_Φ*, and*

$$\begin{pmatrix} c_0 \\ z_0 c_0 \\ \vdots \\ z_0^{2q-1}c_0 \end{pmatrix} \tag{10.20}$$

is an eigenvector of B_Φ *associated with* z_0.

Lemma 4. *If*

$$A = \begin{pmatrix} A_{11} & \cdots & A_{1n} \\ \vdots & & \\ A_{n1} & \cdots & A_{nn} \end{pmatrix}$$

is nonsingular, and each $A_{ij} \in \mathbb{C}^{p\times p}$, *then there is a permutation of the columns of* A *to* $\tilde{A} = (B_{ij})$ *such that* B_{ii} *is nonsingular,* $i = 1,\ldots,n$.

We now begin the main part of the proof of Theorem 1. Let

$$\{z_1,\ldots,z_{pq}\} \subset S_1,\ \left\{\frac{1}{\bar z_1},\ldots,\frac{1}{\bar z_{pq}}\right\} \subset S_0$$

be the determinental roots of Φ. Let $c_1,\ldots,c_{pq}$ be a set of vectors in $\mathbb{C}^p$ such that $c_i^*\Phi(z_i) = 0$, $i = 1,\ldots,pq$. Applying the superscript "*" and subscript "*" respectively we get the following relations:

$$\begin{cases} \Phi(z_i)^* c_i = 0 \\ \Phi_*(z_i) c_i = 0 \end{cases} \tag{10.21}$$

or

$$\begin{cases} \Phi(\frac{1}{\bar z_i}) c_i = 0 \\ \Phi(z_i) c_i = 0. \end{cases} \tag{10.22}$$

We pick the right latent vectors of $\Phi(z_i)$, $\Phi(\frac{1}{\bar z_i})$ to be c_i $i = 1,\ldots,pq$.

We then use Lemma 3 to form a matrix A whose columns are eigenvectors of B_Φ corresponding to $z_1,\ldots,z_{pq}$, $\frac{1}{\bar z_1},\ldots,\frac{1}{\bar z_{pq}}$. We get

$$A = \begin{pmatrix} c_1 & & c_{pq} & c_1 & & c_{pq} \\ z_1 c_1 & & z_{pq}c_{pq} & \frac{1}{\bar z_1} c_1 & & \frac{1}{\bar z_{pq}} c_{pq} \\ \vdots & \cdots & \vdots & \vdots & \cdots & \vdots \\ z_1^{2q-1} c_1 & & z_{pq}^{2q-1} c_{pq} & \frac{1}{\bar z_1^{2q-1}} c_1 & & \frac{1}{\bar z_{pq}^{2q-1}} c_{pq} \end{pmatrix} \tag{10.23}$$

Since the eigenvalues of B are assumed distinct, we see that A is nonsingular. By Lemma 4 there is a permutation of the columns of A such that the upper left corner is a nonsingular matrix in $\mathbb{C}^{p\times p}$. In our case this means there is a subset of $\{c_1,\ldots,c_{pq}\}$, say $c_1,\ldots,c_p$ (after relabeling) that are linearly independent.

We are now ready to specify a factor of Φ. We have

$$G_\Phi(z_i)c_i = 0 \qquad i = 1,\ldots,p \tag{10.24}$$

where $C = (c_1,\ldots,c_p)$ is nonsingular.

Define $D = \mathrm{diag}(-z_1,\ldots,-z_p)$ (where, of course the z_i's have been relabelled along with the c_i's). and $U_1 = CDC^{-1}$. From now on we write G for G_Φ.

We propose to show that

$$G(z) = G_1(z)(Iz+U_1) \tag{10.25}$$

where $\deg G_1 < \deg G$ by using the remainder theorem. Indeed

$$\begin{aligned} &G_{(r)}(-U_1) \\ &= (C_q^T C + C_{q-1}^T C[-D] + \ldots + C_0 C[-D]^q + \ldots + C_q C[-D]^{2q})C^{-1} = 0\cdot I_p, \end{aligned} \tag{10.26}$$

which follows from $G(z_i)c_i = 0 \quad i = 1,\ldots,p$. Therefore

$$G(z) = G_1(z)(Iz+U_1). \tag{10.27}$$

We use a similar argument to establish

$$G_{(\ell)}(-U_1^{*-1}) = 0\cdot I_p, \tag{10.28}$$

which follows from

$$c_i^* G(\tfrac{1}{z_i}) = 0, \quad i = 1,\ldots,p. \tag{10.29}$$

Therefore

$$G(z) = (I+U_1^* z)G_2(Z) \tag{10.30}$$

where $\deg G_2 < \deg G$.

We must now combine (10.27) and (10.30) to get our final result. Define $G_C(z) = C^*G(z)C$, where C is the matrix defined above. Then

$$G_C(z) = C^*G_1(z)C(Iz+D) = G_{1,C}(z)(Iz+D)$$
$$= (I+D^*z)C^*G_2(z)C = (I+D^*z)G_{2,C}(z). \tag{10.31}$$

Since $\{z+z_i\colon\ i = 1,\ldots,p\}$, $\{1+\bar{z}_i z\colon\ i = 1,\ldots,p\}$ have disjoint roots we have

$$G_C(z) = (I+D^*z)G_{3,C}(z)(Iz+D) \tag{10.32}$$

for some appropriate $G_{3,C}$, and so

$$G(z) = C^{*-1}G_C(z)C^{-1} = (I+U_1^*z)G_3(z)(Iz+U_1) \tag{10.33}$$

where $G_3(z) = C^{*-1}G_{3,C}(z)C^{-1}$. Therefore

$$\Phi(z) = (I_p z^{-1}+U_1^*)\Phi_2(z)(I_p z+U_1) \tag{10.34}$$

with

$$\Phi_2(z) = z^{-q+1}G_3(z),\ G_3(z) \text{ of degree } \leq 2(q-1). \tag{10.35}$$

Q.E.D.

The proof of Theorem 1 gives $\Phi_2(z)$ as an extended hermitian QPM of degree $\leq$ q-1. We then simply apply the theorem q times to obtain

$$\Phi(z) = (I_p z^{-1}+U_1^*)\ldots(I_p z^{-1}+U_p^*)A_0(I_p z+U_q)\ldots(I_p z+U_1), \tag{10.36}$$

where, of course, $\det(I_p z+U_i) = 0$ has all roots in S_1, $i = 1,\ldots,p$, and

$$A_0 = (I_p+U_q^*)^{-1}\ldots(I_p+U_1^*)^{-1}\Phi(1)(I_p+U_1)^{-1}\ldots(I_p+U_q)^{-1}. \tag{10.37}$$

Since $\Phi(1)$ is positive definite, we can write $\Phi(1) = \Phi(1)^{1/2}\Phi(1)^{1/2}$ where $\Phi(1)^{1/2}$ is the positive definite square root of $\Phi(1)$. Thus

$$A_0 = U^*_{q+1}U_{q+1}, \quad U_{q+1} = \Phi(1)^{1/2}(I_p+U_1)^{-1}\cdots(I_p+U_q)^{-1}. \tag{10.38}$$

Consequently, on multiplying monomials,

$$\begin{cases} \Phi(z) = \Gamma_*(z)\Gamma(z) \quad \text{or} \\ \Phi(z) = \Gamma(z)^*\Gamma(z) \quad \text{on} \quad |z| = 1 \\ \Gamma(z) = \Gamma_0+\Gamma_1 z + \ldots + \Gamma_q z^q. \end{cases} \tag{10.39}$$

Before relaxing the condition of distinct determinental roots, we give an example. Finding a set of vectors $\{c_1,\ldots,c_{pq}\}$ (right latent vectors) once we have found the distinct determinental roots $\{z_1,\ldots,z_{pq}\} \subset S_1$ is not difficult, as the remarks in Robinson (1967), pages 167-169 explains. Indeed, given $G_\Phi(z)$ with distinct $\{z_1,\ldots,z_{pq}\}$, the adjugate of $G_\Phi(z_i)$

$$\text{adj } G_\Phi(z_i)\cdot G_\Phi(z_i) = \det G_\Phi(z_i)\cdot I_p \tag{10.40}$$

can be expressed as

$$\text{adj } G_\Phi(z_i) = c_i r_i^* \qquad i = 1,\ldots,pq. \tag{10.41}$$

The fact that the c_i can be computed in this way is found in Frazer et al. (1952), pages 61-62.

<u>Example 1</u>. We shall compute a factorization of

$$\begin{cases} \Phi(z) = \begin{pmatrix} 1 & 0 \\ 0 & 1 \end{pmatrix} z^{-1} + \begin{pmatrix} 4 & 1 \\ 1 & 4 \end{pmatrix} + \begin{pmatrix} 1 & 0 \\ 0 & 1 \end{pmatrix} z \\ \quad = \begin{bmatrix} z^{-1}+4+z & 1 \\ 1 & z^{-1}+4+z \end{bmatrix}. \\ \det \Phi(z) = z^{-2} + 8z^{-1} + 17 + 8z + z^2. \end{cases} \tag{10.42}$$

Let $u = z+z^{-1}$, so $\det \Phi(z) = u^2+8u+15$. This has roots at

$$u = \frac{-8\pm\sqrt{64-60}}{2} = -5,\ -3, \tag{10.43}$$

or

$$\begin{cases} z+z^{-1} = -5 \\ z = \dfrac{-5\pm\sqrt{21}}{2} \end{cases} \qquad \begin{cases} z+z^{-1} = -3 \\ z = \dfrac{-3\pm\sqrt{5}}{2} \end{cases}. \tag{10.44}$$

Let $z_1 = \frac{-5-\sqrt{21}}{2}$, $z_2 = \frac{-3-\sqrt{5}}{2}$; the other roots are $\frac{1}{\bar{z}_1}, \frac{1}{\bar{z}_2}$. Furthermore $\{z_1, z_2\} \subset S_1$. Then

$$\begin{cases} \Phi(z_1) = \begin{pmatrix} -1 & 1 \\ 1 & -1 \end{pmatrix} \\ \text{adj}\ \ \Phi(z_1) = \begin{pmatrix} -1 & -1 \\ -1 & -1 \end{pmatrix} = \begin{pmatrix} 1 \\ 1 \end{pmatrix} (-1 \quad -1), \end{cases} \tag{10.45}$$

and

$$\begin{cases} \Phi(z_2) = \begin{pmatrix} 1 & 1 \\ 1 & 1 \end{pmatrix} \\ \text{adj}\ \ \Phi(z_2) = \begin{pmatrix} 1 & -1 \\ -1 & 1 \end{pmatrix} = \begin{pmatrix} 1 \\ -1 \end{pmatrix} (1 \quad -1). \end{cases} \tag{10.46}$$

Note that $\Phi(z_1)\begin{pmatrix} 1 \\ 1 \end{pmatrix} = \Phi(z_2)\begin{pmatrix} 1 \\ -1 \end{pmatrix} = 0$. We can now construct a factorization by letting

$$D = \text{diag}(-z_1, -z_2),\ C = \begin{pmatrix} 1 & 1 \\ 1 & -1 \end{pmatrix},\ U_1 = CDC^{-1}.$$

Then $\Phi(z) = (Iz^{-1}+U_1^*)A_0(Iz+U_1)$, $A_0 = (I+U_1^*)^{-1}\Phi(1)(I+U_1)^{-1}$. Finding a square root of $\Phi(1) = \begin{pmatrix} 5 & 1 \\ 1 & 5 \end{pmatrix}$ will complete the factorization.

Theorem 1 was proven under the assumption that the determinantal roots were distinct. To remove this unappealing condition is not easy but has been done. We shall not go into this, since the details in the construction add little to our understanding of the real problem, but refer the reader to Thrift (1979). Instead we shall apply the factorization theory to obtain an informative representation of the limiting configuration measure induced by $\mathscr{R}$.

5.11. Representation of the random configurations

Combining what we have learnt so far we have found the autoregressive equation

$$\Gamma_0 x_n + \ldots + \Gamma_q X_{n-q} = Z_n. \tag{11.1}$$

Equation (11.1) can be rearranged as

$$\begin{aligned}\Gamma_0 X_n - Z_n &= -[\Gamma_1 X_{n-1} + \ldots + \Gamma_q X_{n-q}] \\ &= \sum_{j=1}^{q} \sum_{k=1}^{p} \gamma_{j,k} X_{n-j,k}\end{aligned} \tag{11.2}$$

where $\gamma_{j,k} = -[k^{th}$ column of $\Gamma_j] \in ¢^p$. We partition the edges corresponding to the bond couples of σ as follows:

$$\begin{cases} E_- = \{(e_{j,k}) \mid j < 0, \quad k \in \{1,\ldots,p\}\} \\ E_0 = \{(e_{0,k}) \mid k \in \{1,\ldots,p\}\} \\ E_+ = \{(e_{j,k}) \mid j > 0, \ k \in \{1,\ldots,p\}\}. \end{cases} \tag{11.3}$$

Recall the definition of $\partial E_-(E_0)$, that is, the boundary of E_- with respect to E_0. We then have the representation of the type we sought due to Thrift:

<u>Theorem 1</u>. *The limiting measure on* $\mathscr{L}(\mathscr{R})$ *leads to a stochastic process* $\{X_j\}$ *that can be expressed as*

$$\Gamma_0 X_n - Z_n = \sum_{(j,k)\in\partial E_-(E_0)} \gamma_{-j,k} X_{n+j,k}. \tag{11.4}$$

<u>Proof</u>:: Let $\{X_n^{(m)}\}$ be the process described in the construction of $\{X_n\}$: $\{X_n^{(m)}\}$ is a process on the circulant

graph with m vertices. If we write (r fixed, m >> r)

$$E(X_n^{(m)} | X_{n-j,k}^{(m)};\quad j \in \{1,\dots,r\},\ k \in \{1,\dots,p\})$$
$$= \sum_{j=1}^{r} \sum_{k=1}^{p} a_{j,k}^{(m)} X_{n-j,k}^{(m)} \qquad a_{j,k}^{(m)} \in \mathbb{R}^p \tag{11.5}$$

and

$$E(X_n | X_{n-j,k};\quad j \in \{1,\dots,r\},\ k \in \{1,\dots,p\})$$
$$= \sum_{j=1}^{r} \sum_{k=1}^{p} a_{j,k} X_{n-j,k} \qquad a_{j,k} \in \mathbb{R}^p. \tag{11.6}$$

Then $a_{j,k}^{(m)} \to a_{j,k}$ (componentwise as $m \to \infty$). This follows from well known properties of normal distributions, and the way we constructed the covariance function of $\{X_n\}$ as a limit of the covariance function of $\{X_n^{(m)}\}$. Then

$$E(\Gamma_0 X_n^{(m)} | X_{n-j,k}^{(m)};\ j \in \{1,\dots,r\},\ k \in \{1,\dots,p\})$$
$$= \sum_{j=1}^{r} \sum_{k=1}^{p} \Gamma_0 a_{j,k}^{(m)} X_{n-j,k}^{(m)}. \tag{11.7}$$

By picking r sufficiently large, we have from (11.5) and the regression theorem in Section 2 that $a_{j,k}^{(m)} = 0$ if $j \in \{1,\dots,q\}$ and $(-j,k) \notin \partial E_-(E_0)$. Consequently, (11.7) implies that

$$E(\Gamma_0 X_n - Z_n | X_{n-j,k};\quad j \in \{1,\dots, r\},\ k \in \{1,\dots,p\})$$
$$= \sum_{j=1}^{q} \sum_{k=1}^{p} \gamma_{j,k} X_{n-j,k} = \sum_{(j,k) \in \partial E_-(E_0)} \Gamma_0 a_{-j,k} X_{n+j,k}$$
$$+ \sum_{j=q+1}^{r} \Gamma_0 a_{j,k} X_{n-j,k}. \tag{11.8}$$

By nondegeneracy, $\Gamma_0 a_{j,k} = 0$ if $j > q$, and so the result follows. Q.E.D.

Theorem 1 relates the autoregression representation of the random configuration with the topological nature of the regularity that we started from.

5.12. Spectral density matrix for Σ = LATTICE(γ)

After the analysis in Sections 8-11 the question arises whether the technique can be applied to other regular structures. We shall see now that this is so, and choose the connection type Σ = LATTICE(γ) in two dimensions. To simplify the notation we shall assume, with some loss of generality, the special form of generators shown in Figure 12.1.

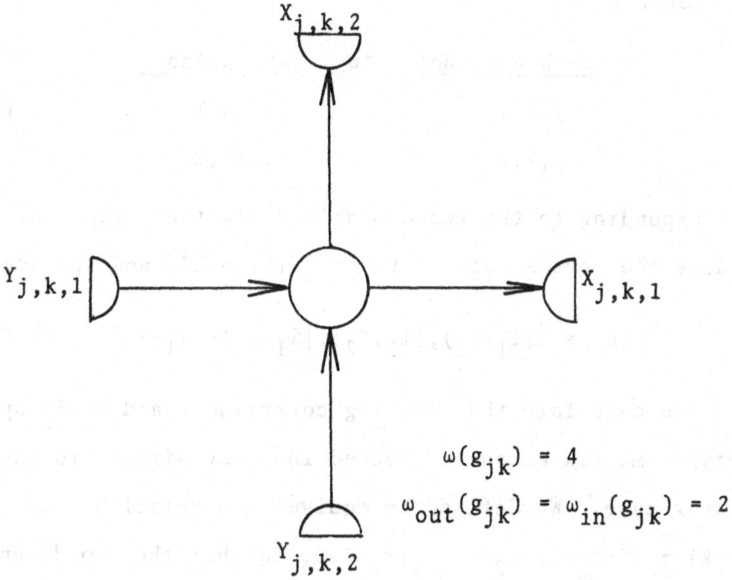

Figure 12.1

The bond values are still real numbers and we shall use the notation as shown in the figure

$$\begin{cases} \text{outbonds} & X_{j,k,1}, X_{j,k,2} \\ \text{inbonds} & Y_{j,k,1}, Y_{j,k,2} \end{cases} \tag{12.1}$$

The Q-measure will be introduced by letting the vector of bond values for g_{jk}

$$\begin{pmatrix} X_{j,k} \\ Y_{j,k} \end{pmatrix} \tag{12.2}$$

have a Gaussian probability measure with mean zero and covariance matrix

$$H^{-1} = \begin{pmatrix} H_{11} & H_{12} \\ H_{12}^T & H_{22} \end{pmatrix}^{-1} ; \quad H_{ij} \in \mathbb{R}^{2\times 2}. \tag{12.3}$$

We keep the bond relation ρ as EQUAL and introduce σ by the connections

outbond label	to	inbond label
j,k,1		j+1,k,1
j,k,2		j,k+1,2

(12.4)

corresponding to the two-dimensional lattice graph L. Of course the vertex set of L is $V(L) = \mathbb{Z}^2$, and the edge set

$$E(L) = \{(i_1,j_1),(i_2,j_2) \,|\, |i_1-i_2|+|j_1-j_2| = 1\}. \tag{12.5}$$

To calculate the limiting covariances and their spectral density matrix we shall proceed in a way similar to the linear case. We define the equivalence relation $(j,k) \sim (j+r_1n, k+r_2n)$ $r_1, r_2 \in \mathbb{Z}$, so that the two-dimensional (infinite) lattice is approximated by the lattice on the two-dimensional (finite) torus. A configuration is given by taking n^2 generators

$$g_{j,k}; \quad 0 \le j \le n-1; \quad 0 \le k \le n-1 \tag{12.6}$$

where $g_{j,k}$ has the outbonds $X_{j,k,1}, X_{j,k,2}$ and inbonds

$Y_{j,k,1}, Y_{j,k,2}$. The generators are positioned on the vertices of the torus graph, with bonds along the respective edges. We now have a finite number of generators, and we want to calculate the distribution of the configuration after conditioning on bond relation EQUALITY.

The bond couples (12.4) are now taken mod n, and state that

$$Y_{j,k} = E_{1,2}X_{j-1,k} + E_{2,2}X_{j,k-1} \bmod n, \tag{12.7}$$

where

$$E_{1,2} = \begin{pmatrix}1 & 0\\ 0 & 0\end{pmatrix} \quad E_{2,2} = \begin{pmatrix}0 & 0\\ 0 & 1\end{pmatrix},\ j,k=0,\dots,n-1. \tag{12.8}$$

This can also be expressed in terms of the fundamental circulant matrix π_n in $\mathbb{C}^{n\times n}$:

$$\begin{cases} Y_j = (\pi_n^{-1}\otimes E_{2,2})X_j + (I_n\otimes E_{1,2})X_{j-1} \\ Y = (I_n\otimes\pi_n^{-1}\otimes E_{2,2} + \pi_n^{-1}\otimes I_n\otimes E_{1,2})X. \end{cases} \tag{12.9}$$

If we assume that the generators $g_{j,k}$ are (initially, before the regularity is controlling them) i.i.d. with covariance matrix

$$H^{-1} = \begin{pmatrix} H_{11} & H_{12} \\ H_{12}^T & H_{22} \end{pmatrix}^{-1} \quad H_{11}, H_{12}, H_{22} \in \mathbb{R}^{2\times 2}, \tag{12.10}$$

then the joint frequency function of $\binom{X}{Y}$ is given by

$$f\binom{X}{Y} = K_0 \exp\{-\tfrac{1}{2} H\binom{X}{Y}\}, \tag{12.11}$$

where K_0 is a normalizing constant, and H is the quadratic form

$$\begin{aligned} &X^T(I_{n^2}\otimes H_{11})X + X^T(I_{n^2}\otimes H_{12})Y \\ &+ Y^T(I_{n^2}\otimes H_{12}^T)X + Y^T(I_{n^2}\otimes H_{22})Y. \end{aligned} \tag{12.12}$$

As before, to get the joint p.d.f. of $X \in \mathbb{R}^{2n^2}$ after conditioning on ρ = EQUAL, we substitute (12.9) into (12.12). The result is a quadratic form in X with matrix

$$(I_{n^2} \otimes H_{11}) + (I_{n^2} \otimes H_{12})(I_n \otimes \pi_n^{-1} \otimes E_{2,2} + \pi_n^{-1} \otimes I_n \otimes E_{12})$$
$$+ (I_n \otimes \pi_n \otimes E_{2,2} + \pi_n \otimes I_n \otimes E_{1,2})(I_{n^2} \otimes H_{12}^T) \tag{12.13}$$
$$(I_n \otimes \pi_n \otimes E_{2,2} + \pi_n \otimes I_n \otimes E_{1,2})(I_{n^2} \otimes H_{22})(I_n \otimes \pi_n^{-1} \otimes E_{2,2} + \pi_n^{-1} \otimes I_n \otimes E_{1,2})$$

or,

$$\begin{aligned} & I_n \otimes I_n \otimes (H_{11} + E_{2,2}H_{22}E_{2,2} + E_{1,2}H_{22}E_{1,2}) \\ & + I_n \otimes \pi_n^{-1} \otimes H_{12}E_{2,2} \\ & + \pi_n^{-1} \otimes I_n \otimes H_{12}E_{1,2} \\ & + I_n \otimes \pi_n \otimes E_{2,2}H_{12}^T \\ & + \pi_n \otimes I_n \otimes E_{1,2}H_{12}^T \\ & + \pi_n^{-1} \otimes \pi_n \otimes E_{2,2}H_{22}E_{1,2} \\ & + \pi_n \otimes \pi_n^{-1} \otimes E_{1,2}H_{22}E_{2,2}. \end{aligned} \tag{12.14}$$

Let

$$\begin{cases} A_{0,0} = H_{11} + E_{1,2}H_{22}E_{1,2} + E_{2,2}H_{22}E_{2,2} \\ A_{0,1} = H_{12}E_{2,2} \\ A_{1,0} = H_{12}E_{1,2} \\ A_{1,1} = E_{2,2}H_{22}E_{1,2}. \end{cases} \tag{12.15}$$

From the Theorem 8.1 we get, inverting the matrix in our quadratic form,

$$R_n(j) = E(X_o X_j^T)$$
$$= \frac{1}{n}\sum_{\ell=0}^{n-1} \{I_n \otimes A_{0,0} + \pi_n^{-1} \otimes A_{0,1} + \pi_n \otimes A_{0,1}^T \tag{12.16}$$
$$+ (I_n \otimes A_{1,0} + \pi_n \otimes A_{1,1})\omega_n^{-\ell}$$
$$+ (I_n \otimes A_{1,0}^T + \pi_n^{-1} \otimes A_{1,1}^T)\omega_n^{\ell}\}^{-1}\omega^{-j\ell},$$

and

$$R_n(j,k) = E(X_{0,0}X_{j,k}^T)$$
$$= \frac{1}{n^2}\Big\{\sum_{\ell=0}^{n-1}\sum_{m=0}^{n-1}(A_{0,0} + A_{1,0}\omega_n^{-\ell} + A_{1,0}^T\omega_n^{\ell} \tag{12.17}$$
$$+ A_{0,1}\omega_n^{-m} + A_{0,1}^T\omega_n^{m}$$
$$+ A_{11},\omega_n^{-\ell+m} + A_{1,1}^T\omega_n^{\ell-m})^{-1}\Big\}\omega_n^{-j\ell-km},$$

where $\omega_n = e^{\frac{2\pi i}{n}}$. $R_n(j,k)$ is a Riemann sum which approximates

$$R(j,k) = \frac{1}{4\pi^2}\int_0^{2\pi}\int_0^{2\pi} e^{-ij\theta_1 - ik\theta_2}$$
$$\Big(A_{0,0} + A_{1,0}e^{-i\theta_1} + A_{1,0}^T e^{i\theta_1} + A_{0,1}e^{-i\theta_2} + A_{0,1}^T e^{i\theta_2} \tag{12.18}$$
$$+ A_{1,1}e^{-i\theta_1 + i\theta_2} + A_{1,1}^T e^{i\theta_1 - i\theta_2}\Big)^{-1} d\theta_1 d\theta_2.$$

Let

$$\Phi(e^{-i\theta_1}, e^{-i\theta_2})$$
$$= A_{0,0} + A_{1,0}e^{-i\theta_1} + A_{1,0}^T e^{i\theta_1} + A_{0,1}e^{-i\theta_2} + A_{0,1}^T e^{i\theta_2}$$
$$+ A_{1,1}e^{-i\theta_1 + i\theta_2} + A_{1,1}^T e^{i\theta_1 - i\theta_2}. \tag{12.19}$$

This is the inverse of the spectral density matrix for the process $\{X_{j,k} | (j,k) \in \mathbb{Z}^2\}$. As before, Φ is positive definite and hermitian at each $(\theta_1, \theta_2) \in [-\pi, \pi)^2$.

We let $z = e^{-i\theta_1}$, $w = e^{-i\theta_2}$, and write

$$\begin{aligned}\Phi(z,w) &= A_{0,0}+A_{1,0}z+A_{1,0}^T z^{-1}+A_{0,1}w+A_{0,1}^T w^{-1} \\ &\quad +A_{1,1}zw^{-1}+A_{1,1}^T z^{-1}w \in \mathbb{C}^{2\times 2}.\end{aligned} \tag{12.20}$$

and get Thrift's result.

Theorem 1. *In the present case the limiting covariances correspond to the spectral density matrix* Φ^{-1}, *where* Φ *is given in* (12.20).

5.13. Factorization of the spectral density matrix in two dimensions

We now meet again the problem of factoring Φ and we shall follow Helson-Lowdenschlager (1958). These authors developed an analytical technique that exactly suits our needs. They discussed the scalar case so that we must extend it to the present situation.

Consider the set $\mathbb{Z}^2$ of lattice vertices. Let S be the following subset of $\mathbb{Z}^2$: $\{(m,n) \in \mathbb{Z}^2 | m > 0$ or both $m = 0$ and $n > 0\}$ as shown in Figure 13.1. The set S has the following properties:

$$\begin{cases} 1. & (0,0) \notin S. \\ 2. & (m,n) \in S \leftrightarrow (-m,-n) \notin S, \text{ unless } (m,n) = (0,0). \\ 3. & (m,n),(m',n') \in S \to (m+m',n+n') \in S. \end{cases} \tag{13.1}$$

Note that S partitions $\mathbb{Z}^2$ into S, $\{(0,0)\}$, -S. We also consider the following Hilbert space. Consider the set of all functions $A\colon T^2 = [-\pi,\pi)^2 \to \mathbb{C}^{2\times 2}$ such that

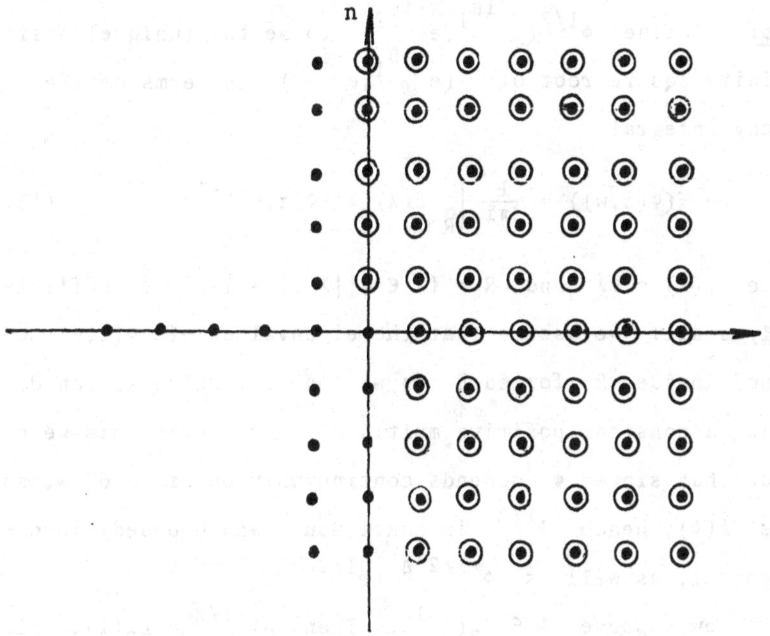

Figure 13.1

$$\frac{1}{4\pi^2} \operatorname{tr} \int_{-\pi}^{\pi}\int_{-\pi}^{\pi} A(\theta_1,\theta_2)\Phi^{-1}(e^{i\theta_1},e^{i\theta_2})A^*(\theta_1,\theta_2)d\theta_1 d\theta_2 < \infty. \tag{13.2}$$

On this space we define the inner product

$$(A,B) = \frac{1}{4\pi^2} \operatorname{tr} \int_{T^2} A\Phi^{-1}B^* d\theta \tag{13.3}$$

and norm $\|A\|^2_{\Phi^{-1}} = (A,A)$. Denote the Hilbert space by $L_2(\Phi^{-1})$. In particular $L_2(I)$ is the Hilbert space when $\Phi \equiv I$, the identity matrix.

Lemma 1. *Let* $A \in L_2(\Phi^{-1})$. *Then* A *can be expressed as*

$$A(\theta_1,\theta_2) = \sum_{(m,n)\in \mathbb{Z}^2} A_{m,n} z^n w^n, \quad z = e^{i\theta_1}, \ w = e^{-i\theta_2}, \tag{13.4}$$

with convergence in $\|\cdot\|_{\Phi^{-1}}$ *norm, and* $A_{m,n} \in ¢^{2\times 2}$.

Proof. Define $\Phi^{1/2}(e^{-i\theta_1}, e^{-i\theta_2})$ to be the (unique) positive definite square root of $\Phi(e^{-i\theta_1}, e^{-i\theta_2})$ in terms of the Cauchy integral

$$f(\Phi(z,w)) = \frac{1}{2\pi i}\int_R f(\lambda)(\lambda I - \Phi(z,w))^{-1} d\lambda \qquad (13.5)$$

where $f(\lambda) = \sqrt{\lambda}$ and $R = \{\lambda \in \mathbb{C} \,|\, |\lambda - 1| = 1-\varepsilon\}$, ε sufficiently small: we assume that the eigenvalues of $\Phi(z,w)$ contained inside R for each (z,w) on T^2, which we can do by taking a constant positive multiple of Φ. From this we conclude that since Φ depends continuously on z and w, so does $f(\Phi)$, hence $\Phi^{1/2}$ is continuous (and bounded) in each component, as well as $\Phi^{-1/2} \triangleq (\Phi^{1/2})^{-1}$.

Now suppose $A \in L_2(\Phi^{-1})$. Then $A\Phi^{-1/2} \in L_2(I)$. Since $\Phi^{1/2}$ is bounded and continuous, we also have $A = (A\Phi^{-1/2})\Phi^{1/2} \in L_2(I)$. By the completeness of the trigonometric system we have

$$A(\theta_1, \theta_2) = \sum_{(m,n)\in \mathbb{Z}^2} A_{m,n} z^m w^n, \quad z = e^{-i\theta_1}, \; w = e^{-i\theta_2}, \qquad (13.6)$$

with covergence in $\|\cdot\|_I$ norm, which follows from the fact that

$$\left\{\begin{pmatrix}1 & 0\\ 0 & 0\end{pmatrix} z^m w^n, \; \begin{pmatrix}0 & 1\\ 0 & 0\end{pmatrix} z^m w^n, \; \begin{pmatrix}0 & 0\\ 1 & 0\end{pmatrix} z^m w^n, \right. \qquad (13.7)$$
$$\left. \begin{pmatrix}0 & 0\\ 0 & 1\end{pmatrix} z^m w^n \;\middle|\; (m,n) \in \mathbb{Z}^2\right\}$$

is a complete orthonormal set in $L_2(I)$.

To show that

$$\sum_{(m,n)\in \mathbb{Z}^2} A_{m,n} z^m w^n \qquad (13.8)$$

converges in $\|\cdot\|_{\Phi^{-1}}$, we consider a partial sum

$$S_{m,n} = \sum_{|m'|\le m, |n'|\le n} A_{m,n} z^m w^n. \tag{13.9}$$

Then

$$\int_{T^2} \operatorname{tr}(S_{m,n}-A)(S_{m,n}-A)^* d\theta \to 0 \tag{13.10}$$

as $m,n \to \infty$. If we let

$$S_{m,n}(\theta_1,\theta_2)-A(\theta_1,\theta_2) = T_{m,n}(\theta_1,\theta_2)$$

$$= \begin{bmatrix} t^{11}_{m,n}(\theta_1,\theta_2) & t^{12}_{m,n}(\theta_1,\theta_2) \\ t^{21}_{m,n}(\theta_1,\theta_2) & t^{22}_{m,n}(\theta_1,\theta_2) \end{bmatrix}, \tag{13.11}$$

then

$$\int_{T^2} |t^{j,k}_{m,n}(\theta_1,\theta_2)|^2 d\theta \to 0 \quad \text{as} \quad m,n \to \infty. \tag{13.12}$$

Let

$$\Phi^{-1/2}(e^{-i\theta_1}, e^{-i\theta_2}) = \begin{bmatrix} \phi_{11}(\theta_1,\theta_2) & \phi_{12}(\theta_1,\theta_2) \\ \phi_{21}(\theta_1,\theta_2) & \phi_{22}(\theta_1,\theta_2) \end{bmatrix} \tag{13.13}$$

Then

$$T_{m,n}(\theta_1,\theta_2)\Phi^{-1/2}(e^{-i\theta_1}, e^{-i\theta_2})$$

$$= \begin{bmatrix} t^{11}_{m,n}(\theta_1,\theta_2) & t^{12}_{m,n}(\theta_1,\theta_2) \\ t^{21}_{m,n}(\theta_1,\theta_2) & t^{22}_{m,n}(\theta_1,\theta_2) \end{bmatrix} \begin{bmatrix} \phi_{11}(\theta_1,\theta_2) & \phi_{12}(\theta_1,\theta_2) \\ \phi_{21}(\theta_1,\theta_2) & \phi_{22}(\theta_1,\theta_2) \end{bmatrix}$$

$$= \left(\sum_{k=1}^{2} t^{ik}_{m,n}\phi_{kj} \right) \tag{13.14}$$

will have the property that

$$\int_{T^2} \left| \sum_{k=1}^{2} t^{ik}_{m,n}\phi_{kj} \right|^2 d\theta \to 0 \quad \text{as} \quad m,n \to \infty, \tag{13.15}$$

since $\Phi^{-1/2}$ is bounded and continuous in each component.

Consequently

$$\int_{T^2} \operatorname{tr} T_{m,n}\Phi^{-1/2}\Phi^{-1/2}T^*_{m,n}d\theta \to 0 \quad \text{as } m,n \to \infty, \tag{13.16}$$

and

$$\sum_{(m,n)\in \mathbb{Z}^2} A_{m,n}z^m w^n \tag{13.17}$$

converge in $\|\cdot\|_{\Phi^{-1}}$.

Therefore we are justified in writing

$$\sum_{(m,n)\in \mathbb{Z}^2} A_{m,n}z^m w^n \tag{13.18}$$

for every $A \in L_2(\Phi^{-1})$. Q.E.D.

We now consider the set $\mathscr{P}$ of trigonometric polynomials

$$P(z,w) = \sum_{(m,n \in S} A_{m,n}z^m w^n, \quad A_{m,n} \in ¢^{2\times 2}, \tag{13.19}$$

and $A_{m,n} = 0 \cdot I_2$ outside some bounded subset of S. The set $\mathscr{P}$ forms a convex subset in $L_2(\Phi^{-1})$. Let $A = \text{closure}(\mathscr{P})$ in $L_2(\Phi^{-1})$. Using a result in Hilbert space theory (Rudin [1973], page 293), there exists a unique $H \in A$ such that

$$\|I+H(z,w)\|^2_{\Phi^{-1}} = \inf_{P\in\mathscr{P}} \|I+P(z,w)\|^2_{\Phi^{-1}}. \tag{13.20}$$

<u>Lemma 2</u>. $H(z,w)$ *can be expressed as*

$$\sum_{(m,n)\in S} \hat{A}_{m,n}z^m w^n, \tag{13.21}$$

converging in $\|\cdot\|_{\Phi^{-1}}$ *norm, and*

$$\|I+H(z,w)\|^2_{\Phi^{-1}} > 0. \tag{13.22}$$

<u>Proof</u>: Using Lemma 1 we write

$$H(z,w) = \sum_{(m,n)\in \mathbb{Z}^2} \hat{A}_{m,n}z^m w^n. \tag{13.23}$$

Let P_n be a sequence of polynomials in $\mathcal{P}$ such that $\|P_n - H\|_{\Phi^{-1}} \to 0$. Now

$$(P_n(z,w)\Phi(z,w), z^m w^n C) = 0; \quad \forall (m,n) \notin S, \qquad \text{and any constant matrix } C. \tag{13.24}$$

Therefore

$$\begin{aligned} &|(P_n(z,w)\Phi(z,w), z^m w^n C) \\ &\quad - (H(z,w)\Phi(z,w), z^m w^n C)| \\ &\quad \leq \|(P_n(z,w) - H(z,w))\Phi(z,w)\|_{\Phi^{-1}} \|C\|_{\Phi^{-1}} \end{aligned} \tag{13.25}$$

is arbitrarily small (as $n \to \infty$). Hence

$$(H(z,w)\Phi(z,w)), z^m w^n C) = 0; \quad \forall (m,n) \notin S, \tag{13.26}$$

and so

$$H(z,w) = \sum_{(m,n)\in S} \hat{A}_{m,n} z^m w^n. \tag{13.27}$$

To show that

$$\|I + H(z,w)\|^2_{\Phi^{-1}} > 0 \tag{13.28}$$

we suppose

$$\|I + H(z,w)\|^2_{\Phi^{-1}} = 0 \tag{13.29}$$

and arrive at a contradiction:

$$\begin{aligned} 0 &= \operatorname{tr} \int_{T^2} (I+H)\Phi^{-1}(I+H)^* d\theta \\ &= \int_{T^2} \operatorname{tr}(I+H)\Phi^{-1/2}\Phi^{-1/2}(I+H)^* d\theta \Rightarrow \end{aligned} \tag{13.30}$$

$I + H(z,w) = 0$ almost surely with respect to Lebesgue measure. But $(P_n(z,w), \Phi(z,w), I) = 0$; $\forall n$; implies $(H(z,w)\Phi(z,w), I) = 0$, or $-(\Phi(z,w), I) = 0$, a contradiction. Q.E.D.

We now follow the construction in Helson and Lowdenslager (1958). We have constructed a Hilbert space $L_2(\Phi^{-1})$ from

Φ and an element $H \in L_2(\Phi^{-1})$ with the property $\|I+H\|_{\Phi^{-1}} = \inf_{P \in \mathscr{P}} \|I+P\| > 0$, and have shown that H can be written

$$H(z,w) = \sum_{(m,n)\in S} \hat{A}_{m,n} z^m w^n. \tag{13.31}$$

Theorem 1. *There exists a nonsingular* $B \in \mathbb{C}^{2\times 2}$ *such that*

$$\Phi(z,w) = \{B^{-1}(I+H(z,w))\}^*\{B^{-1}(I+H(z,w))\} \tag{13.32}$$

on $|z| = |w| = 1$. *Furthermore*

$$(I+H(z,w))^{-1} \in L_2(\Phi^{-1}) \tag{13.33}$$

and can be written

$$(I+H(z,w))^{-1} = I + \sum_{(m,n)\in S} \hat{B}_{m,n} z^m w^n. \tag{13.34}$$

Proof: Let G be a nonsingular matrix, and $\lambda \in \mathbb{C}$. Now

$$\|I+H(z,w) + \lambda G z^m w^n\|^2_{\Phi^{-1}} \tag{13.35}$$

has a unique minimum at $\lambda = 0$ for each $(m,n) \in S$. By Rudin (1973), page 293,

$$(I+H(z,w), G z^m w^n) = 0 \tag{13.36}$$

for every nonsingular matrix G. This can be written

$$\mathrm{tr}\left\{\int_{T^2} (I+H(z,w))\Phi^{-1}(z,w) z^{-m} w^{-n} d\theta \cdot G^*\right\} = 0 \tag{13.37}$$

for every nonsingular G. But if $\mathrm{tr}\, AB^* = 0$ for every nonsingular B (and A a constant matrix), then $A = 0$ (the null matrix). Consequently

$$\int_{T^2} (I+H(z,w))\Phi^{-1}(z,w) z^{-m} w^{-n} d\theta = 0 \cdot I_2 \quad \text{for all } (m,n) \in S. \tag{13.38}$$

Recall that S is closed under group addition (property 3), so for $(m,n) \in S$

$$\|(I+\lambda G z^m w^n)(I+H(z,w))\|^2_{\Phi^{-1}} \tag{13.39}$$

has a unique minimum at $\lambda = 0$, or

$$\|I+H(z,w) + \lambda G z^m w^n (I+H(z,w)\|^2_{\Phi^{-1}} \tag{13.40}$$

has a unique minimum at $\lambda = 0$. Arguing as before we get

$$\int_{T^2} (I+H(z,w))\Phi^{-1}(z,w)(I+H(z,w))^* z^{-m} w^{-n} d\theta = 0\cdot I_2 \tag{13.41}$$

for all $(m,n) \in S$. Taking complex conjugate in (13.41) we get

$$\int_{T^2} (I+H(z,w))\Phi^{-1}(z,w)(I+H(z,w))^* z^m w^n d\theta = 0\cdot I_2, \tag{13.42}$$

for all $(m,n) \in S$.

Combining (13.41) and (13.42), we get

$$(I+H(z,w))\Phi^{-1}(z,w)(I+H(z,w))^* = C \tag{13.43}$$

a constant matrix; that is all other coefficient matrices of $z^m w^n$ $(m,n) \neq (0,0)$ are 0 by orthonormality. C is obviously hermitian, and is nonsingular, which follows from an argument due to Hannan (1970), page 160.

$$\inf_{\substack{\det A_0=1 \\ P\in\mathscr{P}}} \operatorname{tr} \int_{T^2} (A_0+P(z,w))\Phi^{-1}(z,w)(A_0+P(z,w))^* d\theta > 0. \tag{13.44}$$

since we can take the A_0 factor into Φ^{-1}, and define a new $\Phi^{-1}_{A_0} = A_0\Phi^{-1}A_0^*$; at $\hat{A}_0$ (the optimal A_0) we have positivity. Consider

$$\operatorname{tr} A_0 C A_0^* = \operatorname{tr}\int (A_0+A_0H(z,w))\Phi^{-1}(z,w)(A_0+A_0H(z,w))^* d\theta. \tag{13.45}$$

Suppose C is nonsingular. Choose first row a_1^* of A_0 such that $a_1^* C a_1 = 0$. Choose second row a_2^* so that

$$\det\begin{pmatrix} a_1^* \\ a_2^* \end{pmatrix} = 1. \tag{13.46}$$

Then

$$\det\begin{pmatrix} \frac{1}{\varepsilon}\, a_1^* \\ \varepsilon\, a_2^* \end{pmatrix} = 1. \tag{13.47}$$

By letting ε be as small as desired we have

$$\operatorname{tr}\begin{pmatrix} \frac{1}{\varepsilon} & a_1^* \\ \varepsilon & a_2^* \end{pmatrix} C \begin{pmatrix} \frac{1}{\varepsilon} & a_1^* \\ \varepsilon & a_2^* \end{pmatrix}^* \tag{13.48}$$

as small as desired, which leads to a contradiction. Hence

$$(I+H(z,w))\Phi^{-1}(z,w)(I+H(z,w))^* = BB^* \tag{13.49}$$

for some nonsingular B; and

$$\Phi(z,w) = \{B^{-1}(I+H(z,w))\}^*\{B^{-1}(I+H(z,w))\}. \tag{13.50}$$

Furthermore, from (13.38), for all $(m,n) \in S$

$$\begin{aligned} \int_{T^2} z^m w^n \Phi^{-1}(z,w)(I+H(z,w))^* d\theta &= 0\cdot I_2 \\ \int_{T^2} z^m w^n (I+H(z,w))^{-1} C d\theta &= 0\cdot I_2 \\ \int_{T^2} z^m w^n (I+H(z,w))^{-1} d\theta &= 0\cdot I_2. \end{aligned} \tag{13.51}$$

Since $(I+H(z,w))^{-1} \in L_2(\Phi^{-1})$, we use Lemma 1 to get

$$(I+H(z,w))^{-1} = I + \sum_{(m,n)\in S} \hat{B}_{m,n} z^m w^n.$$

Q.E.D.

5.14. Representations of the random configurations in the two dimensional case

The establishment of Theorem 13.1 allows us to represent $\{X_{j,k} | (j,k) \in \mathbb{Z}^2\}$ as an autoregression, just as in the linear case. If we write

$$B^{-1}(I+H(z,w)) = B^{-1} + \sum_{(m,n)\in S} C_{m,n} z^m w^n, \qquad (14.1)$$

then the autoregression is

$$B^{-1}X_{s,t} + \sum_{(m,n)\in S} C_{m,n} X_{s-m,t-n} = Z_{s,t} \qquad (14.2)$$

where the innovation process $\{Z_{s,t} | (s,t) \in \mathbb{Z}^2\}$ is orthonormal and orthogonal to $\{X_{s-m,t-n} | (m,n) \in S\}$. Write (14.2) as

$$B^{-1}X_{s,t} - Z_{s,t} = \sum_{(m,n)\in S} \sum_{j=1}^{2} \gamma_{m,n,j} X_{s-m,t-n,j} \qquad (14.3)$$

where $\gamma_{m,n,j} \in \mathbb{C}^2$ (column vectors). Partition the edges of the lattice graph as follows

$$\begin{cases} E_- = \{(m,n,j) | (m,n) \in -S, \quad j \in \{1,2\}\} \\ E_0 = \{(0,0,j) | j \in \{1,2\}\} \\ E_+ = \{(m,n,j) | (m,n) \in S, \quad j \in \{1,2\}\}. \end{cases} \qquad (14.4)$$

We, of course, have denoted edge $\{(m,n),(m+1,n)\}$ by $(m,n,1)$, and edge $\{(m,n),(m,n+1)\}$ by $(m,n,2)$. Recall the definition of $\partial E_-(E_0)$, that is, the boundary of E_- with respect to E_0. We can then announce the definitive result, also given in Thrift (1979).

<u>Theorem 1</u>. *The random configurations for* Σ = LATTICE(γ) *can be represented by*

$$B^{-1}X_{s,t}-Z_{s,t} = \sum_{(m,n,j)\in\partial E_-(E_0)} \gamma_{-m,-n,j} X_{s+m,t+n,j} \tag{14.5}$$

with convergence in mean square.

Proof: Let

$$E_-[p] = \{(m,n,j) \in E_- \mid |m| \leq p,\ |n| \leq p\}. \tag{14.6}$$

Consider (14.3) at $s = t = 0$

$$B^{-1}X_{0,0}-Z_{0,0} = \sum_{(m,n,j)\in E_-} \gamma_{-m,-n,j} X_{m,n,j}. \tag{14.7}$$

Then

$$\begin{aligned} &E(B^{-1}X_{0,0} \mid X_{(u,v,k)}, (u,v,k) \in E_-[p]) \\ &= E\Big(\sum_{(m,n,j)\in E_- - {}^{o}E_-[p](E_0)} \gamma_{-m,-n,j} X_{m,n,j} \mid X_{(u,v,k)}, \\ &\qquad\qquad (u,v,k) \in E_-[p]\Big) \\ &+ E\Big(\sum_{(m,n,j)\in {}^{o}E_-[p](E_0)} \gamma_{-m,-n,j} X_{m,n,j} \mid X_{(u,v,k)}, \\ &\qquad\qquad (u,v,k) \in E_-[p]\Big), \end{aligned} \tag{14.8}$$

where ${}^{o}E_-[p](E_0) = E_-[p] - \partial E_-[p](E_0)$. Using the results in Section 2 we establish

$$\begin{aligned} &E(B^{-1}X_{0,0} \mid X_{u,v,k}, (u,v,k) \in E_-[p]) \\ &= E(B^{-1}X_{0,0} \mid X_{u,v,k}, (u,v,k) \in \partial E_-[p](E_0)). \end{aligned} \tag{14.9}$$

Also, for each $(m,n,j) \in E_- - {}^{o}E_-[p](E_0)$

$$\begin{aligned} &E(X_{m,n,j} \mid X_{u,v,k}, (u,v,k) \in E_-[p]) \\ &= E(X_{m,n,j} \mid X_{u,v,k}, (u,v,k) \in \partial E_-[p](E_0)). \end{aligned} \tag{14.10}$$

By mean square convergence of the term

$$\sum_{(m,n,j)\in E_- - {}^oE_-[p](E_0)} \gamma_{-m,-n,j} X_{m,n,j}, \tag{14.11}$$

we have from (14.10), using Jensen's inequality for conditional expectation,

$$\begin{aligned} &E\Big(\sum_{(m,n,j)\in E_- - {}^oE_-[p](E_0)} \gamma_{-m,-n,j} X_{m,n,j} \,|\, X_{u,v,k}, \\ &\qquad\qquad (u,v,k) \in E_-[p]\Big) \\ &= E\Big(\sum_{(m,n,j)\in E_- - {}^oE_-[p](E_0)} \gamma_{-m,-n,j} X_{m,n,j} \,|\, X_{u,v,k}, \\ &\qquad\qquad (u,v,k) \in \partial E_-[p](E_0)\Big). \end{aligned} \tag{14.12}$$

Let

$$\begin{cases} U_1 = B^{-1} X_{0,0} \\ U_2 = \sum_{(m,n,j)\in E_- - {}^oE_-[p](E_0)} \gamma_{-m,-n,j} X_{m,n,j}. \end{cases} \tag{14.13}$$

Then

$$\begin{aligned} &E(U_1 | X_{u,v,k} (u,v,k) \in \partial E_-[p](E_0)) \\ &\quad = E(U_2 | X_{u,v,k} (u,v,k) \in \partial E_-[p](E_0)) \\ &\qquad + \sum_{(u,v,k)\in {}^oE_-[p](E_0)} \gamma_{-u,-v,k} X_{u,v,k}. \end{aligned} \tag{14.14}$$

Therefore for some vectors $\eta_{u,v,k}$, $(u,v,k) \in \partial E_-[p](E_0)$, we have

$$\begin{aligned} &\sum_{(u,v,k)\in \partial E_-[p](E_0)} \eta_{u,v,k} X_{u,v,k} \\ &\quad + \sum_{(u,v,k)\in {}^oE_-[p](E_0)} \gamma_{-u,-v,k} X_{u,v,k} = 0. \end{aligned} \tag{14.15}$$

By nondegeneracy (for every finite collection of edges F,

$\sum_{(m,n,k)\in F} a_{m,n,k} X_{m,n,k} = 0$, implying $a_{m,n,k} = 0$ for $(m,n,k) \in F$, since we can express $X_{m,n,k}$ as a sum of orthogonal innovations via Theorem 13.1) we have $\gamma_{-u,-v,k} = 0$ for $(u,v,k) \in {}^{o}E_{-}[p](E_0)$. Since p is arbitrary, $\gamma_{-u,-v,k} = 0$ for all $(u,v,k) \in E_{-} - \partial E_{-}(E_0)$.

Q.E.D.

This concludes our study of the asymptotics of metric pattern theory for large configurations. The results obtained are both elegant and informative but are limited by the restriction to Gaussian Q-measures. At present we do not know if similar asymptotic theory can be constructed for other Q-measures.

5.15. Laws of large numbers in pattern theory

The limit theorems in classical probability theory are usually stated in terms of triangular arrays of random variables

$$\begin{cases} x_{11} \\ x_{21}, x_{22} \\ x_{31}, x_{32}, x_{33} \\ \cdot \ \cdot \ \cdot \ \cdot \end{cases} \tag{15.1}$$

where, in each row all the variables are real-valued and stochastically independent. The partial sums

$$S_n = x_{n1} + x_{n2} + \dots + x_{nn} \tag{15.2}$$

are studied in terms of their probability distributions F_n, and a typical result is that $F_n \to F$ weakly as $n \to \infty$. The limit F could be for example the distribution with all its mass at $x = 0$, the normal distribution, the Poisson distribution, or some other infinitely divisible distribution. To arrive at such results some condition must be imposed on the triangular array guaranteeing that the individual terms in a row be small. Such a condition would be, for example,

$$\lim_{n} \max_{k} P\{|x_{nk}| > \varepsilon\} = 0 \tag{15.3}$$

for any $\varepsilon > 0$.

A complete discussion of these limit theorems can be found in the definitive work by Gnedenko-Kolmogorov (1954).

If the condition that the x_k be stochastically independent is relaxed to (strict) stationarity, the ergodic theorems tell us that the limit of the left side of (15.1), suitable normalized, still exists. The limiting random variable need not be a constant however. It is a constant if the stationary stochastic process is ergodic, but not always otherwise.

If the values of the x_{nk} are not real but form some other algebraic structure than the real line, some limit theorems remain in force. Take for example, the law of large numbers

$$\lim_{n\to\infty} \frac{1}{n} \sum_{k=1}^{n} x_k = m = E(x_1) \quad \text{in probability,} \tag{15.4}$$

for x_k i.i.d. and with their values in a separable Banach space with a mean m as interpreted as a Pettis integral. See Grenander (1963), where extensions to other algebraic structures can also be found.

When we turn to the regular structures appearing in combinatory pattern theory *we must give up stochastic independence* from the very beginning. The reason is that the operations in the corresponding image algebras are usually not entire functions, only partial ones. An exception of limited interest is given by the free image algebras, but otherwise the independence should be replaced by the conditional independence studied in Volume I, Chapter 2.

Given a triangular array of random images from an image algebra $\mathcal{T}$

$$\begin{cases} I_{11} \\ I_{21}, I_{22} \\ I_{31}, I_{32}, I_{33} \\ \dots \end{cases} \tag{15.5}$$

and connections $\sigma_1, \sigma_2, \sigma_3, \ldots$, in accordance with the connection type Σ of $\mathcal{T}$, *when is it possible to prove convergence in distribution of the random images*

$$I_n = \sigma_n(I_{n1}, I_{n2}, \ldots I_{nn}) \tag{15.6}$$

to some limiting distribution P over $\mathcal{T}$?

Reasoning by analogy with the classical case we should assume that the I_{nk} are small in some probabilisitc sense. In an image algebra there will usually be many (partial) unit elements (see Chapter 3) and it is tempting to ask that the I_{nk} should have most of its probability mass close to some unit element if n is large.

The question raised above, which is a fundamental one in metric pattern theory, is at this time almost completely open, and we shall begin to shed some light on it by first examining a couple of special cases.

The first one is very simple. Consider the free image algebra $\mathcal{T}_\infty$ made up of half planes in R^2 as generators, and where the identification rule R identifies intersections of half planes. In other words the (pure) images are convex polygons. In this, as well as in the next case, the measurability questions cause no difficulty: the real difficulties lie in the analytic treatment of the limit problem.

Let us assume that with probability one all the "realizations" of the random set I are contained in the fixed square Q of finite area A. This assumption is not crucial and is introduced only for analytical convenience.

As the criterion of convergence we shall use the expected value of Lebesgue area of the symmetric difference of the two sets involved. We can then announce

Theorem 1. *Under the given conditions we have*

$$\sigma(I_1, I_2, \ldots I_n) \to I_{certain} \tag{15.7}$$

where $I_{certain}$ *consists of all points* z *such that*

$$P(z \in I) = 1. \tag{15.8}$$

Proof: Consider the i.i.d. sequence of random images $I_1, I_2, \ldots$ and with the associated indicator functions $I_1(z), I_2(z), \ldots$ where $z \in R^2$. Recall that since $\mathcal{T}$ is free we have only one type of connector. Also one should note that the present regularity induces no stochastic dependence upon the I_k; this of course simplifies the analysis a good deal. Since R identifies sets by intersection we have simply

$$I^n = \sigma(I_1, I_2, \ldots I_n) = \bigcap_{k=1}^{n} I_k \tag{15.9}$$

or, expressed in indicator functions

$$I^n(z) = \prod_{k=1}^{n} I_k(z). \tag{15.10}$$

Obviously $I^n \supseteq I_{certain}$. The measure of the symmetric difference is then $m(I^n - I)$ and we get for its expected value

$$E[m(I^n - I_{certain})] = \int_{z \in Q} E[I^n(z) - I_{certain}(z)]dz. \tag{15.11}$$

But

$$\begin{aligned} E[I^n(z)] &= E\left[\prod_{k=1}^{n} I_k(z)\right] = \prod_{k=1}^{n} E[I_k(z)] = \{E[I(z)]\}^n \\ &= P^n(z \in I). \end{aligned} \tag{15.12}$$

Since Q has finite Lebesgue measure we can appeal to the theorem of bounded convergence. The integrand in (15.11), expressed as in (15.12), tends to zero iff $P(z \in I) = 1$, which proves the assertion.

Remark 1. In this almost trivial case we have not asked that the individual "terms" I_k be probabilistically close to the unit element, which here is R^2 itself. If we do this we can get more informative limit theorems, one of which was given in Volume I, pp. 213-217, and this deserves to be explored in greater depth.

Remark 2. The limiting element $I_{certain}$ is a convex set but not necessarily a polygon. Hence it need not belong to $\mathcal{T}_\infty$, only to the completion of this space; see the last paragraph of Chapter 4.

We now turn to a more difficult case, still very special, but of greater interest.

Let the generators consist of linear functions over finite intervals $[a,b]$, $a \leq b$, so that it can be completely specified by its in-bond

$$b_{in} = (a,f(a)) \tag{15.13}$$

and out-bond

$$b_{out} = (b,f(b)). \tag{15.14}$$

With Σ = LINEAR, ρ as EQUAL, and with R identifying functions we get the image algebra of continuous linear splines. A conditional unit has $a = b$, $f(a) = f(b)$.

Here we have as connectors $\sigma(\cdot,\cdot)$ concatenation to the right or to the left; say that we choose the first alternative.

In this setting a law of large numbers could assume the following appearance, to mention just one possibility. Consider the triangular array of random images (15.5) where $I_{n\nu}$ has

$$\begin{cases} b_{in} = (\frac{\nu-1}{n}, x_\nu) \\ b_{out} = (\frac{\nu}{n}, x_{\nu+1}) \end{cases} \tag{15.15}$$

with $\nu = 1,2,\ldots n$. Recall that we want the individual $I_{n\nu}$ to contract close to a unit element. We now treat $x_\nu, x_{\nu+1}$ as random, say over the unit square, following the procedure of Volume I, Chapter 2. Say that all $(x_\nu, x_{\nu+1})$ have a Q-measure given by a density

$$q(u,v) = a(u)b_n(v|u) \tag{15.16}$$

which we have written in terms of a marginal density $a(u)$ for u and a conditional density b_n of v given u. Of course we need not have the densities defined relative to Lebesgue measure (as we do here) but relative to some other fixed measure. Also we could have let the first components of b_{in} and b_{out} be random but we wanted as clear cut a case as possible and therefore avoided this here. Also $a(u)$ could have been allowed to depend upon n.

Then the P measure over the vector $x = (x_1, x_2, \dots x_{n+1})$ will be given by a density

$$p_n(x_1, x_2, \dots x_{n+1}) = C_n^{-1} \prod_{\nu=1}^{n} q(x_\nu, x_{\nu+1}) \tag{15.17}$$

where C_n is a normalizing constant.

As $n \to \infty$ we shall let $b_n(v|u)$ contract around u, and we shall assume at this time that this happens in such a way that

$$\lim_{n\to\infty} \frac{1}{\varepsilon} \int_{|v-u|>\varepsilon} b_n(v|u)\,dv = 0 \tag{15.18}$$

uniformly in u and ε.

Also assume $\underline{a}$ to be a positive C_2-function and denote by ξ the x-value for which $a_{max} = \max_x a(x)$ is attained. For the moment ξ shall be assumed to be unique and with $a''(\xi) \neq 0$.

Theorem 2. *Under the given conditions we have*

$$I^n = \sigma(I_{n1}, I_{n2}, \dots I_{nn}) \to I_o \tag{15.19}$$

where $I_o(x) \equiv a_{max}$, *and the convergence is interpreted in expected sup norm.*

Proof: Introduce the function

$$\phi_n(x_1, x_2, \dots x_{n+1}) = a(x_1) \prod_{\nu=1}^{n} b_n(x_{\nu+1}|x_\nu). \tag{15.20}$$

It is a frequency function in R^{n+1} and therefore defines a probability measure Φ_n. The probability measure P_n for all the bonds can be written in terms of its frequency function p_n as

$$p_n(x_1,x_2,\ldots x_{n+1}) = C_n^{-1}\Phi_n(x_1,x_2,\ldots x_{n+1})\prod_{\nu=2}^{n} a(x_\nu) \qquad (15.21)$$

with

$$C_n = \int_{R^{n+1}} \prod_{\nu=2}^{n} a(x_\nu)\Phi_n(dx). \qquad (15.22)$$

The measure Φ_n describes the distribution of all the bonds if we had given x_1 the distribution with frequency function $a(\cdot)$ and all the rest made Markovian with the transition density $b(\cdot|\cdot)$. Given an arbitrary positive c we have for fixed $x_1 \in [0,1]$

$$\begin{aligned}\Phi_n(E_n^c) &\leq \sum_{\nu=1}^{n} \Phi_n[|x_{\nu+1}-x_\nu| > c/n^2] \\ &\leq n \max_u \int_{|v-u|>c/n^2} b_n(v|u)dv\end{aligned} \qquad (15.23)$$

for the event

$$E_n = \{|x_{\nu+1}-x_\nu| \leq c/n^2, \quad \nu = 1,2,\ldots n\}. \qquad (15.24)$$

According to Eq. (15.18), the right hand side of (15.23) tends to zero in such a way that

$$\lim_{n\to\infty} n[1-\Phi^n(E_n)] = 0. \qquad (15.25)$$

Writing, with $a_{max} = \max_{0\leq x\leq 1} a(x)$,

$$\alpha_n(x) = a_{max}^{-n+1} \prod_{\nu=2}^{n} a(x_\nu) \qquad (15.26)$$

it is obvious that $0 \leq \alpha_n(x) \leq 1$. With this function the normalizing constant (15.22) can be expressed via the relation

$$C_n' = \sqrt{n}\, C_n\, a_{max}^{-n+1} = \sqrt{n}\int_{E_n} + \sqrt{n}\int_{E_n^c} \alpha_n(x)\Phi_n(dx). \qquad (15.27)$$

Since the integrand $\alpha_n(x)$ is uniformly bounded (15.25) tells us that the second term tends to zero. To get the limit of the first term let us note that for $x \in E_n$ we have $|x_\nu - x_1| \leq c/n$ so that, since $\underline{a}$ has a bounded derivative

$$a(x_\nu) = a(x_1) + 0(\tfrac{c}{n}); \tag{15.28}$$

hence, if c is small enough, the ratio

$$\frac{\prod_2^n a(x_\nu)}{a^{n-1}(x_1)} \tag{15.29}$$

differs from 1 by an arbitrary small factor. Hence, as $n \to \infty$,

$$C_n' \sim \sqrt{n} \int_{E_n} \left[\frac{a(x_1)}{a_{max}}\right]^{n-1} \Phi^n(dx). \tag{15.30}$$

But the marginal measure of x_1 for Φ^n is given by just the density $a(x_1)$, so that

$$\begin{aligned} C_n' &\sim \sqrt{n} \left\{\int_0^1 \frac{a^n(x_1)}{a_{max}^{n-1}} dx_1 - \int_{E_n^c} \left[\frac{a(x_1)}{a_{max}}\right]^{n-1} \Phi^n(dx)\right\} \\ &\sim \sqrt{n} \int_0^1 \frac{a^n(x_1)}{a_{max}^{n-1}} dx_1 \sim \sqrt{\frac{2\pi\, a_{max}}{A}}; \quad A = -a''(\xi_0). \end{aligned} \tag{15.31}$$

We have thus shown that

$$C_n \sim \sqrt{\frac{2\pi}{nA}}\, a_{max}^{n-1/2}. \tag{15.32}$$

Hence for a fixed $\varepsilon > 0$

$$P\{|x_\nu - \xi| > \varepsilon\} = C_n^{-1} \int_{|x_\nu - \xi| > \varepsilon} \prod_2^n a(x_\nu) \Phi^n(dx). \tag{15.33}$$

But (15.33) can be written as

$$C_n^{-1}(I' + I'') \tag{15.34}$$

where

$$I' = \int_{|x_n - \xi| > \varepsilon \text{ and } F_n} \prod_2^n a(x_\nu) \Phi^n(dx) \tag{15.35}$$

and

$$I'' = \int_{|x_n-\xi|>\varepsilon \text{ and } F_n^c} \prod_2^n a(x_\nu)\Phi^n(dx) \tag{15.36}$$

with

$$F_n = \{|x_\nu - x_{\nu+1}| < cn^{-3};\ \nu = 1,2,\ldots n\}. \tag{15.37}$$

However

$$I' \le \int_{|x_\nu-\xi|>\varepsilon-cn^{-3},\forall\nu} \prod_2^n a(x_\nu)\Phi^n(dx)$$

$$\le (1-\delta)^{n-1} a_{max}^{n-1} \tag{15.38}$$

and

$$I'' \le \int_{F_n^c} a_{max}^{n-1}\Phi^n(dx) \le a_{max}^{n-1} n \max_u \int_{|v-u|>cn^{-3}} b_n(v|u)du \tag{15.39}$$

so that the left side of (15.33) tends to zero.

However the sup norm of the difference

$$||I^n - I_0|| = \max_\nu |x_\nu - \xi| \tag{15.40}$$

so that the stated convergence (15.19) holds in the sense specified. Q.E.D.

Now let us assume instead that the maximum of $a(x)$ is not unique and $a(x)$ attains a_{max} at $x = \xi_1, \xi_2, \ldots \xi_m$ with $a''(\xi_i) \ne 0$, $i = 1,2,\ldots m$. We can then verify, with the same method as above, that (15.25) holds, but it is no longer possible to show that x_1 will be close to a particular value with large probability. Instead of (15.32) we now get

$$C_n \sim a_{max}^{n-1/a} \sqrt{\frac{2\pi}{n}} \sum_{i=1}^m \frac{1}{\sqrt{-a''(\xi_i)}} \tag{15.41}$$

and x_1 will no longer have a one-point limiting distribution. Instead its limiting distribution will be given by

$$\lim_{n\to\infty} P(x_1=\xi_j) = \frac{\dfrac{1}{\sqrt{-a''(\xi_j)}}}{\sum_{i=1}^{m} \sqrt{-a''(\xi_i)}} = P_j \tag{15.42}$$

or we get, after a short argument,

Theorem 3. *Under the same conditions as in Theorem 2, except that* $a(x)$ *achieves its maximum at* $\xi_1,\xi_2,\ldots\xi_r$ *with* $a''(\xi_i) \neq 0$, *we have for an arbitrary but fixed* $x \in [0,1]$

$$\lim_{n\to\infty} P[I(x) = \xi_j] = p_j \tag{15.43}$$

where p_j *is given in* (15.42).

Hence we cannot claim that the random images I^n settle down close to a fixed image I_0, but rather that they behave distribution-wise as the random image $I(x) = \xi$ where ξ is a random variable as above. This is similar to the non-ergodic situation for stationary stochastic processes.

Remark 3. If $a(x)$ attains a_{max} on an interval in $[0,1]$ we conjecture that the limiting distribution of x_1 will be uniform over this interval. We have not proved this.

What happens when the condition (15.18) does not hold? We shall only make some heuristic remarks. In other words, let us examine the case when v conditioned by u is not that close to u. One possibility of some interest is when v is very close, not to u, but to $u+g(u)/n$, where $g(u)$ is a given function, so that $|v-u|$ in (15.18) is replaced by $|v-u-g(u)/n|$. Let us assume that g is continuous and positive for $0 \leq x < 1$, $g(1) = 0$, and with

$$\int_0^1 \frac{du}{g(u)} = +\infty, \tag{15.44}$$

a condition whose role will be clear later on. With the same proof we can show that with large probability and for any $c > 0$

$$x_{\nu+1} - x_\nu - g(x_\nu)/n \tag{15.45}$$

will be smaller than c/n^2. Hence we can expect that

$$I^n(t) \sim \text{solution of the equation } I'(t) = g(I). \tag{15.46}$$

Let us therefore introduce the function

$$\mu(x) = \int_0^x \frac{du}{g(u)} \tag{15.47}$$

which is monotonically increasing from 0 to $+\infty$ as x varies from 0 to 1. Hence μ^{-1} is well defined and we would expect

$$I^n(t) \sim \mu^{-1}\{t+\mu[I^n(0)]\}. \tag{15.48}$$

Note that the inverse function μ^{-1} takes $t+\mu[I^n(0)]$ back to the interval $[0,1]$ as required in the description of the image algebra. That is why (15.44) was introduced.

Here $I^n(0)$ would play the role of x_1 previously, and we can make a guess about its limiting distribution. We observe that the method of Laplace, that we used to derive (15.32), may be used with some modification. Indeed, we would expect to get a one-point (or several-points when the maximum of $a(x)$ is attained at several points) distribution at a value ξ for which

$$\prod_2^n a\,\{\mu^{-1}[(\frac{\nu-1}{n} + \mu(\xi)]\} \tag{15.49}$$

is a maximum. Hence, asymptotically for $n \to \infty$, we should look for a ξ that maximizes

$$\int_0^1 \ell n\, a\{\mu^{-1}[t+\mu(\xi)]\}dt$$
$$= \int_\xi^{\mu^{-1}[1+\mu(\xi)]} \ell n\, a(u) \frac{du}{g(u)} = \int_\xi^{h(\xi)} \ell n\, a(u) \frac{du}{g(u)} \tag{15.50}$$

where $h = h(\xi)$ is the root of the equation

$$1 = \int_\xi^h \frac{du}{g(u)} . \tag{15.51}$$

Whether this can be proved rigorously is another matter that we shall not go into here.

Instead we shall study a variation of the same problem but using an analytic method that promises to be of greater scope than just the special case to which it is applied. Let the generators be linear functions g: $a+b$; $x \in [0,1]$. To make their support tend to zero requires that we shrink the function, say to g: $anx+b$; $x \in [0, \frac{1}{n}]$.

The comparison between the present case and the classical limit theorems in probability has been summarized in Table 15.1.

Table 15.1

	classical case	present case
ρ	TRUE	continuity condition
Σ	LINEAR	LINEAR
g	real number	linear function
$\sigma(I_1,I_2)$	add real numbers	concatenate functions
S	multiply by constant	translate function
unit	zero on R	one point function
normalization	divide by n	shrink function

Let all generators have supporting intervals of length 1 before normalization as described. Say that the bond values $\beta_1 = y_1$ and $\beta_2 = y_2$ of g have a joint probability distribution Q over B × B. If B is finite we can represent Q by a square matrix

$$Q = \{g_{k\ell};\ \ k,\ell = 1,2,\ldots r\}. \tag{15.52}$$

Since we deal with symmetric regularity it seems natural to assume that Q is symmetric, and this will be done.

The assumption that B be finite is of course very restrictive. However, it is believed that the method to be developed can be applied to the case when B is only assumed to be compact, for example a finite interval.

Note that Q *is not a stochastic matrix* but the sum of all its entries is equal to 1. If B is finite it will mean no significant loss of generality to let the bond values be the integers: B = N = (1,2,...r). With $c = \sigma(g_1,g_2,\ldots,g_n) \in \mathscr{C}_n(\mathscr{R})$ and $\beta_1(g_\nu) = y_\nu$, $\beta_2(g_\nu) = y_{\nu+1}$ we get the probability over $\mathscr{C}_n(\mathscr{R})$

$$P(c) = Z_n^{-1} q_{y_1y_2}\cdot q_{y_2y_3}\cdot q_{y_3y_4}\ \cdots\ q_{y_ny_{n+1}} \tag{15.53}$$

with the partition function

$$Z_n = \sum_{y's} q_{y_1y_2}\cdot q_{y_2y_3}\cdot q_{y_3y_4}\ \cdots\ q_{y_ny_{n+1}}. \tag{15.54}$$

The resulting image $\hat{I}_n$ represents a linear spline function on [0,n], and after normalization we get an image $\hat{I}_n$ defined on [0,1]. We cannot just apply Markov chain theory to get the limit theorem: Q does not represent a transition probability matrix.

What can be claimed about the limiting probabilistic properties of $\hat{I}_n$? Since $\hat{I}_n$ will have a more chaotic form as n increases we cannot hope to prove that it settles down to any fixed image as was the case earlier in this section. Instead we shall show that the probability measure P_n of $\hat{I}_n$ converges, in some sense, to some measure P_∞.

To make this statement precise we must specify the mode of convergence employed. We shall identify $\hat{I}_n$ with the density of a signed bounded measure on $[0,1]$. In other words we operate in $BV([0,1])$. We shall show that P_n converges in probability according to the weak* topology, so that for any continuous f we have

$$\int_0^1 f(x)\hat{I}_n(x)dx \to \int_0^1 f(x)I_\infty(x)dx \tag{15.55}$$

in probability where the random image $I_\infty(\infty)$ will be defined below.

To gain some intuition into the problem let us first consider two simple special cases representing the extreme situations.

<u>Case 1</u>. If the bonds of any generator are exactly equal with probability one, $\beta_1 = \beta_2 = \beta$, then if $P(\beta=k) = p_k$ we get only constant images over $[0,1]$, so that

$$P\{\hat{I}_n \equiv k\} = Z_n^{-1}\, p_k^n, \quad k = 1,2,\ldots\,. \tag{15.56}$$

Hence

$$Z_n = \sum_{k=1}^{r} p_k^n \tag{15.57}$$

which decreases exponentially as $n \to \infty$ in such a way that we can expect

$$P\{I_\infty \equiv k\} = \begin{cases} \frac{1}{\#(K)} & \text{if } k \in K \\ 0 & \text{else} \end{cases} \tag{15.58}$$

where

$$K = \{k \mid \max_{\ell} p_\ell = p_k\}. \tag{15.59}$$

In this case

$$Q = \text{diag}[p_1, p_2, \ldots, p_r] \tag{15.60}$$

and the $\hat{I}_n$ are indeed constant. Since the p_ν are the eigenvalues of Q it will become clear that their largest value corresponds to λ_1 in the statement of the next theorem.

Case 2. The opposite extreme is when the two bond values of a generator are stochastically independent. The images I_n will then become highly irregular as n tends to infinity. We then get, if $P(\beta=k) = g_k$ for both in- and out-bonds,

$$P\{\hat{I}_n(\tfrac{\nu-1}{n}) = y_\nu;\ \nu = 1,2,\ldots n+1\} = Z_n^{-1} q_{y_1} q_{y_2}^2 q_{y_3}^2 \cdots q_{y_n}^2 q_{y_{n+1}} \tag{15.61}$$

so that

$$Z_n = \left(\sum_1^r q_k^2\right)^{n-1}. \tag{15.62}$$

Hence the ordinates $I_n(\frac{\nu-1}{n})$; $\nu = 2,3,\ldots n$; will be stochastically independent and with the marginal distribution given by the probabilities

$$q_k^2 / \left(\sum_1^r q_k^2\right). \tag{15.63}$$

For $\nu = 1$ and $n+1$ the marginal probabilities remain q_k. In this case

$$Q = \{q_k q_\ell;\ \ k,\ell = 1,2,\ldots r\} \tag{15.64}$$

which is singular with the only non-zero eigenvalue Σq_k^2. Since the ordinates $\hat{I}(\frac{\nu-1}{n})$ are i.i.d. for $\nu = 2,\ldots n$ it is reasonable to expect I_∞ to take the value

$$\left(\sum_1^r k q_k^2\right) / \left(\sum_1^r q_k^2\right) \tag{15.65}$$

with probability one; recall the mode of convergence given in Eq. (15.55).

After considering these two special instances of Q let us introduce the vectors $e = \mathrm{col}[1,1,\ldots 1] \in R^r$ and e_k as unit column vectors along the kth coordinate axis. We shall prove

Theorem 4. *Let the largest eigenvalue* λ_1 *of* Q *be associated with the projection operator* P *and assume that all other eigenvalues are smaller in absolute value. Then* $\hat{I}_n$ *converges in probability according to the weak* distribution to* $I_\infty(x)$

$$P\{I\ (x) \equiv \mu_\nu\} = (e^T P z_\nu)^2 / \|Pe\|^2 \tag{15.66}$$

where z_ν *is an eigenvector of* PNP *with eigenvalue* μ_ν *and* $N = \mathrm{diag}[1,2,3,\ldots r]$.

Proof: We shall plan the proof, which is long and technical, as follows. First assume that λ_1 is simple. Classical Frobenius theory tells us that λ_1 must be positive as well as one of its eigenvectors, appropriately normalized. We shall show that the probabilisitic behavior of the ordinates $\hat{I}_n(\frac{\nu-1}{n})$ converges and use this to prove the theorem; in this case by a perturbation argument. We then let λ_1 be multiple but shall study the ordinates by a different method. After that we show that $\hat{I}_n$ as a whole converges in the mode stated above.

It is easy to find the asymptotic expression for the partition function Z_n. Indeed

$$Z_n = \Sigma q_{y_1y_2}q_{y_2y_3} \cdots q_{y_ny_{n+1}} = e^TQ^ne \tag{15.67}$$

where the summation is over all combinations of y-values with subscripts from 1 to r. Writing the spectral decomposition of Q as

$$Q = O^T\Lambda O \tag{15.68}$$

in terms of the orthogonal $r \times r$ matrix O and the diagonal matrix $\Lambda = \text{diag}[\lambda_1,\lambda_2,\ldots\lambda_r]$, we shall assume that the eigenvalues have been enumerated so that $\{|\lambda_k|\}$ is a non-increasing sequence. Remember that $O^Te_k = v_k$ is an eigenvector of Q.

Consider the largest eigenvalue λ_1 which is assumed to be simple for the moment. There is an associated eigenvector v_1, all whose components are non-negative so that $c = (e,v_1) > 0$. Then

$$\begin{aligned} Z_n &= e^TQ^ne = e^TO^T\Lambda^nOe \\ &= \lambda_1^ne^TO^T \text{diag}\left[1,\left(\frac{\lambda_2}{\lambda_1}\right)^n,\left(\frac{\lambda_3}{\lambda_1}\right)^n,\ldots\right]Oe \\ &\sim \lambda_1^ne^TO^Te_1e_1^TOe = \lambda_1^n\, e^Tv_1v_1^Te \\ &= c^2\lambda_1^n \quad \text{as} \quad n \to \infty,\ c = e^Tv_1. \end{aligned} \tag{15.69}$$

Introduce now the *partial sum of bond values* for fixed $\alpha \in (0,1]$

$$Y_n = \frac{1}{n}\sum_{k=1}^{m} y_k, \quad m = [\alpha n] \tag{15.70}$$

and its characteristic function

$$\phi_n(z) = E\left[e^{iY_n z}\right]. \tag{15.71}$$

Using (15.53) this expected value can be expressed

$$\phi_n(z) = Z_n^{-1} \sum_{y's} e^{i\frac{z}{n}y_1} q_{y_1y_2} e^{i\frac{z}{n}y_2} q_{y_2y_3} \cdots e^{i\frac{z}{n}y_m} q_{y_my_{m+1}} \cdot \tag{15.72}$$

$$\cdot\, q_{y_{m+1}y_{m+2}} q_{y_{m+2}y_{m+3}} \cdots q_{y_ny_{n+1}}$$

This can be rewritten as

$$\phi_n(z) = Z_n^{-1} \sum_{y's} p^{(n)}_{y_1y_2} p^{(n)}_{y_2y_3} \cdots p^{(n)}_{y_my_{m+1}} q_{y_{m+1}y_{m+2}} \cdots q_{y_ny_{n+1}} = Z_n^{-1}\sum \tag{15.73}$$

with

$$p^{(n)}_{k\ell} = e^{i\frac{z}{2n}k} q_{k\ell} e^{i\frac{z}{2n}\ell}; \quad P(n) = \{p^{(n)}_{k\ell};\ k,\ell = 1,2,\ldots r\}. \tag{15.74}$$

except for the first and mth factor in the sum where

$$\begin{cases} p^1_k(n) = e^{i\frac{z}{n}k} q_{k\ell} e^{i\frac{z}{n}\ell}; \ P^1(n) = \{p^1_{k,\ell}(n);\ k,\ell=1,2,\ldots r\} \\ p^m_{k\ell}(n) = e^{i\frac{z}{n}k} q_{k\ell} e^{i\frac{z}{n}}; \ P^m(n) = \{p^m_{k,\ell}(n);\ k,\ell=1,2,\ldots r\} \end{cases} \tag{15.75}$$

To evaluate the sum in (15.73) we express it as a matrix-vector product

$$\sum = e^T P^1(n) P^{m-2}(n) P^{(m)}(n) Q^{n-m}, \tag{15.76}$$

or

$$\Sigma = \sum_{k=1}^{r} a_k^{(n)} b_k^{(n)}. \tag{15.77}$$

Recalling that we introduced the vector e_k with zeroes everywhere except in the kth position (15.77) can be expressed as

$$\begin{cases} a_k(n) = e^T P^1(n) P^{m-2}(n) P^{(m)}(n) e_k \\ b_k(n) = e_k^T Q^{n-m} e \end{cases} \tag{15.78}$$

The $a_k(n)$ and $b_k(n)$ can be evaluated asymptotically as follows. The second one is easily obtained with exactly the same method as for Z_n, and we get

$$b_k(n) = e_k^T O^T \Lambda^{n-m} O e = \lambda_1^{n-m} e_k^T O^T e_1 e_1^T O e [1+\text{exponentially decreasing terms}]. \tag{15.79}$$

The leading term in (15.79) is then

$$\lambda_1^{n-m}(e_k, v_1)(e, v_1). \tag{15.80}$$

For $a_k(n)$ a closer examination is required. We have

$$P(n) = D(n)QD(n);\ D(n) = \operatorname{diag}\left[e^{i\frac{z}{2n}}, e^{i\frac{z}{2}2}, e^{i\frac{z}{2}3}; \ldots e^{i\frac{z}{2}r}\right]. \tag{15.81}$$

This implies, for large n,

$$D(n) = I + \frac{iz}{2n} N + O(n^{-2}), \tag{15.82}$$

so that in terms of a new matrix A

$$P(n) = Q + \frac{iz}{2n}[NQ+QN+O(n^{-2})] = Q + \frac{iz}{2n}A + O(n^{-2}),\ n\to\infty. \tag{15.83}$$

Similarly we have

$$\begin{cases} P^{(1)}(n) = Q + O(\frac{1}{n}) \\ P^{(m)}(n) = Q + O(\frac{1}{n}). \end{cases} \tag{15.84}$$

These relations lead naturally to employ classical perturbation technique. Let the absolutely largest eigenvalue of $P(n)$ be $\lambda_1(n)$ with an appropriately normed eigenvector $v_1(n)$ and with similar notation for the smaller eigenvalues, and associated eigenvectors. Then using a well known formula for perturbation calculations

$$\begin{cases} \lambda_1(n) = \lambda_1 + \frac{iz}{2n} v_1^T A v_1 + O(n^{-2}) \\ v_1(n) = v_1 + O(n^{-1}) \end{cases} \tag{15.85}$$

where the second relation could easily be made sharper, but this is not required. Hence

$$a_k(n) = e^T P^{(1)}(n) O^T(n) \Lambda^{m-2}(n) O(n) P^{(m)}(n) e_k \tag{15.86}$$

with

$$\Lambda(n) = \text{diag}[\lambda_1(n), \lambda_2(n), \ldots \lambda_r(n)] \tag{15.87}$$

and

$$O^T(n) e_k = v_k(n). \tag{15.88}$$

Then

$$\lambda_1^{2-m} a_k(n) = e^T P^{(1)}(n) O^T(n) \text{diag}\left[\left(\frac{\lambda_1(n)}{\lambda_1}\right)^{m-2}, \left(\frac{\eta_2(n)}{\lambda_1}\right)^{m-2}, \ldots\right] O(n) P^{(m)}(n) e_k \tag{15.89}$$

and with (15.85) substituted we obtain

$$\left[\frac{\lambda_1(n)}{\lambda_1}\right]^{m-2} = \left[1 + \frac{iz}{2n} v_1^T A v_1 + O(n^{-2})\right]^{m-2}. \tag{15.90}$$

Making n tend to infinity and using (15.90) this gives us

$$\lim_{n\to\infty} \lambda_1^{2-m} a_k(n) = e^T Q O^T \operatorname{diag}\left[e^{\frac{i\alpha v_1^T}{2\lambda_1}Av_1}, 0, 0, \ldots\right] O Q e_k$$

$$= e^{\frac{i\alpha vT}{2\lambda_1}Av_1} e^T Q O^T e_1 e_1^T O Q e_k \tag{15.91}$$

$$= e^{\frac{i\alpha v_1^T}{2\lambda_1}Av_1} e^T Q v_1 v_1^T Q e_k .$$

But $Qv_1 = \lambda_1 v_1$ so that the above reduces to

$$\lim_{n\to\infty} \lambda_1^{2-m} a_k(n) = e^{\frac{i\alpha v_1^T}{2\lambda_1}Av_1} \lambda_1^2 (e_k, v_1)(e, v_1). \tag{15.92}$$

Inserting into (15.77) together with (15.80) this leads to

$$\lim_{n\to\infty} \lambda_1^{-n} \sum_{k=1}^{r} a_k(n) b_k(n)$$

$$= \sum_{k=1}^{r} \left[\lim_{n\to\infty} \lambda_1^{-m} a_k(n)\right]\left[\lim_{n\to\infty} \lambda_1^{m-n} b_k(n)\right] \tag{15.93}$$

$$= e^{i \frac{\alpha}{2} v_1^T A v_1} \sum (e_k, v_1)^2 (e, v_1)^2 .$$

Using Parseval's relation the above sum can be evaluated to be $(e, v_1)^2 = c^2$.

Now we go back to the asymptotic expression for Z_n given in (15.69) and obtain directly from (15.73), (15.93), (15.77) and (15.70)

$$\lim_{n\to\infty} \phi_n(z) = e^{\frac{i\alpha v_1^T}{2\lambda_1} Av_1} \tag{15.94}$$

which is the Fourier transform of the measure concentrated in the constant $\frac{\alpha}{2\lambda_1} v_1^T A v_1$. Using again the fact that v_1 is an eigenvector of Q and relation (15.83) defining A we get the constant

$$\alpha v_1^T N v_1 = \alpha \sum_{k=1}^{r} k\, v_{1k}^2 = \alpha V, \quad V = \sum_{k=1}^{r} k v_{1k}^2 \tag{15.95}$$

where $v_1 = \text{col}[v_{1k};\ k = 1,2,\ldots r]$.

We have thus shown that

$$Y_n \to \alpha \sum_{k=1}^{r} k v_{1k}^2 \tag{15.96}$$

in probability as $n \to \infty$.

Let us now consider an arbitrary real valued continuous function f on $[0,1]$ and form the integral corresponding to the weak* topology in $B([0,1])$ as described above

$$\hat{I}_n(f) = \int_0^1 \hat{I}_n(x) f(x)\, dx. \tag{15.97}$$

Approximate f from above by ϕ and from below by ψ

$$\psi(x) \le f(x) \le \phi(x) \tag{15.98}$$

where ϕ and ψ are finite functions (constant on intervals). We can then write

$$\begin{cases} \phi(x) = \sum_{\nu} \phi_\nu 1_{\alpha_\nu}(x) \\ \psi(x) = \sum_{\beta} \psi_\nu 1_{\alpha_\nu}(x) \end{cases} \tag{15.99}$$

using the indicator function 1_α for the set $[0,\alpha] \subseteq [0,1]$. But we know that (15.96) holds, which implies

$$\hat{I}_n(1_\alpha) \to \alpha V \tag{15.100}$$

and hence

$$\begin{cases} \hat{I}_n(\phi) \to V \sum_{\nu} \phi_\nu \alpha_\nu = \hat{I}_\infty(\phi) \\ \hat{I}_n(\psi) \to V \sum_{\nu} \psi_\nu \alpha_\nu = \hat{I}_\infty(\psi) \end{cases} \tag{15.101}$$

with the image

$$\hat{I}_\infty(x) \equiv V. \tag{15.102}$$

Of course V need not be an integer so that $\hat{I}_\infty$ need not be in $\mathcal{T}$: we have to extend the image algebra. This is just as in the classical law of large numbers: the limit need not be in the support of the original random variables. Since f is continuous we can choose ϕ and ψ such that $\phi(x)-\psi(x) \leq \varepsilon$ uniformly in x for any given $\varepsilon > 0$. This implies that for any $f \in C[(0,1)]$

$$\hat{I}_n(f) \to I_\infty(f), \quad n \to \infty \quad \text{in distribution.} \tag{15.103}$$

In this case λ_1 is simple with the projection operator $P = v_1 v_1^T$ so that

$$PNP = v_1 v_1^T N v_1 v_1^T \tag{15.104}$$

is singular with the only non-zero eigenvalue

$$\mu_1 = v_1^T N v_1 = V \tag{15.105}$$

associated with the eigenvector v_1. The corresponding probability given by (15.66) reduces to

$$\frac{(e^T P v_1)^2}{\|Pe\|^2} = \frac{(e^T v_1)^2}{(e^T v_1)^2} = 1 \tag{15.106}$$

so that the theorem has been shown to hold in the case λ_1 being simple.

To proceed to multiple λ_1 we first note that (15.69) is immediately generalized to $\lambda_1^{-n} Z_n \sim \|Pe\|^2$ by essentially the same argument.

To deal with the case when λ_1 has multiplicity $s > 1$, perturbation methods are less convenient since the eigenvalue λ_1 can split up under perturbation, introducing a complication that can be avoided by using the following idea instead which offers an alternative approach to our problem.

We need high powers of $P(n)$ to calculate $a_k(n)$ in (15.86). But this leads to an expression like

$$\begin{aligned}(Q + \frac{B}{n})^n &= Q^n + \frac{1}{n}(Q^{n-1}B+Q^{n-2}BQ +\dots+ BQ^{n-1})\\ &\quad + \frac{1}{n^2}(Q^{n-2}B^2+Q^{n-3}BQB +\dots)\\ &\quad + \dots\\ &= S_0 + \frac{1}{n} S_1 + \frac{1}{n^2} S_2 + \dots\end{aligned} \tag{15.107}$$

It is clear that

$$\lambda_1^{-n} S_0 \to P. \tag{15.108}$$

Similarly "most" of the terms in S_1 are of the form $Q^s B Q^t$ with s and t large so that

$$\frac{1}{n} \lambda_1^{-n} S_1 \to PBP/\lambda_1. \tag{15.109}$$

Proceeding this way we get

$$\frac{1}{n} \lambda_1^{-n} S_\nu \to P(BP)^\nu/(\nu!\lambda_1^\nu). \tag{15.110}$$

We can also dominate the norms of the terms in (15.107) by using

$$\frac{\lambda_1^{-n}}{n^\nu} \|S_\nu\| \le \frac{\|B\|^\nu}{\lambda_1^\nu} \binom{n}{\nu} \frac{1}{n^\nu} \le \frac{\|B\|^\nu}{\nu!\lambda_1^\nu} \tag{15.111}$$

so that a standard uniformity argument leads to (see Notes B) the useful relation

$$\lim_{n\to\infty} (Q + \frac{B}{n})^n = P \exp \frac{(BP)}{\lambda_1} = P \exp \frac{(PBP)}{\lambda_1} P. \tag{15.112}$$

In the present case $B = \frac{iz}{2} A + O(n^{-1})$ so that

$$\begin{aligned}\lim_{n\to\infty} \lambda_1^{-m} a_k(n) &= \lambda_1^{-2} e^T QP \exp(i \frac{\alpha z}{2}{}_1 PAP) PQe_k\\ &= e^T P \exp(i \frac{\alpha z}{2\lambda_1} PAP) Pe_k.\end{aligned} \tag{15.113}$$

For $b_k(n)$ we get as before

$$\lim_{n\to\infty} \lambda_1^{m-n} b_k(n) = e_k^T Pe = e^T Pe_k \tag{15.114}$$

so that, again using Parseval's relation and (15.73), (15.69), and the fact that P is self-adjoint,

$$\begin{aligned}\phi(z) &= \lim_{n\to\infty} \phi_n(z) = \|Pe\|^{-2}\cdot(Pe, Q \exp(i \tfrac{\alpha z}{2\lambda_1} PAP)Pe) \\ &= \|Pe\|^{-2}\cdot(e,P \exp(i \tfrac{\alpha z}{2\lambda_1} PAP)Pe).\end{aligned} \tag{15.115}$$

This expression can be simplified if we express the symmetric matrix $A = NQ + QN$ in spectral form via the relation

$$PAP = \sum_{\nu=1}^{r} \mu_\nu' A_\nu;\ A_\nu = \text{projection operator} \tag{15.116}$$

so that

$$\phi(z) = \|Pe\|^{-2} \sum_{\nu=1}^{r} e^T PA_\nu Pe\ e^{i \frac{\alpha z}{2\lambda_1} \mu_\nu'} \tag{15.117}$$

with

$$\begin{aligned}\pi_\nu &= \|Pe\|^{-2}\cdot e^T PA_\nu Pe = \|Pe\|^{-2}\cdot e^T PA_\nu^T A_\nu Pe \\ &= \|Pe\|^{-2}\cdot \|A_\nu PE\|^2 \geq 0\end{aligned} \tag{15.118}$$

we have

$$\sum_{\nu=1}^{n} \pi_\nu = \|Pe\|^{-2}\ e^T PPe = 1 \tag{15.119}$$

so that the π's are probabilities. We have thus shown that the limiting characteristic function is the Fourier transform of a probability measure with the probabilities π_ν at the points $\frac{\alpha}{2\lambda_1} \mu_\nu'$.

But

$$PAP = P(QN+NQ)P = PQNP + PNQP = 2PNP\lambda_1. \tag{15.120}$$

The possible values of I_∞ are then the eigenvalues of μ_ν of PNP. We can disregard eigenvalues $\mu_\nu = 0$ since N is bounded from below by I with the usual partial ordering of operators so that

$$x^T PNPx = 0 \Rightarrow Px = 0 \tag{15.121}$$

and from (15.118) we then get $\pi_\nu = 0$. If $\mu_\nu \neq 0$ we must have any eigenvector x_ν in the range of P.

The rest of the argument can be carried out just as for λ_1 simple. The proof of the theorem is complete. Q.E.D.

Remark 4. It is tempting to try to simplify the proof by reducing the problem to one for ordinary Markov chains. This is not easily done. Consider the transition probability matrix for y_{k+1} given y_k. It will depend upon k and n. It can be shown however, see Grenander (1978b), that this matrix converges as $n \to \infty$. Perhaps this could be used as the starting point for a simpler proof.

Remark 5. We do not know if the theorem also holds for the weak topology.

Remark 6. Laws of large numbers for other regular structures have not been studied yet; the next case could be Σ = SQUARE LATTICE.

5.16. Random dynamics for configurations

So far we have only studied static randomness in configuration space and the induced measures over image algebras. The perspective changes when we let the measure P_t on $\mathscr{L}(\mathscr{R})$ depend upon time t.

A configuration can be described by its content(c) = $\{g_1, g_2, \ldots g_n\}$ and its connector σ. The dynamics can, for example, let the content be fixed *while the connections are opened and closed over the set*

$$\mathscr{C}[g_1,g_2,\dots g_n] \subset \mathscr{C}(\mathscr{R}) \tag{16.1}$$

of all regular configurations that $\mathscr{R}$ admits built from the generators $g_1,g_2,\dots g_n$.

On the other hand we will sometimes encounter configurations where the connector σ is fixed, applicable to given generator skeletons $\gamma_1,\gamma_2,\dots\gamma_n$, but where the *choice of generators varies with time.* Of course one can also have combinations of these two cases: both content and connector vary dynamically.

All cases that have been studied so far, with continuous or discrete time, have formed a Markov process over the state space $\mathscr{C}(\mathscr{R})$. Even more: the process has usually been of birth and death type. This applied to generators so that on G there are given two intensities

$$\begin{cases} M : G \to [0,\infty) \\ \Lambda : G \to [0,\infty) \end{cases} \tag{16.2}$$

Here $M(g)dt$ denotes the probability that a particular generator $g \in \text{content}(c)$ will disappear from c in a time interval of length dt assuming that $\mathscr{R}$ admits this. Similarly $\Lambda(g)dt$ denotes the probability that a new generator g will be added to c during $(t,t+dt)$, but we must also describe if and how g should be connected to the old generators in c. For example we may specify that no new bond couple shall be closed. Or a particular connector shall be used to join g to c. This will be stated explicitly in each special case.

Some examples will make clear how this works. With G finite, Σ = LINEAR and ρ = EQUAL let us put $M \equiv 0$ and arbitrary Λ. Consider the regular configuration $c = \sigma(g_1, g_2, \ldots g_n)$, and let σ denote the usual linear connector from left to right. Then we select a g with probability $\Lambda(g)dt$ and join it on the right, with the result $\sigma(g_1, g_2, \ldots g_n, g)$, *if* this is regular. In other words we pick g at random but conditioned by the constraint

$$\beta_{in}(g) = \beta_{out}(g_n). \tag{16.3}$$

A special case of this is when G has a special element, say #, which symbolizes termination of the dynamics. For example, in syntax controlled finite state languages this means end of the derivation of a sentence; see Volume I, 2.10.

The same set up, but with Σ = TREE, needs a specification to which out-bond (in the old configuration) the in-bond of the new generator shall be attached. We may specify this with a random choice with probabilities depending upon ext(c), and of course again conditioned by $\mathcal{R}$.

Just as we used birth and death intensities for the introduction and deletion of generators we shall introduce *birth and death intensities* for bond couples. They shall be functions of the bond values:

$$\begin{cases} \mu : B \times B \to [0,\infty) \\ \lambda : B \times B \to [0,\infty). \end{cases} \tag{16.4}$$

The resulting probabilities will again be understood as conditioned by $\mathcal{R}$.

For example on $\mathscr{L}[g_1,g_2,\ldots g_n]$ we can let (16.4) generate the birth and death process c_t. All the values of c_t are in the given state space and have identical content. Their topology varies dynamically at random, however, and is governed by μ and λ together with the rules of regularity $\mathscr{R}$.

We postpone the analysis of random dynamics to Chapter 7, where a detailed study will be presented of a special case typical for this problem.

CHAPTER 6
PATTERNS OF SCIENTIFIC HYPOTHESES

6.1. Hypotheses as regular structures

The question of how scientific hypotheses are formed is an intriguing problem that has puzzled psychologists and philosophers of science for a long time. More recently it has been studied from the point of view of artificial intelligence (see Notes A) and this may eventually lead to a better understanding of how humans create hypotheses. The way hypotheses are suggested in the sciences may have more in common with the mental processes of artistic creation than with the strict schemes of the computer programs used in artificial intelligence.

Whatever the case may be, we shall argue that hypotheses within one and the same area of scientific discourse form regular structures that can be understood in terms of pattern theory. In this section we shall discuss this in general and rather vague terms, while the remaining sections in the chapter will narrow down and make precise what we arrive at in the present section by restricting it to a specific scientific discipline.

In Chapter 1 we referred to Poincaré's phrase "... atoms of thought hooked together..." and this is indeed a suggestive formulation helping us to approach this difficult question. We shall view hypotheses as combinations of certain primitives - the atoms of thought - connected together by logical operations. What the primitives are depends upon the domain of discourse and the same is true of the logical operations.

To exemplify this we consider Figure 1.1 that is intended to illustrate the hypothesis that water boils at 100°. The additional inputs to the third atom "of thought" could

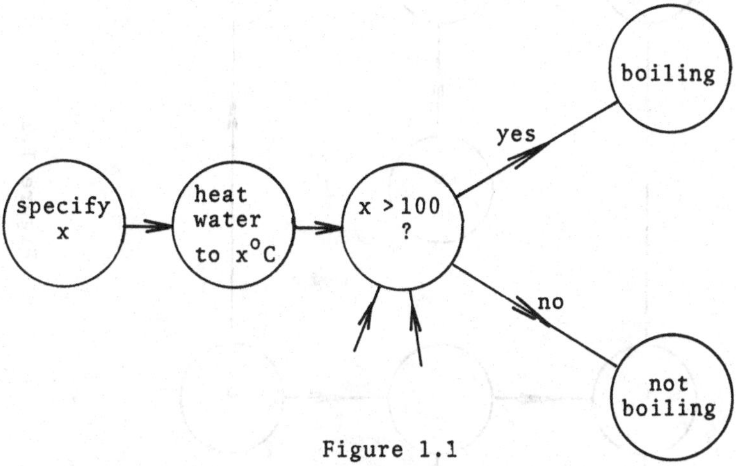

Figure 1.1

specify at what latitude the experiment takes place or at what barometric pressure and so on. Anyway, we get a causal chain of "atoms", a sort of flow chart.

Or, for the synthesis of hypotheses for a Galilean experiment, see Figure 1.2, synthesizing the hypothesis

$$z = a + bt + ct^2 \tag{1.1}$$

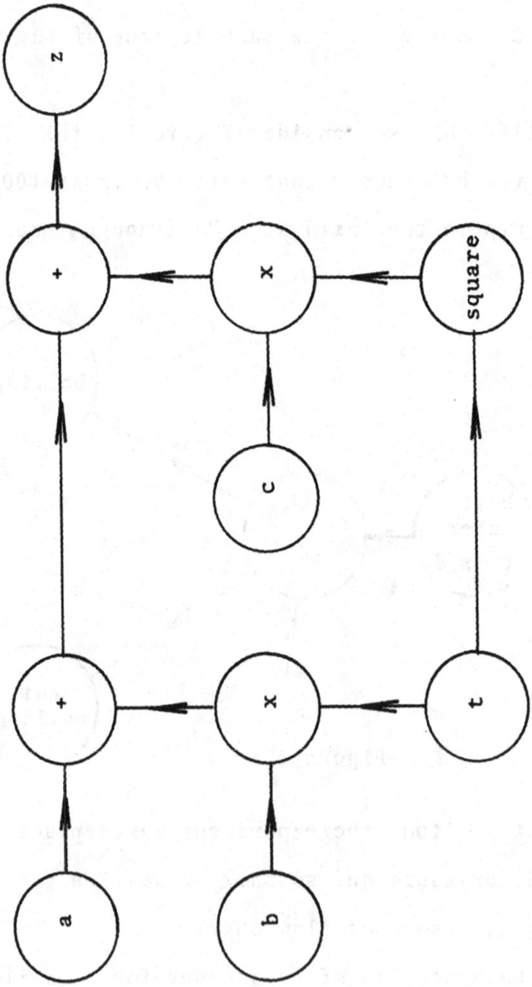

Figure 1.2

for the motion of a material body subject only to the gravitational force. Here z is the height of the body at time t, and a,b,c are constants. In the diagram we use five assignment "atoms", giving the values of a,b,c,t, and z as well as three computational "atoms", namely dyadic addition "+", dyadic multiplication "×" and the monadic "square".

In the two figures the truth or falsehood of the hypotheses do not appear. They are not relevant for the present discussion, only the logical form of the hypotheses matters so far.

Of course the same hypothesis could be described by other "atoms". A more sophisticated version is shown in Figure 1.3. There the first atom represents the function z(t), while "D" stands for differentiation and "g" is the gravitational acceleration.

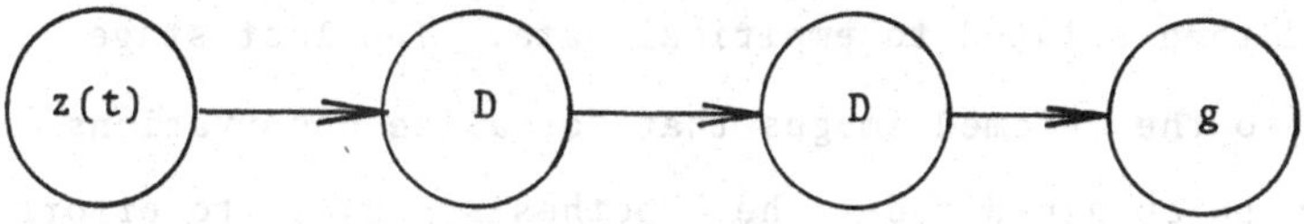

Figure 1.3

Starting from certain "atoms" we can certainly not combine them at will. In a diagram like the one in Figure 1.2, for example, the atom "+" has to be connected to three other atoms, two should be connected with arrows directed inwards and one outwards. Here it is assumed tacitly that the "atoms" represent real numbers (or real variables) and simple arithmetic operations on them.

In a more complicated diagram we must make sure that the logic of the diagram is meaningful so that the successive operations can be carried out. To do this we can indicate the domain sets and range sets of the operations, just as has been done repeatedly in Volume I and II. In other words the bond values should be specified.

If this is done we can *interpret* the formal diagram, if it is regular w.r.t. some $\mathcal{R}$, as a meaningful hypothesis. The formula in Figure 1.2 *means* the function $z = a+bt+ct^2$ for real t-values, and we are naturally led to identify all formulas that mean this function. In other words an *identification rule* R must be introduced to produce equivalence classes, the images, of equivalent configuration diagrams. In this way a hypothesis is an element of an image algebra $\mathcal{J}$.

It is now becoming clear how the synthesis of patterns of hypotheses will be structured, at least in general terms. A regular structure will generate sets of (well-formed) configurations; they will be interpreted via an identification rule and then related to empirical data. The last stage leads us to the defomed images that formalize observations which can be obtained under the hypothesis subject to errors of measurement, finiteness of the data and other experimental restrictions.

It goes without saying that for a given domain of discourse we can have more than one image algebra of hypotheses. The relations between such image algebras will play an important role for our understanding of how scientific hypotheses are synthesized.

To make the discussion more precise we shall look at a particular domain of discourse and choose that of statistical hypotheses. This has been studied by the author and R. Ristow and the following is closely related to this work.

6.2. Patterns of statistical hypotheses

To synthesize statistical hypotheses we should first specify the "atoms", or, in pattern theoretic terminology, the set G of generators. What should they be?

We could always start from a very fundamental level, say predicate calculus, and build from this upwards, but this is not what we have in mind. We want a way of synthesizing the pattern of hypothesis formation that expresses their main features in a way close to our intuition. Therefore, in the present context, the generators should be the basic constructs in probability theory, such as stochastic variables, frequency and distribution functions, and so on.

It is natural to start from the stochastic variable concept. Say that we consider the following statistical hypothesis: we have six observations, independent of each other, the three first of which are $N(0,\sigma^2)$ and the remaining three are $N(0,\sigma^2)$. Or, written out in detail,

$$\begin{cases} x_1 = m_1 + N(0,\sigma^2) = m_1+e_1 \\ x_2 = m_1 + N(0,\sigma^2) = m_1+e_2 \\ x_3 = m_1 + N(0,\sigma^2) = m_1+e_3 \\ x_4 = m_2 + N(0,\sigma^2) = m_2+e_4 \\ x_5 = m_2 + N(0,\sigma^2) = m_2+e_5 \\ x_6 = m_2 + N(0,\sigma^2) = m_2+e_6. \end{cases} \tag{2.1}$$

The symbols in (2.1) would usually be understood as stochastic variables and addition of stochastic variables. The general hypothesis would be arbitrary real m_1 and m_2, arbitrary positive σ, so that it is three-dimensional. The null hypothesis could be its restriction to $m_1 = m_2$ so that it is two-dimensional.

It would be possible to use as generators in (2.1) and similar systems arithmetic operators together with certain simple stochastic variables. However the stochastic variable, i.e. measurable function on some reference space, is not really what we describe in the typical statistical hypothesis. We are not concerned how this function looks, only how its values are distributed.

Although the generators will be introduced in terms of stochastic variables with given probability distributions the identification rule will *identify distributions*, not stochastic variables.

When we speak of a generator being a probability distribution (there are also other generators) we actually mean a stochastic variable $x = x(\omega)$, $\omega \in \Omega$, a measurable function on some reference space Ω, and with the given distribution. This will be understood tacitly in the following.

Going back to (2.1) consider the distribution $N(0,\sigma^2)$ of the e_i's. Note that this is not a completely specified distribution: σ is an arbitrary positive number. This is quite common for statistical hypotheses where we are used to encounter compositie hypotheses. We should therefore allow such a generator to consist of a *set of probability distributions* (actually stochastic variables representing them; see above).

Similarly for assignment generators (see Volume I, p. 18). They do not have to specify a variable completely, and this leads us to let such generators *specify such a variable by restricting it to a set.*

But this is not enough; some rather unexpected generators will also be needed. To show that this is so let us think of the system (2.1) generalized to general sample sizes for the two sample problem

$$\begin{cases} x_i = m_1 + N(0,\sigma^2); & i = 1,2,\ldots n_1 \\ y_i = m_2 + N(0,\sigma^2); & i = 1,2,\ldots n_2 \end{cases} \tag{2.2}$$

We can still synthesize the hypotheses with the same generators as long as n_1 and n_2 are fixed natural numbers. Suppose however that they are not predetermined but found by a pilot experiment or in some other way learning their values unspecified to begin with. Then the above pattern synthesis will not do the job: the number of generators in the configuration is not fixed.

One could of course get out of this dilemma by saying that a hypothesis will be synthesized by many configurations but this is really avoiding the issue. Instead we shall introduce *another generator that samples from a given distribution* and has as one input, namely the sample size. This will be described more carefully later on.

For a similar reason we shall need still *another generator that copies one object* and produces a number of identical copies of it. For example, if g is a generator with one output, taking a real number (or set of real numbers) as values then the generator X (for Xerox) will take as output values an n-vector, all whose components are equal to the output of g. This is indicated by the configuration diagram in Figure 2.1. The out-bond value of X will be a space of N-vectors formed over the space represented by the in-bond

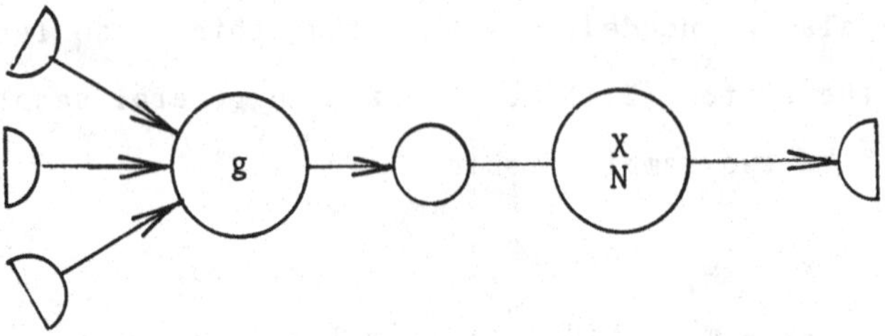

Figure 2.1

value. A few more generators of this somewhat unconventional appearance will be introduced in Section 3.

We shall show in the following sections that these generators suffice for the synthesis of the hypothesis patterns of a wide class of statistical hypotheses. It should be pointed out that the latter do not include hypotheses for sequential designs (see Notes A) for which "looping" generators would also be required.

When we have completed the pattern synthesis we will arrive at some image algebra. To make this concrete we shall carry out all the steps, and aim for an image algebra that corresponds to the usual hypotheses one would find in an elementary introduction to mathematical statistics. As a matter of fact we shall study several image algebras of this sort and briefly examine relations between them.

6.3. Generators for statistical hypotheses

Let us first introduce distribution generators. We have listed some obvious candidates to be included in G in Table 3.1 (see Notes A). In the column headed B_{in} we give the set in which the in-bonds take as values subsets. For example the "normal" generator in the fourth row can have its first in-bond (corresponding to the mean value) take the value $[0,\infty) \subset \mathbb{R}$, or the second one (corresponding to the variance) take as the value the set with a single point $\{1\} \subset \mathbb{R}^+$. Similarly the out-bonds take as values subsets of the sets listed in the last column. The rationale behind making in-bonds values take sets as values was given in the previous section.

Note that the eight generators listed are naturally divided into two classes

$$G_{dist} = G_{dist}^{par} \cup G_{dist}^{nonpar} \qquad (3.1)$$

where G_{dist}^{par} includes the six first (and perhaps several others) while G_{dist}^{nonpar} includes the last two (and perhaps others). G_{dist}^{par} contains the distributions with a finite dimensional parameter space, and G_{dist}^{nonpar} the others. This distinction will be of interest later on.

We now continue by mentioning some of the arithmetic operators that seem required. See Table 3.2. Some more should probably be added.

Note that all the arithmetic generators in Table 3.2 operate on and produce scalar values. At least some of them should be extended to operate on vectors, for example addition, subtraction and multiplication by scalars. A convenient way of formalizing this would be to allow the typical operations in linear algebra as arithmetic generators. They should then include the ones that do inner product, matrix-vector and matrix-matrix multiplication as well as matrix inverse. To simplify the discussion, however, we shall not assume that such generators are in G_{arith} (see Notes B).

Table 3.1

Distribution Generators: G_{dist}				
name	ω_{in}	ω_{out}	B_{in}-set	B_{out}-set
Bernoulli = b(p)	1	1	[0,1]	{0,1}
Binomial = B(n,p)	2	1	$\mathbb{N}$, [0,1]	{0,1,2,...n}
Poisson = P(m)	1	1	$\mathbb{R}^+$	$\{0\} \cup \mathbb{N}$
Normal = $N(m,\sigma^2)$	2	1	$\mathbb{R}$, $\mathbb{R}^+$	$\mathbb{R}$
Rectangular = R(0,1)	0	1	-	$\mathbb{R}$
Chi-square = $\chi^2(n)$	1	1	$\mathbb{N}$	$\mathbb{R}^+$
. . . .				
general continuous = C	0	1	-	$\mathbb{R}$
general absolutely continuous = AC	0	1	-	$\mathbb{R}$
. . . .				

Table 3.2

Arithmetic Operators: G_{arith}				
name	ω_{in}	ω_{out}	B_{in}-set	B_{out}-set
addition +	2	1	$\mathbb{R}, \mathbb{R}$	$\mathbb{R}$
subtraction -	2	1	$\mathbb{R}, \mathbb{R}$	$\mathbb{R}$
multiplication ×	2	1	$\mathbb{R}, \mathbb{R}$	$\mathbb{R}$
division ÷	2	1	$\mathbb{R}, \mathbb{R}-\{0\}$	$\mathbb{R}$
exponentiation a^x	2	1	$\mathbb{R}^+, \mathbb{R}$	$\mathbb{R}^+$
logarithm ln x	1	1	$\mathbb{R}^+$	$\mathbb{R}$
. . . .	. .			

Now to the assignment operators. A few are given in
Table 3.3. They all have in-arity zero and out-arity one.
The out-bond value is a subset of the set mentioned in the
last column.

The reader may notice that the generators can be thought
of as *computational modules if this term is taken to mean
that random elements are permitted.* This will also be true
for the generators to be introduced later. An in-bond set
should be thought of as describing the largest possible support of any distribution in the class associated with this
particular bond site. Similarly for the out-bonds.

If two generators have the same bonds and if they compute
results whose *probability distirubtions are the same,*

Table 3.3

Assignment Generators: G_{assign}				
name	ω_{in}	ω_{out}	B_{in}-set	B_{out}-set
real line $\mathbb{R}$	0	1	-	$\mathbb{R}$
positive real line $\mathbb{R}^+$	0	1	-	$\mathbb{R}^+$
unit interval [0,1]	0	1	-	[0,1]
truth value set {0,1}	0	1	-	ϕ,{0},{1},{0,1}
natural numbers $\mathbb{N}$	0	1		$\mathbb{N}$
. . . .				

conditioned by the same fixed input, then they will be considered *similar*. Although we use a single symbol to denote such a computational module, there can be several others similar to it. If so they ought to have been denoted by different symbols, but this has not been done explicitly in the tables (see Notes C).

The similarities form a group of permutations that relate computational modules equal in terms of (conditional) distributions of outputs. The similarities should be thought of as forming a group of bijective transformations of the background spaces.

Local regularity will be chosen by taking ρ as INCLUSION; *global regularity* by using the connection type Σ = POSET.

6.4. Examples of configurations

To familiarize the reader with the regular structure $\mathscr{L}(\mathscr{R})$ that we have constructed so far and to bring out more clearly what is still missing let us look at a few simple examples of statistical hypotheses analyzed in terms of configuration diagrams.

Trying to synthesize the linear model in (2.1) we immediately encounter one difficulty in that all the three first x's should have the same mean value and the remaining ones some other (possibly the same) mean value. The generators introduced so far do not allow this and we therefore introduce the copying generators $X_n(\beta)$ briefly mentioned in Section 2. One is shown in Figure 4.1 for $n = 3$. Note that

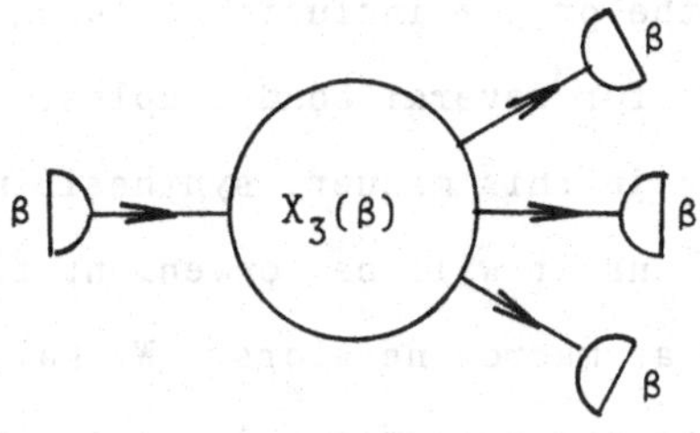

Figure 4.1

for general n we have $\omega_{in}[X_n(\beta)] = 1$, $\omega_{out}[X_n(\beta)] = n$ and that all bond-values are identical and equal to β.

With the aid of the copying generator the two sample hypotheses (2.1) can be synthesized as in Figure 4.2. We have indicated bond values close to the bonds in the diagram. Three assignment generators are used, one arithmetic, and six distribution generators.

The distribution generator $N \in G^{par}_{dist}$ has two in-bonds which must be separated by a bond structure parameter, see discussion in Chapter 3, taking for example the value 1 for the mean value and 2 for the variance.

The usual null hypothesis in the sign test situation is synthesized in Figure 4.2 (a) and (b) for the sample size $n = 4$ and employing Bernoulli generators "b". The symbol "int" is used to denote the real interval $[0,1]$. For the usual alternative hypothesis with p arbitrary in $[0,1]$ we need one of the copying generators, here X_4(int); without it we could not specify an arbitrary p common for all four Bernoulli distributions.

In Figure 4.3 we have synthesized the statistical hypothesis of a chi-square distribution with three degrees of freedom. Note the proper inclusion between the bond values $\mathbb{R}^+$ and $\mathbb{R}$ for several bond couples.

As we go along in this manner, synthesizing many of the standard distributions it will be convenient to use some of the configurations as macrogenerators. We may, for example have macrogenerators as in Figure 4.4(a), a single chi-square distribution, or as in Figure 4.4(b), all chi-square distributions. If we do this we can use very simple configuration diagrams for, say, the F-distribution hypothesis in Figure 4.5(a) and (b). In (b), where we synthesize a composite statistical hypothesis we need a copying generator.

It is now becoming apparent that the diagrams will become clearer and easier to read if we introduce the generators "sample n" with its obvious interpretation: from the set of distributions inputted we select an i.i.d. sample of

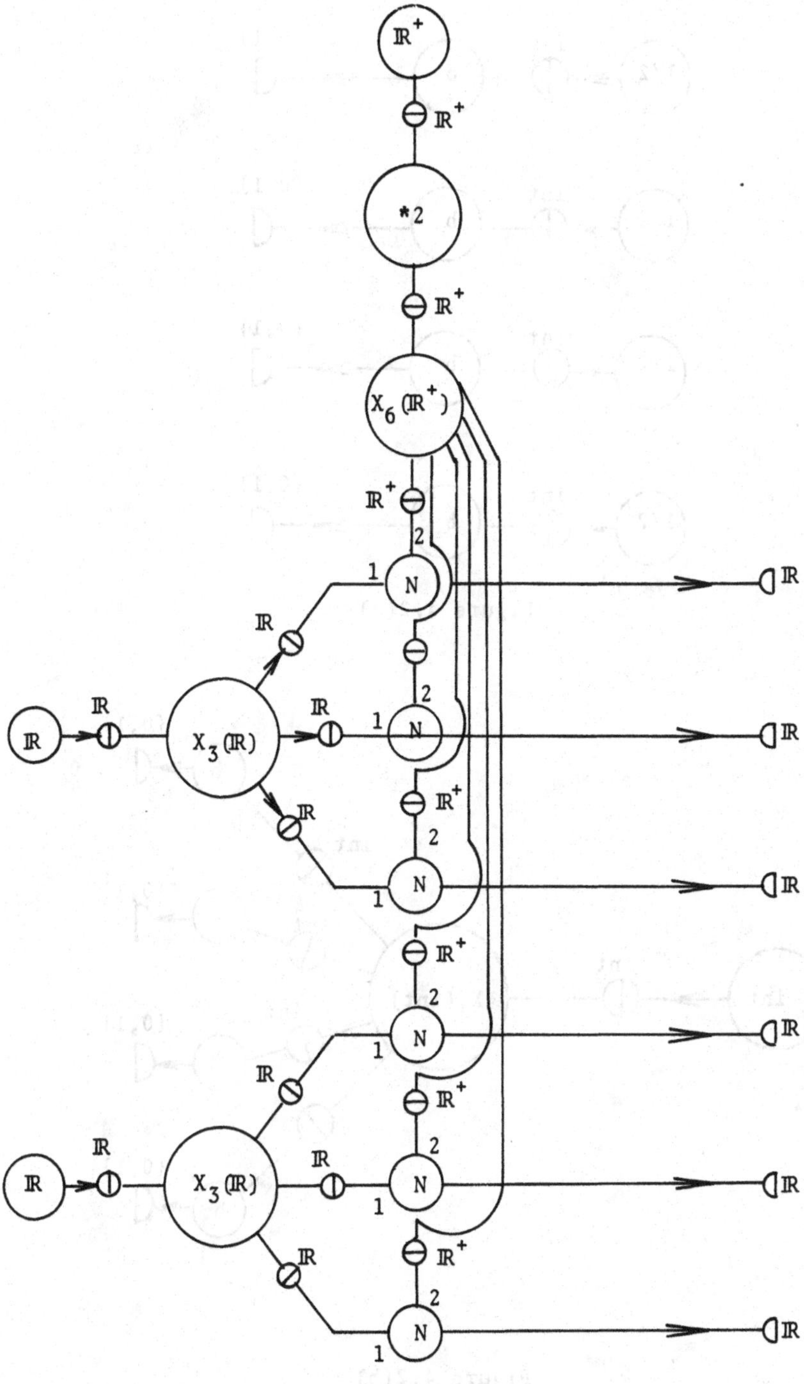

Figure 4.2

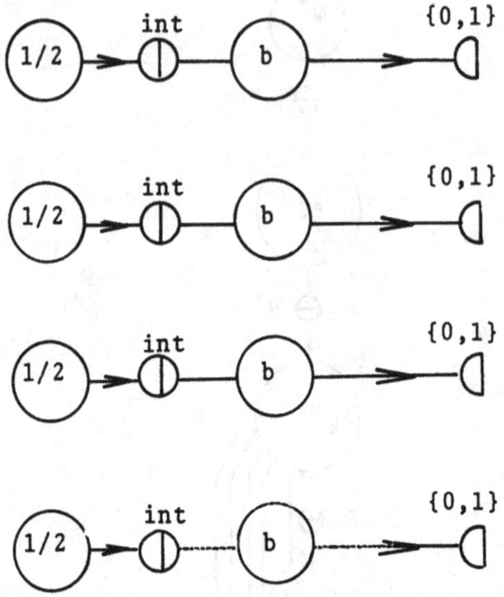

Figure 4.2(a)

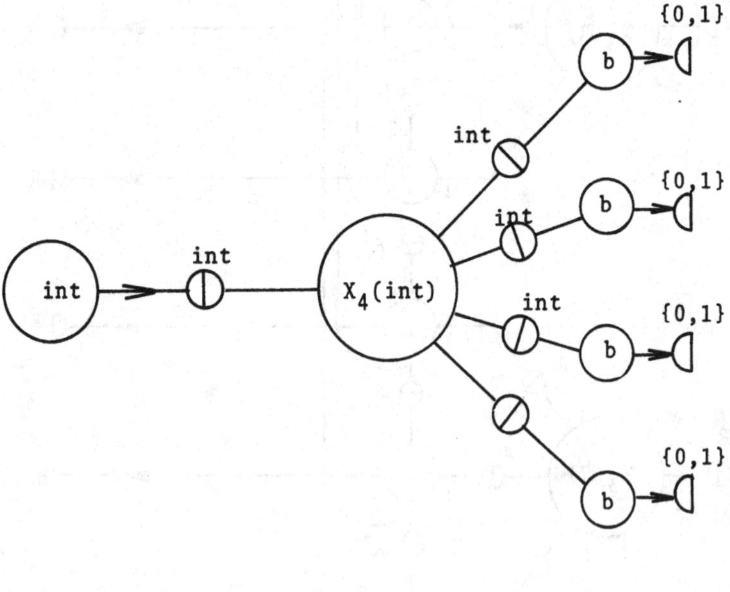

Figure 4.2(b)

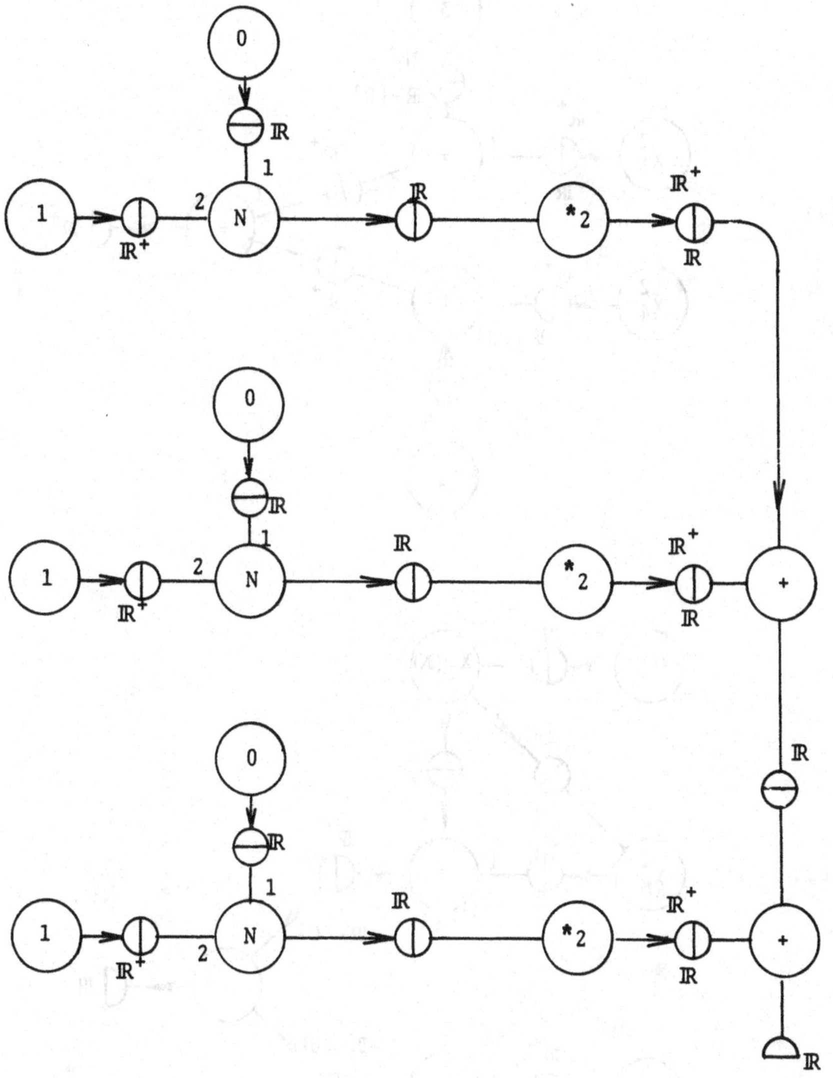

Figure 4.3

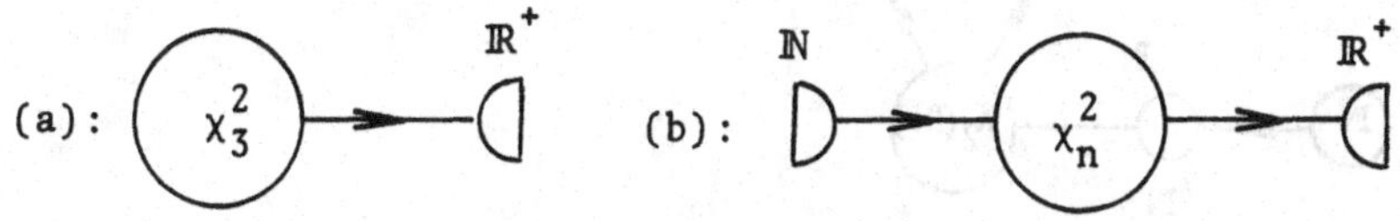

Figure 4.4

(a):

(b):

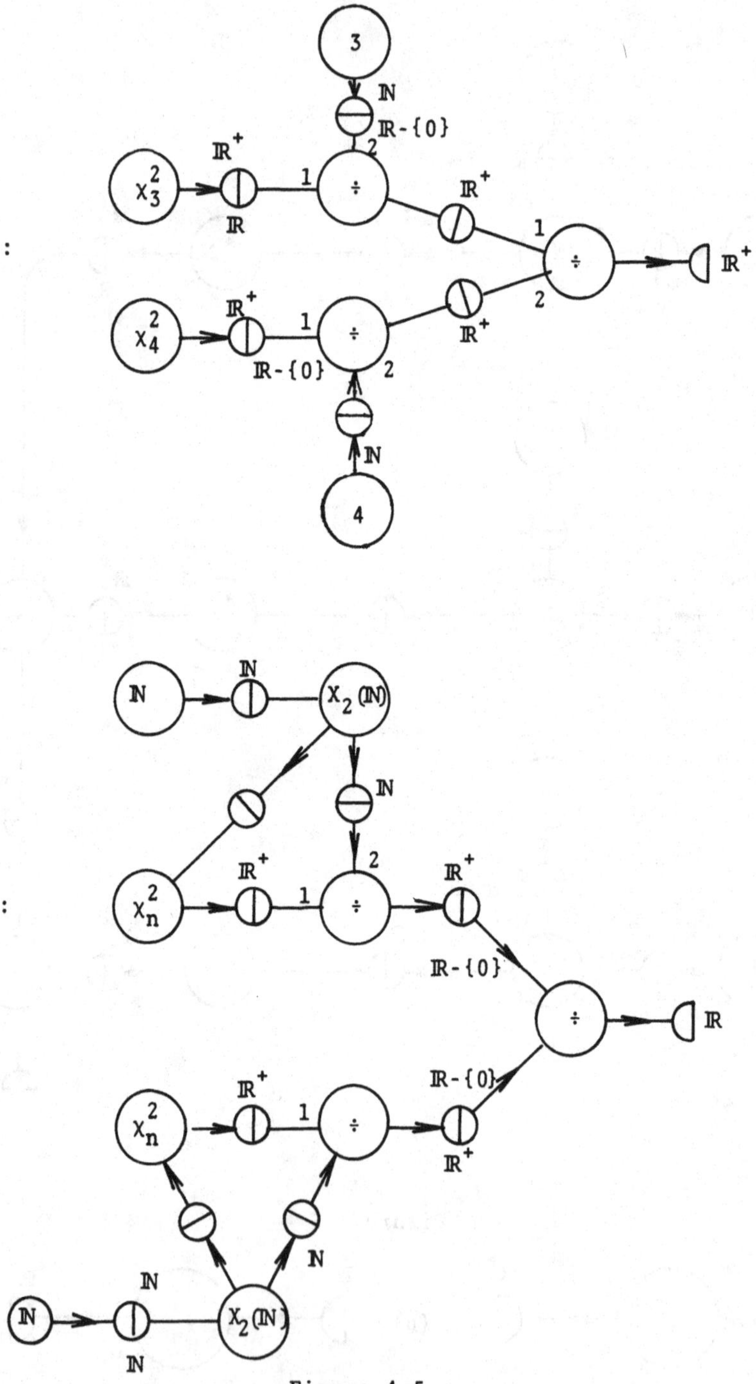

Figure 4.5

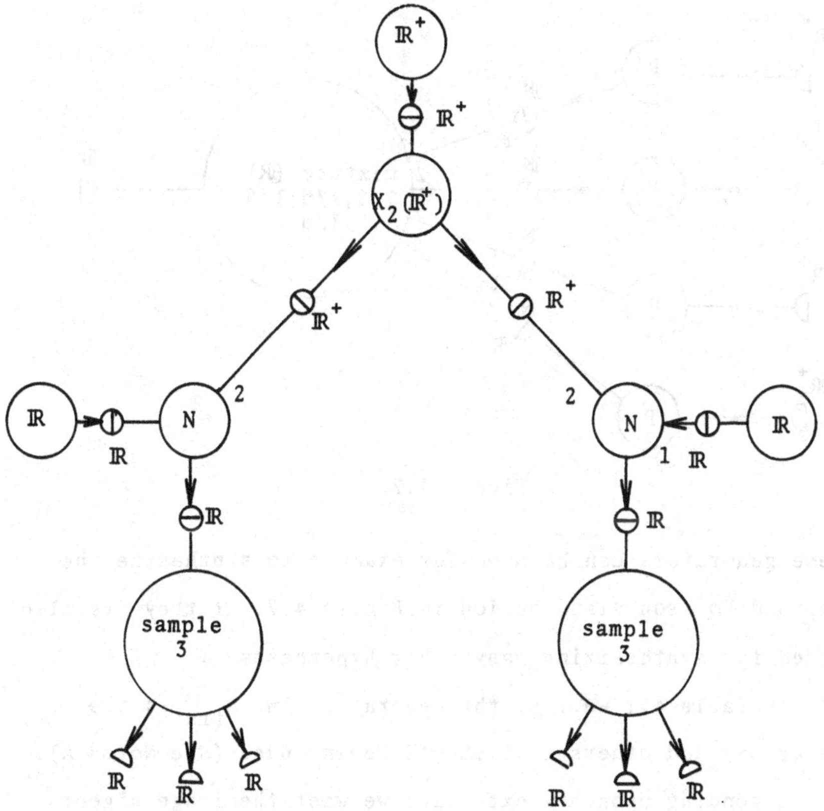

Figure 4.6

size n. Then we could synthesize the two sample hypotheses in Figure 4.2 for example by the more transparent configuration diagram in Figure 4.6. It may be remarked that all randomness synthesized is tacitly assumed to be independent, conditioned by the inputs from the preceding levels in the POSET connection type.

Still another type of generator, "mixture", has in-arity n and out-arity one. An attribute $p = (p_1, p_2, \ldots p_n)$ of probabilities summing to one describes how one of the inputs is selected according to the probabilities in p.

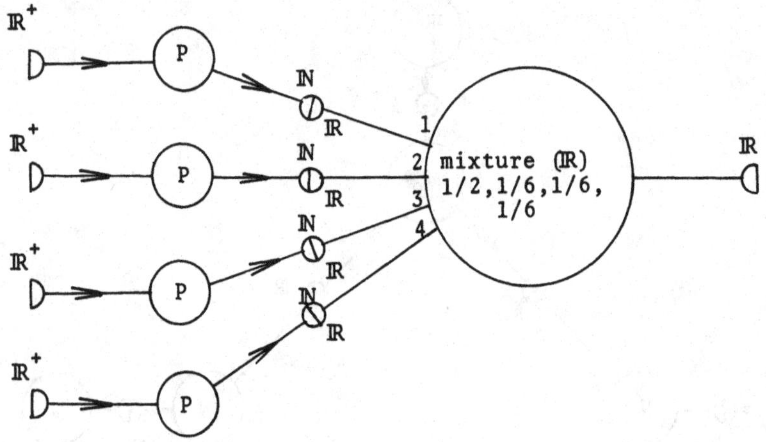

Figure 4.7

These generators can be used for example to synthesize the compound Poisson distribution in Figure 4.7 but they are also needed for synthesizing many other hypotheses.

In Table 4.1 we list the generators in G_{trans}; the reader may add others that should be included (see Notes A).

Depending upon how extensive we want the image algebra of hypotheses to be, the set of generators will be deemed sufficient or not. Whatever the case may be we are now sufficiently familiar with the approach to synthesizing configurations, meaning statistical hypotheses, to go ahead to discuss the resulting image algebra.

Table 4.1

Transformation Generators: G_{trans}				
name	ω_{in}	ω_{out}	B_{in}-set	B_{out}-set
$X_n(\beta)$	1	n	$\{\beta\}$	$\{\beta\}$
$\text{mix}(p_1,p_2,\ldots p_n;\beta)$	n	1	$\{\beta\}$	$\{\beta\}$
$\text{sample}_n(\beta)$	1	n	$\{\beta\}$	$\{\beta\}$
. . . .				

6.5. Hypotheses as images

We shall now attribute a meaning to the formulae (regular configurations) that represent hypotheses and also identify them according to their meaning.

Theorem 1. *In* $\mathcal{L}(\mathcal{R})$, *with* G *and* S *as described and with the regularity* $\mathcal{R} = \langle\text{INCLUSION},\text{POSET}\rangle$, *consider the relation* $c_1 R c_2$ *meaning that* $B_{ext}(c_1) = B_{ext}(c_2)$ *and that the two configurations compute the same set of joint conditional probability distributions at their out-bond sites. Then* R *is an identification rule so that* $\mathcal{L}(\mathcal{R})/R$ *is an image algebra.*

Proof: Let c be a configuration with $\mathcal{R}$-regularity. Since we have no generators with out-arity zero we have $\omega_{out}(c) > 0$. Recalling that all bond values are sets we shall think of configurations as representing a set of probability distributions over the respective sets. Constants are thought of as degenerate probability distributions, having all their

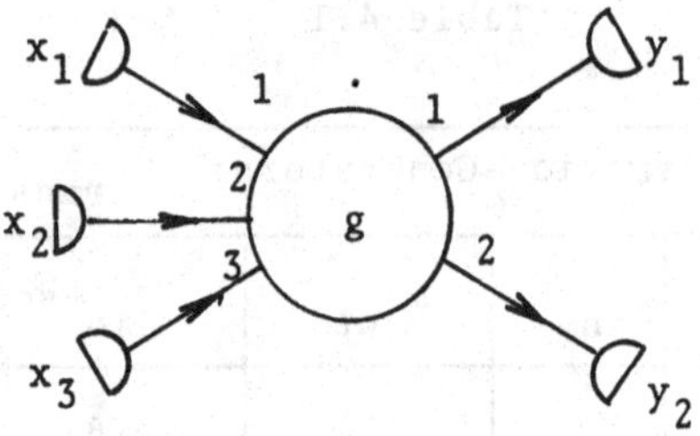

Figure 5.1

mass at a single point.

If c contains a generator g with in-arity r and out-arity s let us write $x_1, x_2, \ldots x_r$ with x_i being the symbol for the stochastic variables associated with the ith in-bond. Similarly, $y_1, y_2, \ldots y_s$ for the out-bonds; see Figure 5.1 where $r = 3$ and $s = 2$.

Since, at an arbitrary bond site, we have a class of distributions, rather than a single one, we should think of $\{x_i\}$ and $\{y_j\}$ as variables representing a set of stochastic vector variables. When we select one representative for each x_i the generator "computes" a result consisting of certain stochastic variables, s in number. This is so whether g is deterministic or random: inspect the generators in Tables 3.1, 3.2, 3.3, and 4.1. As the selected representatives for the x's range over their sets the computed results range over some sets denoted by the variables y_j.

On the lowest level of the POSET structure of c consider all the x's. Recalling the conditional independence assumption from Section 4, their marginal conditional distribution determines their joint conditional distribution as a product measure (there are no preceding generators in the POSET). This is true for each representative in the class of distributions.

Now move up in the levels of the POSET and define the successive results computed by the generators. Since we have no cycles and each generator has a well defined set of preceding generators this construction is unique and leads to a well defined set of conditional distributions for the stochastic variables associated with the out-bonds of c. Therefore the out-bond distributions are defined and the definition of R makes sense.

It remains to show that it satisfies the four conditions in Definition 3.1.1 of Volume I. Condition (i), that R be an equivalence relation is obvious since R is defined via equality of certain sets characterizing the configurations partially. Condition (ii) holds since Definition 1 required it. The condition (iii) can be verified by following each step in the previous construction of resulting out-bond distribution, and noting that similarity means same (conditional) resulting distribution at each step. Condition (iv), finally, also follows from the construction and for the same reason. Hence R is an identification rule and $\mathcal{T} = \mathcal{L}(\mathcal{R})/R$ is well-defined. Q.E.D.

Theorem 1 enables us to build up systems of hypotheses in a combinatory manner restricted only by the choice of G and by the rules of regularity $\mathcal{R}$. We shall illustrate this by a few examples (see Notes A).

In Figure 5.2(a) we have shown one image, indicated by a rectangular box, consisting of all regular configurations identified mod(R) with the one shown inside the box (the inner one). This image combined with another one, POL, are

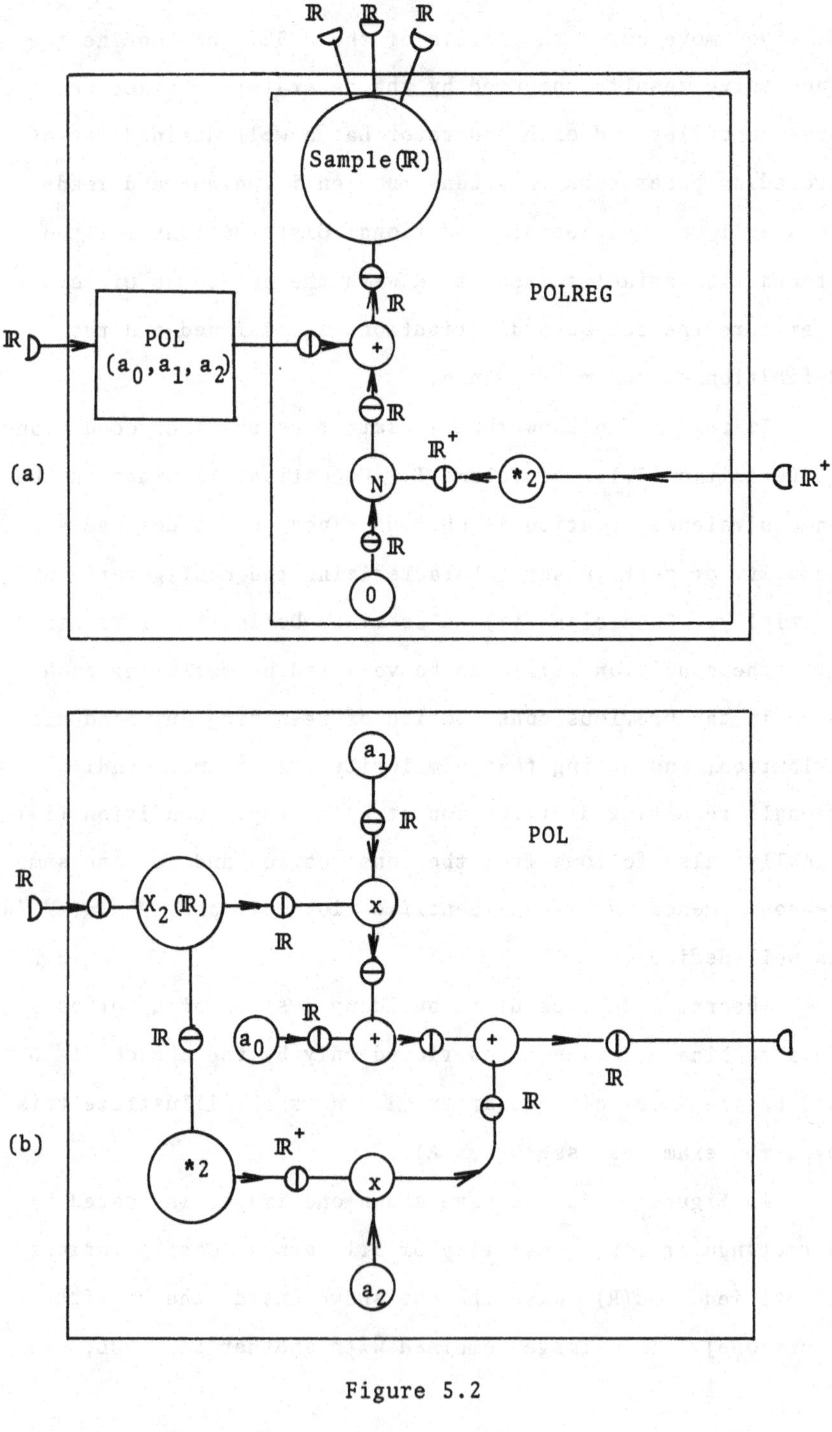

Figure 5.2

in turn combined to a large one that will be denoted POLREGR. It has $\omega_{in} = 2$, $\omega_{out} = 3$. The image POL, for polynomial, is synthesized in (b). It has $\omega_{in} = \omega_{out} = 1$, and "means" a second order polynomial evaluated at a single (arbitrary) point.

To synthesize the hypothesis image corresponding to a second order polynomial evaluated at three arbitrary points with the result disturbed by Gaussian additive noise we use three copies of POLREG. This is shown in Figure 5.3. This image has $\omega_{in} = 4$, $\omega_{out} = 3$.

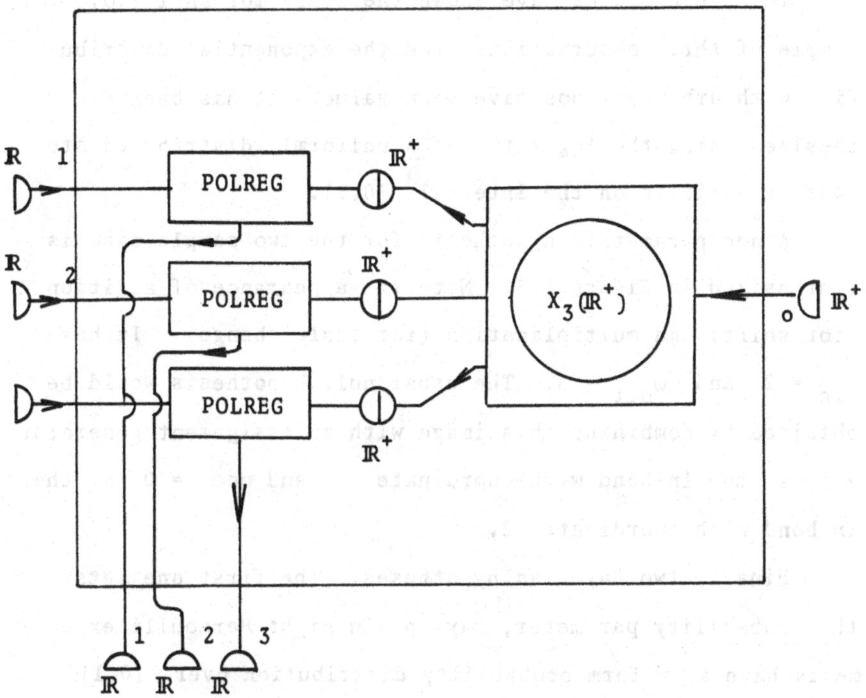

Figure 5.3

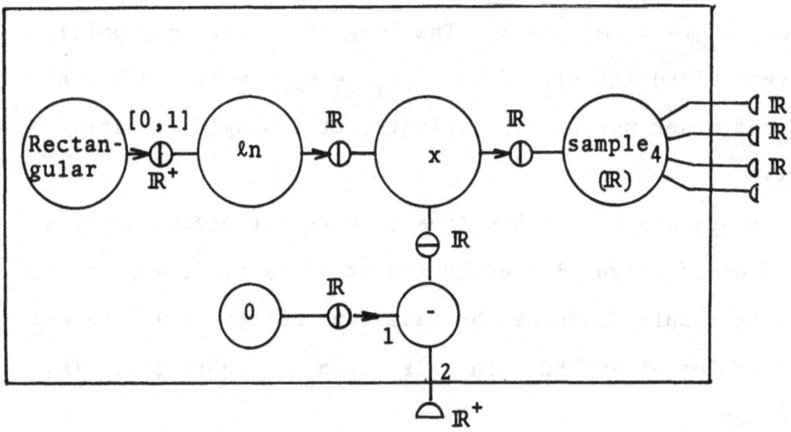

Figure 5.4

In Figure 5.4 we have shown the image for an i.i.d. sample of three observations from the exponential distribution with arbitrary positive mean value. It has been synthesized using the logarithm of a uniformly distributed stochastic variable on the interval [0,1].

A non-parametric hypothesis for the two sample case is synthesized in Figure 5.5. Note the appearance of addition (for shift) and multiplication (for scale change). It has $\omega_{in} = 2$ and $\omega_{out} = 5$. The usual null hypothesis would be obtained by combining this image with an assignment generator $= 1$ at the in-bond with coordinate 1 and one $= 0$ at the in-bond with coordinate 2.

Finally two Bayesian hypotheses. The first one lets the probability parameter, say p, in eight Bernoulli experiments have a uniform probability distribution over [0,1]. It produces four values of the corresponding binomial distribution $B(p,8)$. This is shown in Figure 5.6.

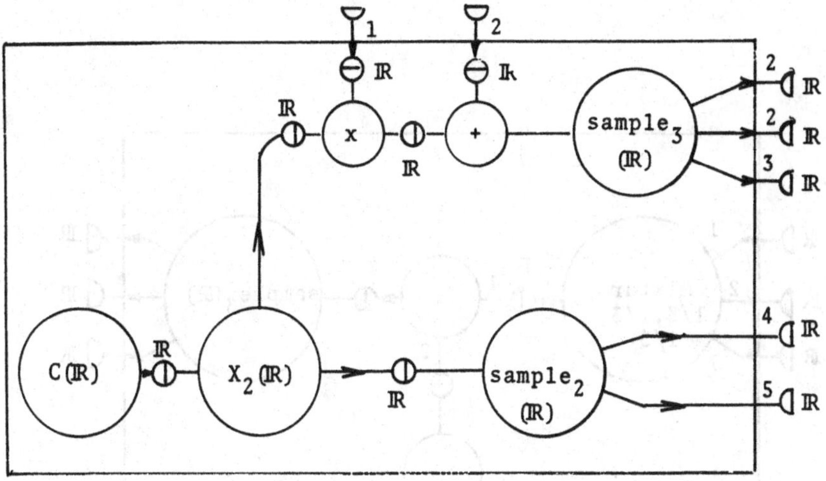

Figure 5.5

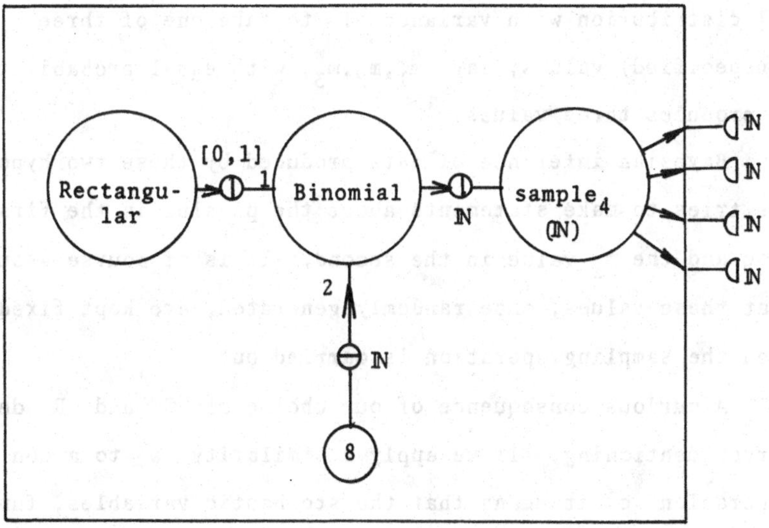

Figure 5.6

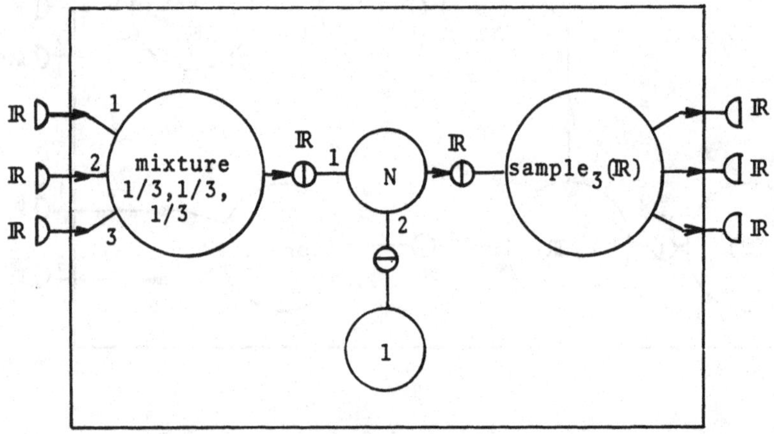

Figure 5.7

The last one, in Figure 5.7, allows the mean of a normal distribution with variance 1 to take one of three (unspecified) values, say m_1, m_2, m_3, with equal probabilities. It produces three values.

Bayesian inference of data produced by these two hypotheses tries to make statements about the p-value in the first case and the m_i-value in the second. It is of course assumed that these values, once randomly generated, are kept fixed when the sampling operation is carried out.

A curious consequence of our choice of S and R deserves mentioning. If we apply a similarity s to a configuration c it means that the stochastic variables, functions on Ω, that c represents will be changed by a mapping s between reference spaces. This mapping preserves P-measure however so that c and sc have the same distribution (for fixed inputs). But then they are equivalent modulo R, (sc)Rc, which implies

$$sI = I; \quad \forall s \in S; \quad \forall I \in \mathcal{T}. \tag{5.1}$$

The similarities, which are non-trivial on $\mathcal{C}(\mathcal{R})$, thus degenerate on $\mathcal{T}$ to the identity operation.

6.6. Image algebras of hypotheses

The generators given in Sections 3 and 4, together with the regularity $\mathcal{R}$ = <INCLUSION,POSET> lead to a configuration space $\mathcal{C}(\mathcal{R})$; modulo the identification rule of the last section we have a well defined image algebra $\mathcal{T}$ of statistical hypotheses.

If we restrict G, but keep $\mathcal{R}$ and R, we obtain sub-image algebras and we mention as an illustration the one we get with G' consisting only of

$$\begin{cases} \text{assignment operators in } G_{assign} \\ \text{unary addition, e.g. "+3"} \\ \text{unary multiplication, e.g. "×5"} \\ \text{binary multiplication} \\ \text{normal distributions} \end{cases} \tag{6.1}$$

Let us call the resulting image algebra

$$\mathcal{T}_{normal} = \langle G', S, \mathcal{R}, R \rangle. \tag{6.2}$$

On the other hand let the generator space G" be equal to G' except that the normal distributions are replaced by D($\mathbb{R}$), the set of distributions on the real line. This last generator shall be treated as having two in-bonds, the first

specifying the mean value, the second specifying the standard deviation. Its "meaning" is the set of all distributions
on the real line with specified first two moments. Introduce similarly to (6.2)

$$\mathcal{T}_{linear} = <G'',S,\mathcal{R},R>. \qquad (6.3)$$

Consider the mapping h: $\mathcal{T}_{normal} \rightarrow \mathcal{T}_{linear}$ where h
replaces N in $\mathcal{C}_{normal}$ by D in $\mathcal{C}_{linear}$ as will be discussed in more detail below.

<u>Theorem 1</u>. *It is a homomorphism if we let* S = S', Σ = Σ'
and extend the definition by defining hs = s, hσ = σ; *see
Section* 3.7.

<u>Proof</u>: To see that h is homomorphic we first note that
h is well defined. If c ∈ $\mathcal{C}_{normal}$ and

$$c = \sigma(g_1, g_2, \ldots g_n); \quad g_i \in G_{normal} \qquad (6.4)$$

we put

$$hc = \sigma(hg_1, hg_2, \ldots hg_n); \quad hg_i \in G_{linear}$$

where hN(m,σ) → hD(m,σ) and hg = g if g is not a normal
distribution generator. In other words we start from a
generator map and then try to extend it to all of $\mathcal{C}_{normal}$.
We can now apply Theorem 3.4.1 if the generator map is covariant, i.e. if h(sg) = sh(g). This relation is trivially
satisfied if g is not a distribution generator. If it is a
distribution generator g is associated with a normal stochastic variable x(ω), ω on a reference space Ω. As discussed sx(ω) means "changing the randomness" so that, with
some abuse of notation, sx(ω) = x(sω). In the same way hg

is a class of distributions (all with specified moments of order 1 and 2)

$$hg = \{y_\rho(\omega);\ \text{all } \rho \text{ in some indexing set}\} \qquad (6.5)$$

so that

$$h(sg) = \{y_\rho(s\omega);\ \text{all } \rho\} = s\{y_\rho(\omega);\ \text{all } \rho\} = shg. \qquad (6.6)$$

Hence h is covariant.

Also note that hg and g have the same bonds so that the assumption about homologous bonds in Theorem 3.4.1 is satisfied. The theorem guarantees that h is a configuration homomorphism.

On the other hand if $c_1 R c_2$ holds for $c_1, c_2 \in \mathscr{T}_{\text{normal}}$ it means that the "results" of c_1 and c_2 have the same probability distributions for given inputs. But the "formulas" hc_1 and hc_2 look the same except that normal distributions are replaced by sets of general distributions with the same first moments. Recalling that G' and G'' have only linear arithmetic it is clear that hc_1 and hc_2 lead to the same set of probability distributions for given inputs. Hence $(hc_1)R(hc_2)$ holds and we can apply Theorem 3.7.3. It tells us that h is an image homomorphism as stated. Q.E.D.

Arguing in the same way, but in the opposite direction we see that $h^{-1}: \mathscr{T}_{\text{linear}} \to \mathscr{T}_{\text{normal}}$ is well defined and homomorphic between the two image algebras (see Notes A).

Here we are dealing with two scientific theories, expressed as the two image algebras $\mathscr{T}_{\text{normal}}$ and $\mathscr{T}_{\text{linear}}$. *The two theories are seen to be isomorphic* via the mapping h: the two theories are not identical since they do not deal with exactly the same set of logical objects, but *their logical structure is the same.*

One may believe that the same map would be an image homomorphism when the arithmetic generators include all the ones listed in Table 3.2. This is not true, however, as can be seen when we try to prove that c_1Rc_2 imply $(hc_1)R(hc_2)$. We can only state in this case that h is a configuration homomorphism.

For fixed Σ and a given family Γ of generator skeletons consider the set <u>Conf</u> of configuration spaces together with their homomorphisms. On this category consider the functor $\mathcal{MTX}$, see Section 3.5. What does it represent in the present case?

Recalling that $\mathcal{MXT}$ operates by forming finite unconnected unions of arbitrary regular configurations. The new connection type is therefore formed by unions of σ's, all σ's in the original Σ. If Σ = POSET, as has been assumed throughout this chapter we have $\mathcal{MTX}$POSET = POSET so that nothing is changed by applying the functor. This is the case mentioned in the discussion following Eq. (5.6) in Section 3.5.

If, however, Σ is not closed w.r.t. this functor, we get an operation that has an interesting statistical interpretation. Say, for example, that we only allow connected POSET connectors in $\mathcal{R}$ = <INCLUSION,CONNECTED POSET>. Then we do not allow the configuration in Figure 6.1 as regular. On the other hand it belongs to the regularity <INCLUSION, POSET>. It "means" that the statistical experiment in the upper part of the configuration diagram is replicated, perhaps with other parameter values.

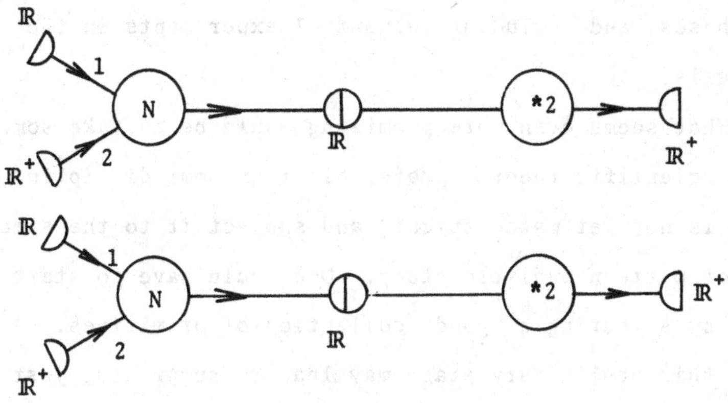

Figure 6.1

Note that this could not be obtained by combining the sampling generator with the upper half of Figure 6.1 since the latter would lead to equal parameter values. The $\mathcal{MXT}$ functor represents, in the present context, hypotheses corresponding to combinations of statistical experiments.

6.7. Conclusions

We have shown how a certain scientific theory can be understood in terms of regular structures: hypotheses are synthesized in a combinatory manner from simple building blocks, the generators, and with local and global regularity as described. This was done for the special case of statistical theories.

The pattern synthesis was not carried very far but it is sufficiently detailed to show the feasibility of this approach. It should be continued and deepended, establishing other homomorphisms, for example relating Bayesian to non-Bayesian hypotheses, studying covariant mappings of statistical hypotheses, and including sequential experiments in the synthesis.

What seems even more promising would be to take some other scientific theory, preferably from some discipline which is not yet mathematical, and subject it to the same sort of pattern analytic study. One would have to start again by selecting a "good" collection of primitives. Already this preliminary stage may lead to surprises, just as we were forced to introduce some rather surprising generators to synthesize statistical hypotheses. One would then go on to study particular hypotheses and try to find a natural identification rule. Once the image algebras have been fixed one should search for homomorphic relations between them. In the humanities these may perhaps be given historical interpretations, or be related to intellectual pressures in the idealogical climate of the period.

Analysis of a scientific theory in such terms, carried out with the full rigor that is required, may increase our understanding of how theories are synthesized.

CHAPTER 7
SYNTHESIS OF SOCIAL PATTERNS OF DOMINATION

7.1. Patterns in mathematical sociology

Sociology studies systems of interacting individuals or groups of individuals, and mathematical sociology creates the mathematical tools needed for such studies. We suggest that pattern theory - the mathematical theory of regular structures - can be used to create such tools.

The crucial fact from which we shall start is that the units interact with each other. Typically this leads to non-linear models with the ensuing mathematical difficulties. Interactions can conveniently be described in terms of bonds, local and global regularity, and other fundamental notions of pattern theory. Of course this will not reduce the mathematical difficulties, just formalize them in a precise form.

In our approach - which is completely non-empirical - we shall illustrate what we have in mind for a particular type of social linkage, namely domination. Many research workers have tried to study systems of domination by mathematical methods. One of the earliest was H. G. Landau (1951), (1965), whose work has been followed up by others, see in particular I. Chase (1974). An idea that seems to have occurred to many social scientists is to exploit the analogy to statistical mechanics, where one also deals with units that interact with each other. A major difficulty is that no obvious social parameters correspond to mass, acceleration, and force.

In Section 2 we shall describe the pattern theoretic set-up on which the following synthesis is based (see Notes A). The social dynamics that results is admittedly difficult to analyze mathematically. For that reason we start to study it by computer simulations, some numerical results of which are presented in Sections 3 and 4. This leads to some conjectures that are dealt with later.

The resulting probabilistic set up turned out to be, to our surprise, a special case of one we have advocated in many other contexts - the regularity controlled probability model. It will be examined analytically in Section 4.

In this connection we have to analyze the concept of a "typical configuration". Since our mathematical structure is not linear in any obvious sense - the basic algebraic operations are here represented by connectors - we do not have access to the concept of an average. Instead we shall argue in favor of a concept *typical set* in configuration space.

Returning to the analogy with statistical mechanics one would hope that as the size of the configuration increases it would, in some sense, tend to a macroscopically deterministic limit: the thermodynamic limit. Again employing simulations we shall see however in Section 5 that an important modification is needed in the model, one that could be neglected for configurations of constant size.

The basic analytical result is given in Section 6. It tells us how the social interaction assumed will indeed lead to a deterministic limit, although one that would be difficult to guess at the beginning. We believe that this will give us a handle on the problem of pattern analysis of these patterns although this question will not be treated here.

The computer programs used for the simulations are collected in an Appendix at the end of the chapter together with comments for their use.

We make no empirical claims. We do now know whether real systems of domination behave like the patterns we have synthesized. That is not what we are after. The main point is that if we start from certain simple assumptions about bonding - and they could certainly be varied in many ways - we have the mathematical capability of deducing non-intuitive consequences about the patterns of domination that result. Without such capability it seems a hopeless task to carry out inference from observed domination structures.

In the last section we mention some possibilities of deepening the mathematical analysis as well as of extending the model to cover more flexible domination behavior.

One could criticize the model used below: it is much simpler than anything one would expect to encounter in real social systems. This is certainly a valid objection, but misses the point of the paper. We argue that, even with this simple model, *an analysis of some depth is needed to explain the tendency to the limit* given by Eq. (6.7) of Section 6. For more complex models other, also counter-intuitive, phenomena should be expected.

7.2. Domination regularity

The generators represent the individuals or groups of individuals in the social structure under consideration. The arities may vary over G, but the in- and out-arities, $\omega_{in}(g)$ and $\omega_{out}(g)$, of any fixed generator g remain constant in time.

We shall assume throughout this discussion that we have only a single species: *the social structure is homogeneous.* There will then be only a single value of the generator index $\alpha(g)$. To deal with heterogeneous social structures, for example expressed in terms of sex, occupation, or class, G should be made up of several generator index classes. In the following $\alpha(g)$ will not be mentioned since its value is constant.

To each g we associate, in addition to other possible attributes, a scalar attribute $x(g) \geq 0$ which expresses g's social influence and degree of activity. The value of $x(g)$ may vary over time.

The similarity group shall consist, as in many other regular structures, of all permutations that leave the bonds unchanged: $g' = sg$ if $B(g') = B(g)$. The bonds will be discussed below. It will be clear that two generators can be similar although their x-attributes differ.

The directed regularity formalized as a connector $\sigma \in \Sigma$ expresses the way the generators connected via σ dominate or do not dominate each other. If g_1 has an in-bond connected to an outbond of g_2 this means that g_1 dominates g_2. In Figure 2.1 g_3 dominates g_2 who dominates g_4 who dominates g_1. The generator g_5 is isolated from the rest.

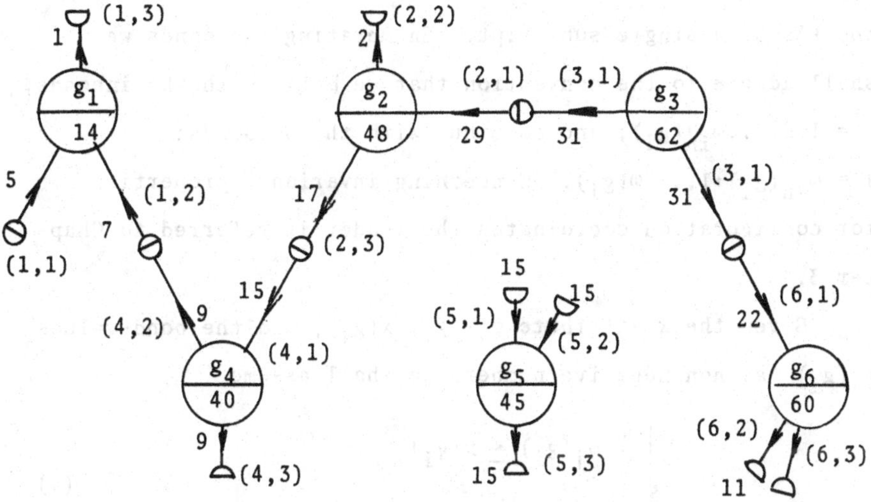

Figure 2.1

The values of the x-attributes are shown below the identifier in the circle symbolizing each generator; they are also given in the second column of Table 2.1.

Table 2.1

g	x	$\omega_{in}(g)$	$\omega_{out}(g)$	β_{in1}	β_{in2}	β_{out1}	β_{out2}
g_1	14	2	1	5	7	1	-
g_2	48	1	2	29	-	2	17
g_3	62	0	2	-	-	31	31
g_4	40	1	2	15	-	9	9
g_5	45	2	1	15	15	15	-
g_6	60	1	2	22	-	11	10

For any regular configuration c the generators will be enumerated by the generator coordinates i; $i = 1,2,\ldots n=\#(c)$;

and the corresponding bonds by the bond coordinates $k = (i,j);\ j = 1,2,\ldots\omega(g_i)$. We shall sometimes enumerate the k's by a single subscript. Enumerating the bonds we shall adhere to the convention that we begin with the inbonds; $j = 1,2,\ldots\omega_{in}(g_i)$; and continue with the outbonds; $j = \omega_{in}(g_i)+1,\ldots\omega(g_i)$. Concerning invariance properties for configuration coordinates the reader is referred to Chapter 3.

Given the x-attribute of g_i, $x(g_i)$, and the bond-values $\beta_j(g_i)$ as non-negative numbers we shall assume

$$\begin{cases} \sum_j \beta_j(g_i) \le x(g_i) \\ \beta_{j'}(g_i) \ge \beta_{j''}(g_i) \end{cases} \tag{4}$$

where $1 \le j'' \le \omega_{in}(g_i)$, $\omega_{in}(g_i)+1 \le j'' \le \omega(g)$. The interpretation of the two conditions in (4) is that the first one expresses *a limit for the ability of* g_i *to ward off other individuals as they try to dominate* g_i (this is expressed in terms of the inbond values) and *to the ability of* g_i *to dominate others* (this is expressed in terms of its outbond values).

This condition appears natural in any domination patterns. Individuals have two aims, to stay independent of others and to increase their influence. The condition just says that these aims can only be attained in a limited way that may vary from individual to individual depending upon their x-attribute.

The second condition in (4) says that for any individual *the drive for independence takes precedence over the wish for influence.* One can imagine situations where this will not be a natural assumption. As a matter of fact the author believes that if this assumption is changed some remarkable domination patterns will result; this deserves to be studied but will not be done here.

We now define the regularity $\mathscr{R} = \langle\rho,\Sigma\rangle$ by letting the *local regularity* be

$$\rho = \text{'GREATER THAN'}. \tag{2.1}$$

Hence, if g' has an outbond β' with value b' connected to an inbond β'' with value b'' of the generator g'', we can use (2.1) to give us the inequality $\beta_{in}(g') \geq \beta' > \beta'' \geq \beta_{out}(g'')$. This holds for any inbond value $\beta_{in}(g')$ of g' and any outbond value $\beta_{out}(g'')$ of g''.

Also, if the generators are linked by a chain

$$g_{i_1} \to g_{i_2} \to \cdots \to g_{i_r} \tag{2.2}$$

so that an outbond of g_{i_ν} connects to an inbond of $g_{i_{\nu+1}}$ we see that the bond-values are ordered monotonically. No cycles can occur, and the ordering is transitive. If, in a given regular configuration c, a chain can be found from some g' to some g'', or if $g' = g''$, we write $g' \succ g''$ and it is easily seen that the relation '$\succ$' induces a partial order on c.

Hence the *global regularity* Σ must lead to a partially ordered set. We shall not add any further requirement so that we have simply

$$\Sigma = \text{POSET}. \tag{2.3}$$

To proceed with the pattern synthesis of $\mathscr{L}_n$ we shall select n generators $g_1, g_2, \ldots g_n$ i.i.d. according to a probability measure Q over the generator space G. Once the g_i's have been selected we form the set $\mathscr{C}[g_1, g_2, \ldots g_n]$

of all regular configurations that can be obtained from $\{g_1,g_2,\ldots g_n\}$ by connecting its bonds in accordance with the regularity $\mathcal{R}$. Note that $\mathcal{L}[g_1,g_2,\ldots g_n]$ is a finite set although its cardinality can be large.

For fixed c those g's that are not dominated will be called *the rulers of* c. All the other generators in c, except the isolated ones, are dominated by at least one ruler.

If n is small it is convenient to illustrate the structure of domination by a configuration diagram such as Figure 2.1. For large n this is not practical, and we shall have to be satisfied by descriptive statistics that summarize some of the relevant features of our structure.

One such statistic is the *bond frequency*

$$bf(c) = \#\text{bond couples connected in } c \tag{2.5}$$

as well as the *normalized bond frequency* $nbf(c) = bf(c)/n$. When $bf(c) = 0$ no one dominates anyone else, *complete independence*. The opposite extreme is when

$$\begin{aligned} bf(c) &= \min\left(\sum_{g\in c} \omega_{in}(g),\ \sum_{g\in c}\omega_{out}(g)\right) \\ &= \min[\Omega_{in},\Omega_{out}]. \end{aligned} \tag{2.6}$$

Then no more bound can be connected: *saturated domination*. Saturated domination can usually be realized in many ways. Saturation is the same as to say that the external bond set $B_e(c)$ consists of only in-bonds, only out-bonds, or is empty.

Some other statistics of interest are the normalized numbers of in-bonds $\Omega_{in}(c)/n$, and of out-bonds, $\Omega_{out}(c)/n$, as well as the related quantities for external bonds only. When two regular configurations c' and c'' are given the two latter quantities give us some idea of how $\sigma(c',c'')$ may appear.

We also need $ns(c)$, the number of connected subcomponents of c. In Table 2.1 for example $ns(c) = 2$. This number, as well as the normalized one, $ns(c)/n$, tells us how the domination structure is made up of substructures isolated from each other.

Let c be regular and consider one of its generators g, for example g_6 in Figure 2.1. The *independence ratio*

$$ir(g) = \frac{\min \beta_{in}(g)}{\max \beta_{out}(g)} \geq 1 \tag{2.7}$$

and equality holds if all bonds are equal. The common bond value is then at most $x(g)/\omega(g)$. A large value of $ir(g)$ means that g uses most of its power, expressed by $x(g)$, to ward off domination rather than to dominate others. The value $ir(g) = +\infty$ in particular occurs when all out-bond values are zero (and some in-bond value is positive). In the figure we have $ir(g_6) = 2$, a value that was not big enough to avoid domination by g_3. However if g_6 had changed its *strategy* to, $\beta_{6,1} = 40$, $\beta_{6,2} = 5$, $\beta_{6,3} = 5$, which is consistent with $x(g_6) = 60$, we would get the independence ratio $ir(g_6') = 8$ and g_3 could not dominate the modified generator g_6'.

The strategy of any generator can be divided into two parts. First the independence ratio is selected, as discussed. After that the bond values are chosen without violating (2.4) or (2.7). The way this is done determines how the maximum power $x(g)$ is distributed over the repelling bonds, the in-bonds, and over the attracting bonds, the out-bonds. The *uniform strategy*, that will be assumed in some of the simulation experiments, makes all in-bond values equal, say $= \beta_{in}$, and all out-bond values equal, say β_{out}. Then we must have $\beta_{in} = ir(c)\beta_{out}$ and

$$\omega_{in}\beta_{in} + \omega_{out}\beta_{out} \leq x(g) \tag{2.8}$$

so that

$$\beta_{out} \leq \frac{x(g)}{\omega_{out}(g) + ir(c)\omega_{in}(g)} . \tag{2.9}$$

Another, more drastic, change of strategy would be to change the arities, but this will not be allowed in what follows.

Before we begin to examine the configuration dynamics we must consider closeness relations in $\mathscr{L}[g_1,g_2,\ldots g_n]$. From now on k will enumerate out-bonds, $k = 1,2,\ldots\Omega_{out}$ and ℓ the in-bonds, $\ell = 1,2,\ldots\Omega_{in}$. Consider a regular configuration c and a possible bond couple (k,ℓ).

Let $C_{k\ell}: \mathscr{L}[g_1,g_2,\ldots g_n] \to \mathscr{L}[g_1,g_2,\ldots g_n]$ be the configuration operation that closes the bond $k \to \ell$ if this is allowed by $\mathscr{R}$ and if they were not connected, and leaves them otherwise. Similarly $O_{k\ell}: \mathscr{L}[g_1,g_2,\ldots g_n] \to \mathscr{L}[g_1,g_2,\ldots g_n]$ opens the connection (this is allowed by $\mathscr{R}$ if it was closed) and leaves it unchanged else. All other connections are left as they are. Let T stand for any $C_{k\ell}$ or $O_{k\ell}$ when k and ℓ vary over their possible values.

Starting with an initial configuration $c(0) \in \mathscr{L}[g_1,g_2,\ldots g_n]$ and defining iteratively

$$c(t+1) = T(t)c(t), \quad t = 1,2,\ldots N \tag{2.10}$$

where t represents time, and $T(t)$ is one of $\{C_{k\ell}\}$ or $\{O_{k\ell}\}$ for each t. The Eq. (2.10) gives a history of the way our domination structure has developed. In the next section we shall analyze the probabilistic properties of such a history.

Given $c = \sigma(g_1,g_2,\ldots g_n) \in \mathscr{C}[g_1,g_2,\ldots g_n]$ consider the *neighborhood* of c in configuration space $\mathscr{C}[g_1,g_2,\ldots g_n]$

$$N(c) = \{c' \mid c' = \sigma'(g_1,g_2,\ldots g_n) \text{ and } \sigma = \sigma' \text{ except for one bond couple}\}. \tag{2.11}$$

We shall split up the neighborhood in a positive and a negative part $N(c) = N^+(c) \cup N^-(c)$

$$\begin{cases} N^+(c) = \{c' \mid c' = \sigma'(g_1,g_2,\ldots g_n) \text{ and } \sigma' = \sigma \\ \qquad \text{except that } \sigma' \text{ has one bond couple} \\ \qquad \text{connected that is open in } \sigma\} \\ N^-(c) = \{c' \mid c' = \sigma'(g_1,g_2,\ldots g_n) \text{ and } \sigma' = \sigma \\ \qquad \text{except that } \sigma' \text{ has one bond couple} \\ \qquad \text{unconnected that is closed in } \sigma\}. \end{cases} \tag{2.12}$$

Using the T-operators we see that

$$\begin{cases} C_{k\ell}c \in N^+(c) \cup \{c\} \\ O_{k\ell}c \in N^-(c) \cup \{c\}\ . \end{cases} \tag{2.13}$$

Hence two successive configurations $c(t)$ are either equal or $c(t+1)$ belongs to the positive or negative neighborhoods of $c(t)$. The $c(t)$ function takes only "small" steps.

It should be remarked that the "closeness" relation expressed by these neighborhoods is symmetric. Indeed the statement $c \in N(c')$ is equivalent to the statement $c' \in N(c)$; both mean that one configuration can be obtained by one $C_{k\ell}$ or one $O_{k\ell}$.

We also need the concept of a *marginal set*. Given a set of natural numbers $1 \leq i_1 < i_2 < \ldots i_r \leq n$ we shall denote by

$$\text{marginal}_{i_1,i_2,\ldots i_r} \mathscr{L}[g_1,g_2,\ldots g_n] \tag{2.14}$$

the set of subconfigurations obtained from any $c \in \mathscr{L}[g_1,g_2,\ldots g_n]$ by deleting all g_i's except for $i = i_1,i_2,\ldots i_r$, *and* at the same time leaving bonds from any g_{i_ν} connected that were connected in c.

This construction defines a natural projection mapping

$$\text{project}_{i_1,i_2,\ldots i_r}: \mathscr{L}[g_1,g_2,\ldots g_n] \rightarrow \text{marginal}_{i_1,i_2,\ldots i_r} \mathscr{L}[g_1,g_2,\ldots g_n]. \tag{2.15}$$

Note that all members in the marginal set are regular with respect to $\mathscr{R}$; this follows from the fact that Σ is a monotonic connection type.

Also, when a probability measure P is given on $\mathscr{L}[g_1,g_2,\ldots g_n]$, it induces another one, $P_{i_1,i_2,\ldots i_r}$, on the marginal set by the natural definition

$$P_{i_1,\ldots i_r}(c') = P\{c|\text{projection}_{i_1,\ldots i_r}(c) = c'\} \tag{2.16}$$

for any c' in the marginal set.

7.3. Configuration dynamics

Our dynamical system shall be characterized by a *time-homogeneous probabilistic set up of Markov type.* As all biological systems, domination structures can be expected to have a good deal of randomness in them. This randomness will be specified below in such a way that it does not depend upon absolute time; only relative time plays a role in the dynamics.

If the system changes drastically, for example through births, deaths, emigration, immigration, or through changes of strategy among its members, then *the model will only be applicable during the constant regimes between times of drastic change.* We must then supplement our time-homogeneous model by a stochastic mechanism describing the transition from one regime to another.

The Markov condition means that the development at time t (now to be taken as continuous) will depend only upon the present state, not on the past.

Dynamics: *We shall assume that in the time interval* $(t,t+h)$ *we have*

1) $P[c(t+h) = c(t)|c(t) = c] = 1-h\,q_c + o(h)$

2) $P[c(t+h) \notin N(c) \cup c|c(t) = c] = o(h)$

3a) $P[c(t+h) = c'|c(t) = c] = \lambda_{k\ell}h + o(h)$

if $c' = C_{k\ell}c \neq c$

3b) $P[c(t+h) = c'|c(t) = c] = \mu_{k\ell}h + o(h)$

if $c' = O_{k\ell}c \neq c$.

In 3a) we have a transition to the positive neighborhood $N_+(c)$ and in 3b) to the negative neighborhood $N_-(c)$. The sample functions of this stochastic process (or rather of a separable version) will be piecewise constant and jumps will be "small" as described in the previous section.

To be logically consistent we must have

$$q_c = \sum_{A(c)} \lambda_{k\ell} + \sum_{B(c)} \mu_{k\ell} \tag{3.1}$$

where

$$\begin{cases} A(c) = \{(k,\ell) | C_{k\ell}c \neq c\} \\ B(c) = \{(k,\ell) | O_{k\ell}c \neq c\}. \end{cases} \tag{3.2}$$

The intensities $\lambda_{k\ell}$ and $\mu_{k\ell}$ of creating or annihilating a domination relation will be functions of the out-bond value β' and out-bond value β'', involving only the bonds k and ℓ respectively

$$\begin{cases} \lambda_{k\ell} = \lambda(\beta',\beta'') \\ \mu_{k\ell} = \mu(\beta',\beta''). \end{cases} \tag{3.3}$$

The λ and μ functions should be continuous and λ positive for $\beta' > \beta''$, otherwise arbitrary, while μ should be positive everywhere. We need never consider $\beta' \leq \beta''$, since this offends against local regularity, but in the simulations it will be convenient to define $\lambda(\beta',\beta'')$ to be zero for $\beta' \leq \beta''$.

Starting with an initial regular configuration $c(0)$ at $t = 0$, the above dynamics leads to a probability measure P_t over $\mathscr{L}[g_1,g_2,\ldots g_n]$. What happens as t tends to infinity? The answer is given by

<u>Theorem 1</u>. *The probability measure* P_t *tends to a limit* P, *as* $t \to \infty$, *which is the unique equilibrium measure over* $\mathscr{L}[g_1,g_2,\ldots g_n]$.

Proof: Since $\mathscr{C}[g_1,g_2,\ldots g_n]$ is finite we are dealing with a Markov chain with a finite number of states and with a continuous time parameter. Given any pair of configurations c' and c'' we can find a chain $c' = c_1,c_2,\ldots,c_m = c''$ such that

$$c_{\nu+1} = T_\nu c_\nu, \text{ where } T_\nu \text{ is either an } O_{k\ell} \text{ or a } C_{k\ell}. \tag{3.4}$$

Note that when content(c) is given all the $\lambda_{k\ell},\mu_{k\ell}$ are strictly positive and constant. This means that for any $h > 0$ the transition $c' \to c''$ in $(t,t+h)$ has positive probability. Indeed, the probability that $c(t)$ will remain constant $= c_\nu$ and then jump to $c_{\nu+1}$ and again remain constant in $(t,t+h)$ is easily calculated to be

$$\frac{e^{-hq_{c_\nu}} - e^{hq_{c_{\nu+1}}}}{q_{c_{\nu+1}} - q_{c_\nu}} \lambda_{k\ell} \tag{3.5}$$

if $q_{c_{\nu+1}} \neq q_{c_\nu}$, and to be

$$e^{-hq_{c_\nu}} \lambda_{k\ell} \tag{3.6}$$

if $q_{c_{\nu+1}} = q_{c_\nu}$. Since these probabilities are strictly positive the finite state Markov chain is ergodic so that P_t tends to a limit which is automatically an equilibrium distribution. In an ergodic chain the equilibrium is uniquely determined. Q.E.D.

This result does not tell us anything about the speed of convergence, only that it takes place. To gain some feeling for the time that can be expected to be needed before the system has settled down reasonably close to statistical equilibrium we shall carry out simulation experiments on the computer.

The APL-code for the simulation can be found in the Appendix at the end of the chapter. The function GEN generates a number N = right argument, of generators. The arities are stored in a (N,2) matrix, the first column in-arities, the second one out-arities. The arities are selected at random (uniformly) between 1 and MAXIN and MAXOUT. The x-attributes are given values calling the function XDIST, see below. An empty (0,4)-matrix BOND is set up that will later be updated to contain the closed bond in σ. In BONDLOOP the out and in bond values are stored in a matrix OUT of N rows and MAXOUT columns, similarly for IN. The values are computed as described in Section 2.

The function XDIST generates all x's as i.i.d. from a triangular distribution from 0 to 20. This form is quite arbitrary but is easy to change if so desired. The same is true for the following forms in (3.8) and (3.9).

The function LAM has left argument β_{out} and right argument β_{in} and computes

$$\lambda = \begin{cases} 1-\exp\left[-\left(\frac{\beta_{out}-\beta_{in}}{A2}\right)^{P2}\right] & \text{if } \beta_{out} > \beta_{in} \\ 0 & \text{else} \end{cases} \tag{3.7}$$

The function MU computes

$$\mu = \begin{cases} \exp\left[-\left(\frac{\beta_{out}-\beta_{in}}{A1}\right)^{P1}\right] & \text{if } \beta_{out} > \beta_{in} \\ 1 & \text{else} \end{cases} \tag{3.8}$$

and NU gives

$$\nu = \ell n\, \lambda - \ell n\, \mu\ . \tag{3.9}$$

The main function in this library is DYNAMICS whose right argument is the number of iterations in the simulation for given content(c). It selects an out-bond and an in-bond at random, tests whether it is already closed or not. An open bond is closed with some probability λ, a closed one opened with probability μ as described in Section 3. The BOND array is updated iteratively. It has four columns and NBOND rows. In each row the first element is the number of the generator, the second the number of its out-bond, the third the number of the generator to be dominated, and the fourth one its in-bond number. The NUVALUE and NUVECTOR contain data on the entropy of the configuration; see (3.9).

In ANALYSIS the number of connected subconfigurations, NCS, is computed calling on CONNECT and ELIST, see below. The components themselves are also computed and printed out, together with their sizes. The average size and the standard deviation is also calculated.

In ELIST the right argument is the matrix BOND already mentioned. It forms a list of connected generators and the result is a (NBOND,2)-matrix containing in each row the identifying numbers of two generators connected by a closed bond couple.

This function is needed before executing CONNECT whose right argument is the result of ELIST. It calculates the connected components using a depth-first search algorithm; see Tarjan (1972) and Notes A.

We now execute GENERATE 10 with MAXIN=1, MAXOUT=2 and with full utilization of the power of each generator and uniform strategy over bonds as in (2.9). We get the content(c) shown in Figure 3.1. The numerical values are rounded off.

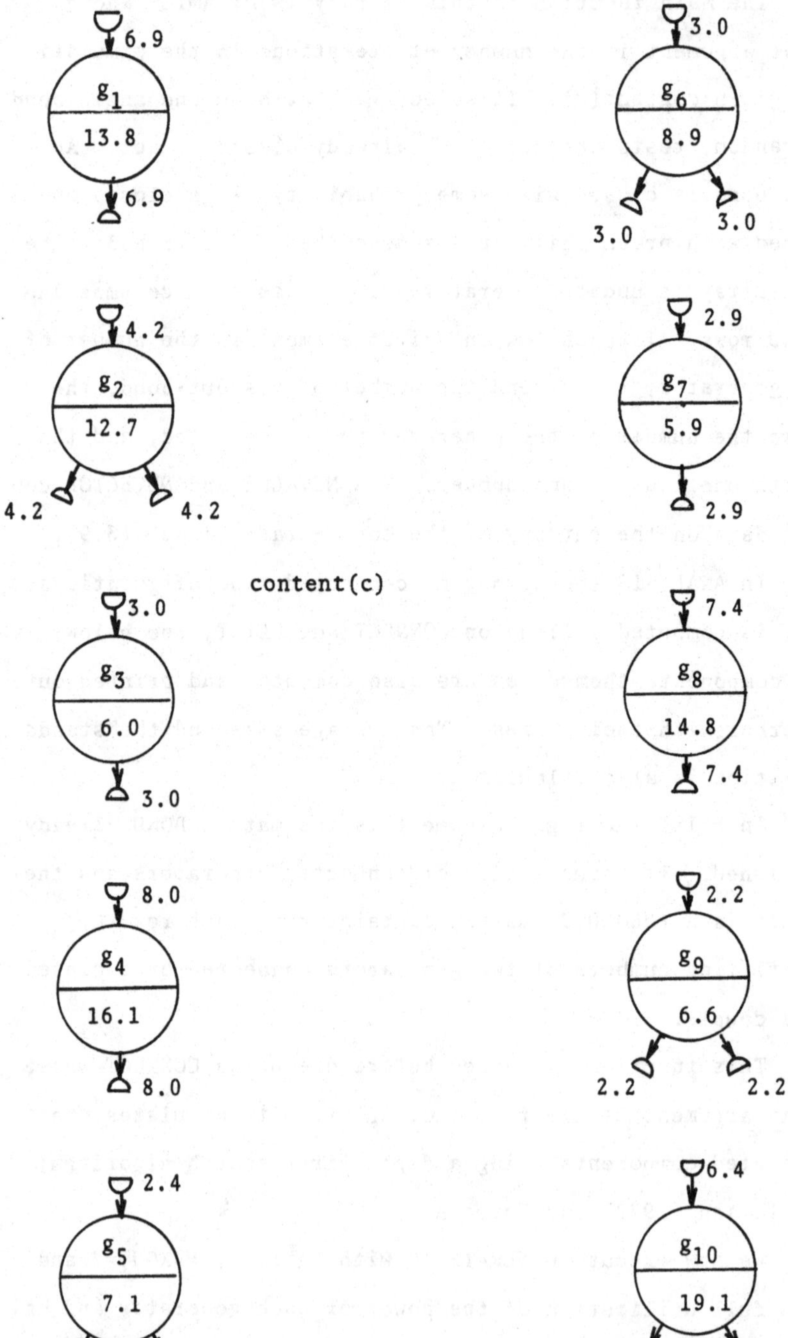

content(c)
g1
13.8
6.9
6.9
g2
12.7
4.2
4.2
4.2
g3
6.0
3.0
3.0
g4
16.1
8.0
8.0
g5
7.1
2.4
2.4
2.4
g6
8.9
3.0
3.0
3.0
g7
5.9
2.9
2.9
g8
14.8
7.4
7.4
g9
6.6
2.2
2.2
2.2
g10
19.1
6.4
6.4
6.4

Figure 3.1

To simulate this dynamical system we begin by executing DYNAMICS 10 followed by ANALYSIS. This gives us the configuration shown in Figure 3.2. Only 3 bonds have been established, two of them from generators with out-arity 1. The latter is reasonable since greater out-arity leads to lower out-bond values (for uniform strategy) and hence less power to dominate other generators.

Now 10 more iterations; the result is shown in Figure 3.3. Two more bonds have been closed, resulting in the components (1,4), (2,6,10), (7,8,9) and the rest of the generators isolated. The not-isolated rulers are 4, 10, 8.

We now iterate 30 times more and get c(50) in Figure 3.4. One more bond has been closed, namely the bond couple (3,1) → (5,1). We have opened (8.1) → (7.1).

After 80 more iterations we get c(130) in Figure 3.5. We have now opened more bonds, namely the bond couples (7,1) → (9,1) and (1,1) → (10,1). On the other hand we have closed the bond couples (2,1) → (7,1), (8,1) → (9,1), (1,1) → (3,1). We have now only 3 components, namely (1,3,4,5) with the ruler 4, (2,6,7,10) with the ruler 10, and (8,9) with the ruler 8. There are no isolated generators left. Note that the second component has a tree connector. We are now close to equilibrium, as far as the number of bonds is concerned, bonds will continue to open and close indefinitely, but with the C and O operations tending to balancing each other.

Another 100 iterations gives c(230) displayed in Figure 3.6. We have opened the bond couples (4,1) → (1,1), (3,1) → (5,1), (2,1) → (7,1) and closed (7,1) → (5,1) and (4,1) → (8,1). We now have 4 components, namely (1,3) with ruler

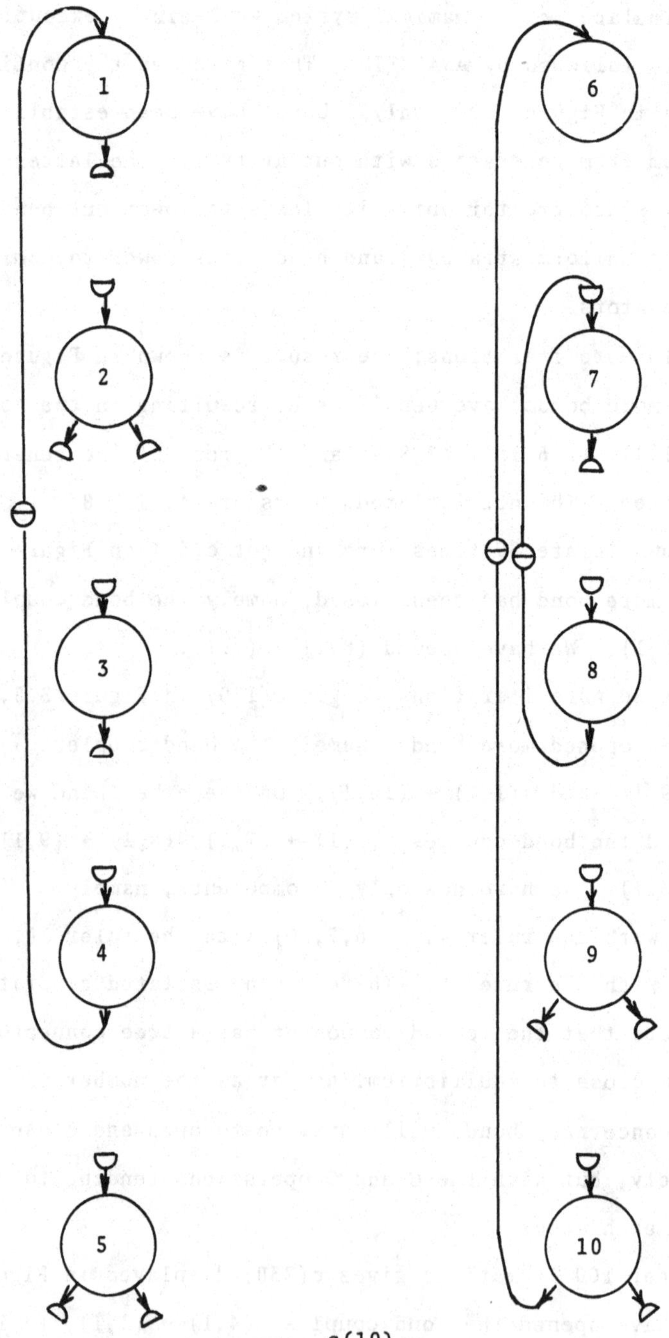

c(10)

Figure 3.2

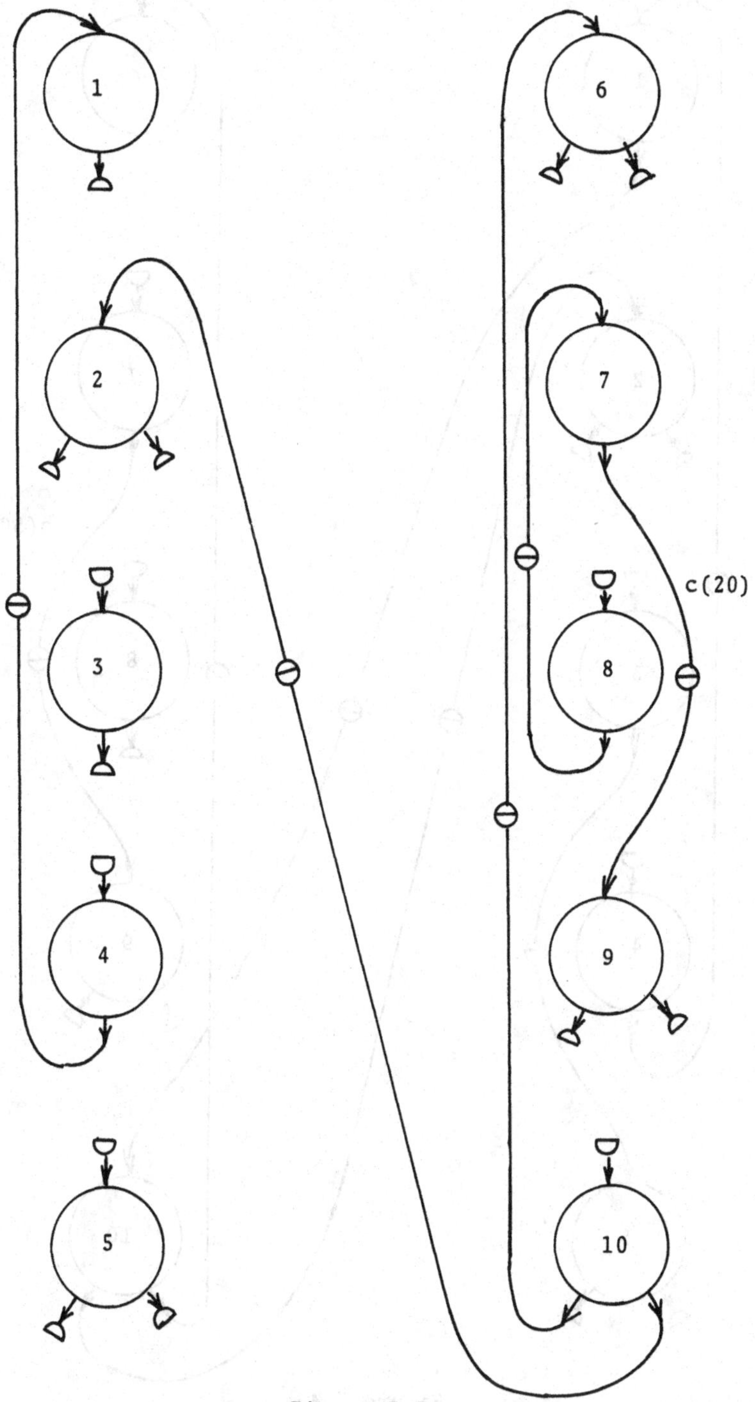

Figure 3.3

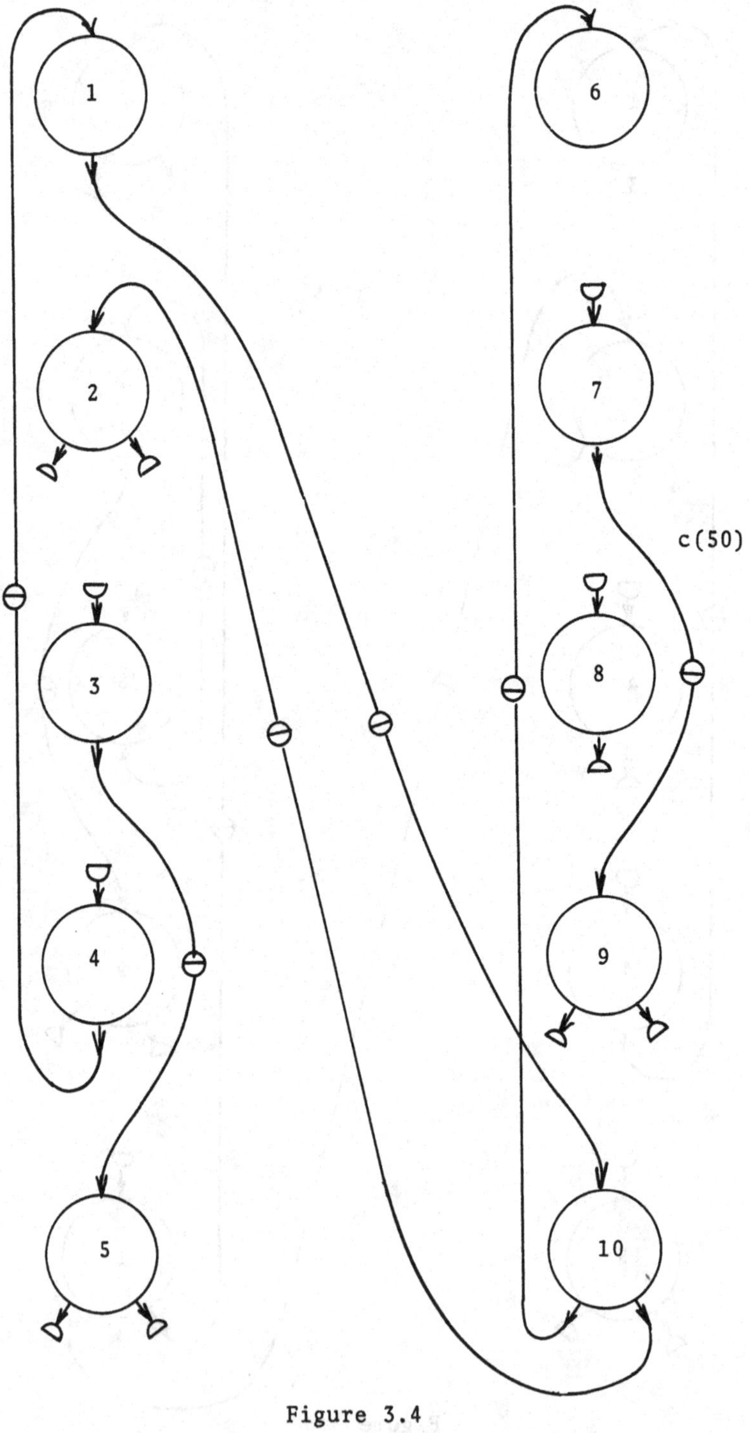

Figure 3.4

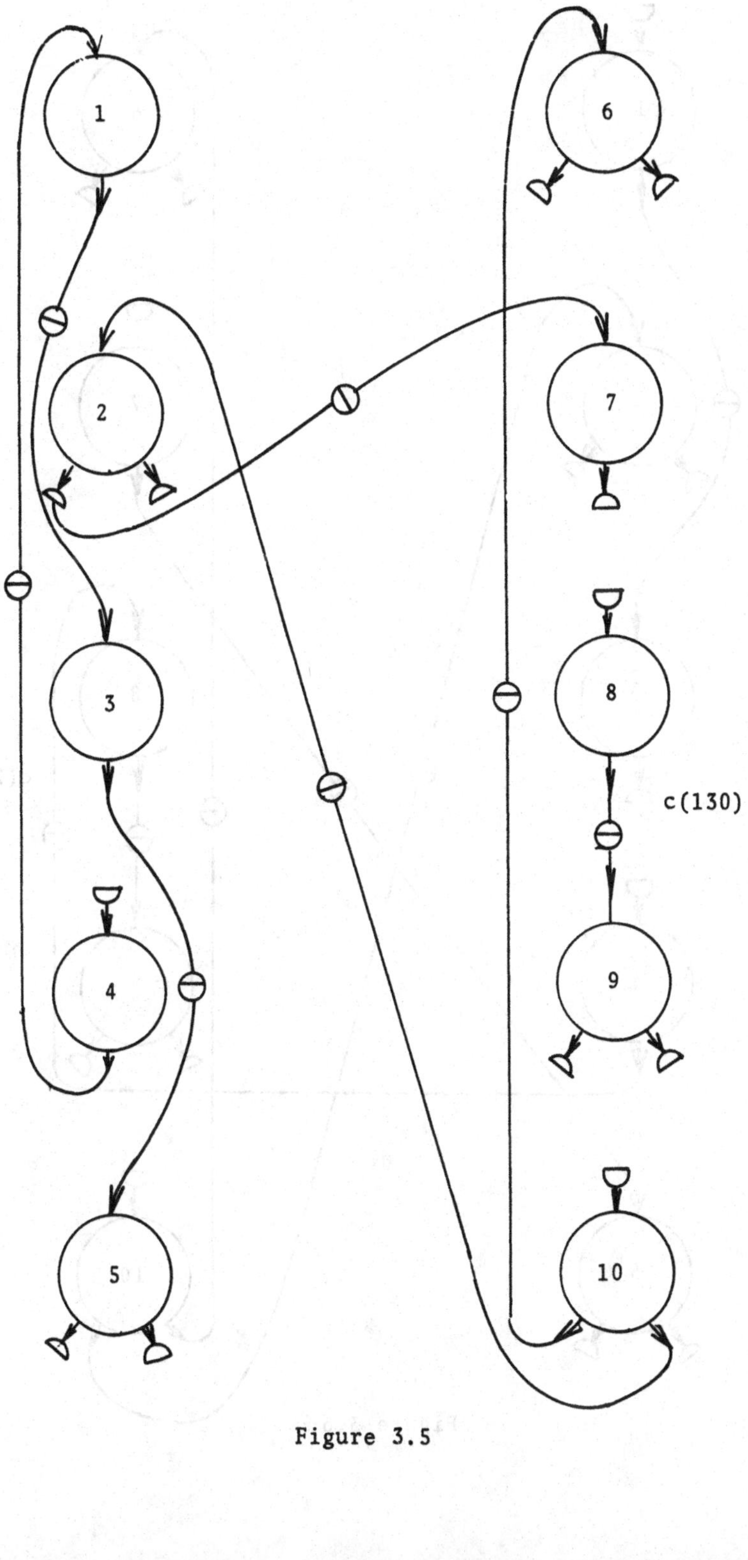

Figure 3.5

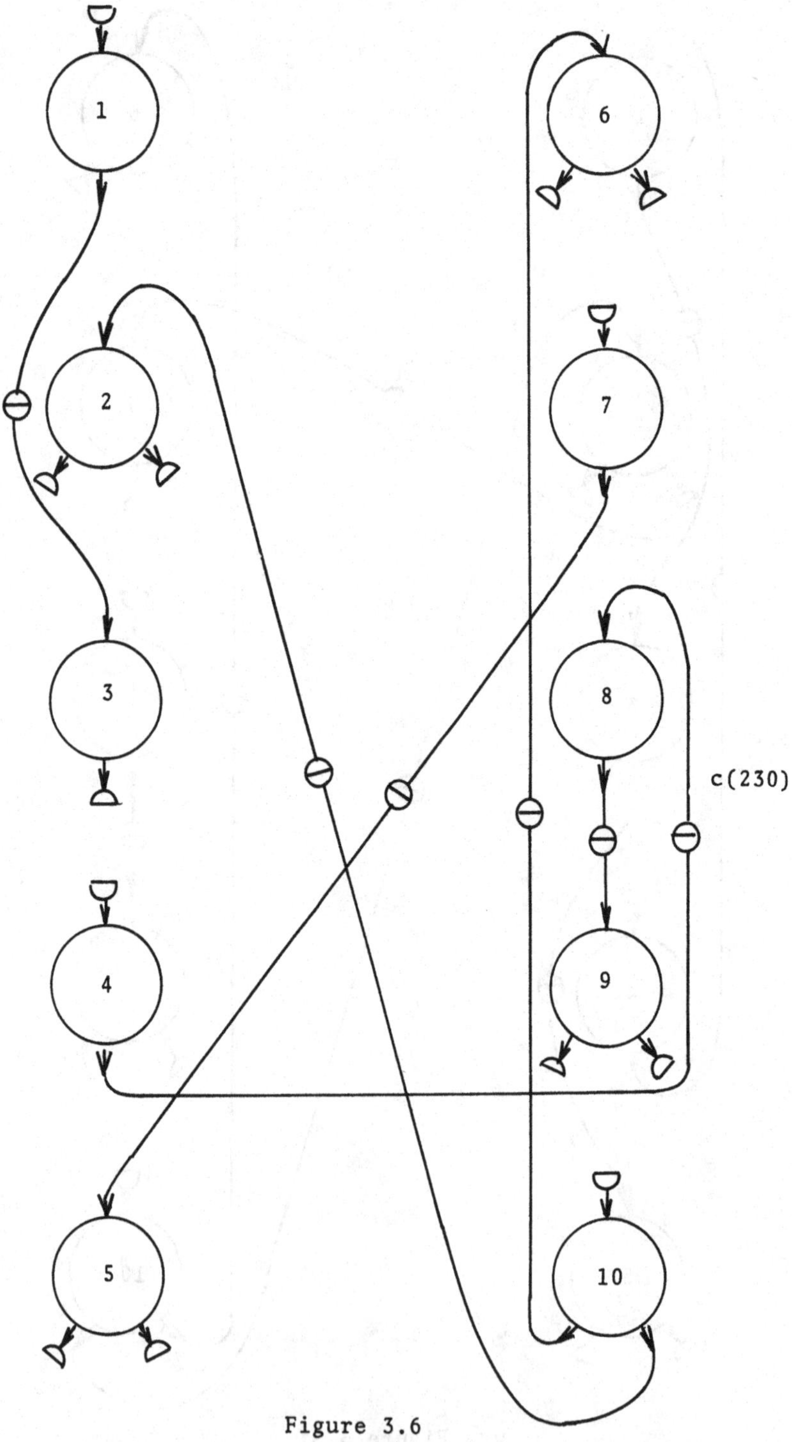

Figure 3.6

1,(2,6,10) with ruler 10, (4,8,9) with ruler 4, and (5,7) with ruler 7.

In Figures 3.7-3.9 we have graphed some configuration statistics as functions of time. In 3.7 the number of closed bonds is shown, in 3.8 the number of components, and in 3.9 a quantity proportional to $\ell n\ p(c)$ in the equilibrium distribution (to be discussed in the next section). They give approximately the same impression: equilibrium is reached to a fair degree after around $t = 100$.

This is of course only a single experiment with fixed content(c). Similar experiments with other realizations of content(c), but with parameters of the same order of magnitude give approximately the same impression.

The time it takes to reach equilibrium depends upon what initial configuration has been chosen. In all our experiments we have started with all generators isolated, NB = 0.

Larger configurations probably take longer to reach statistical equilibrium. It would be useful to get analytic bounds on the *time constant* of the system, even crude ones. At present we do not have any such bounds.

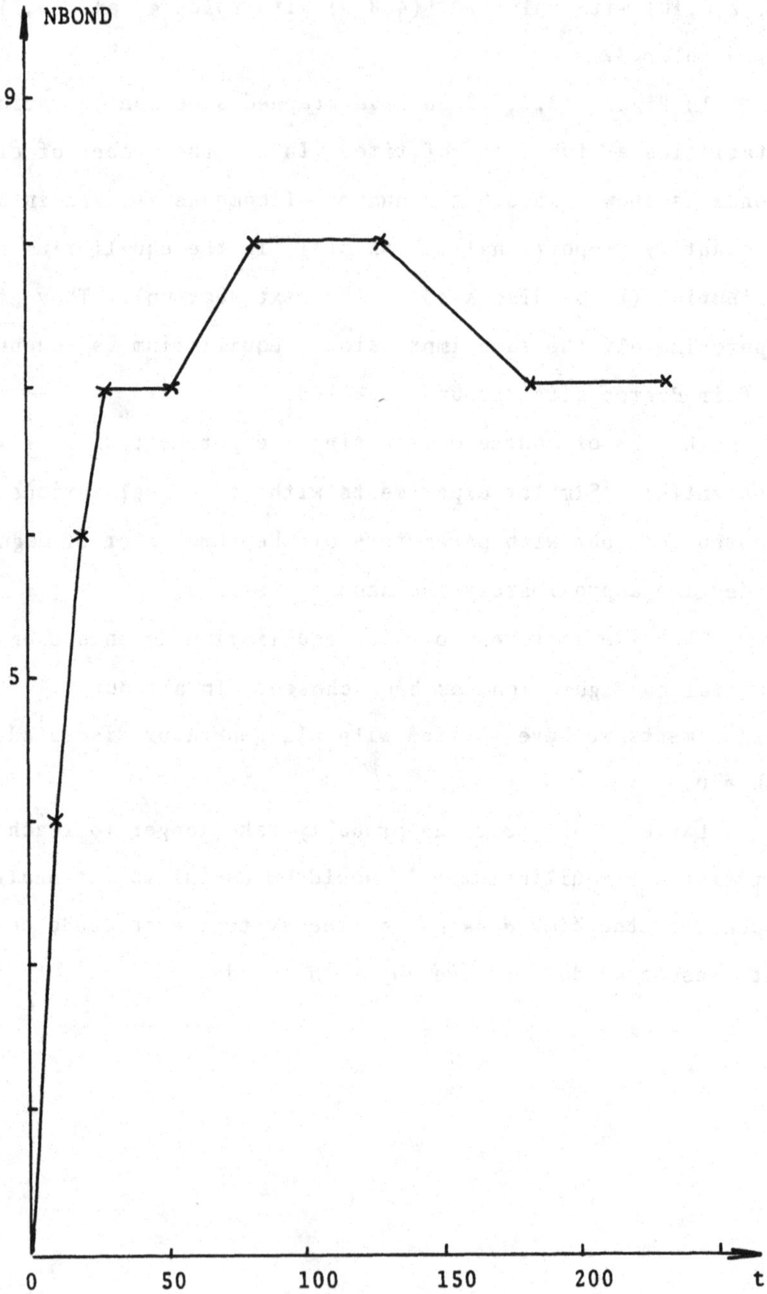

Figure 3.7

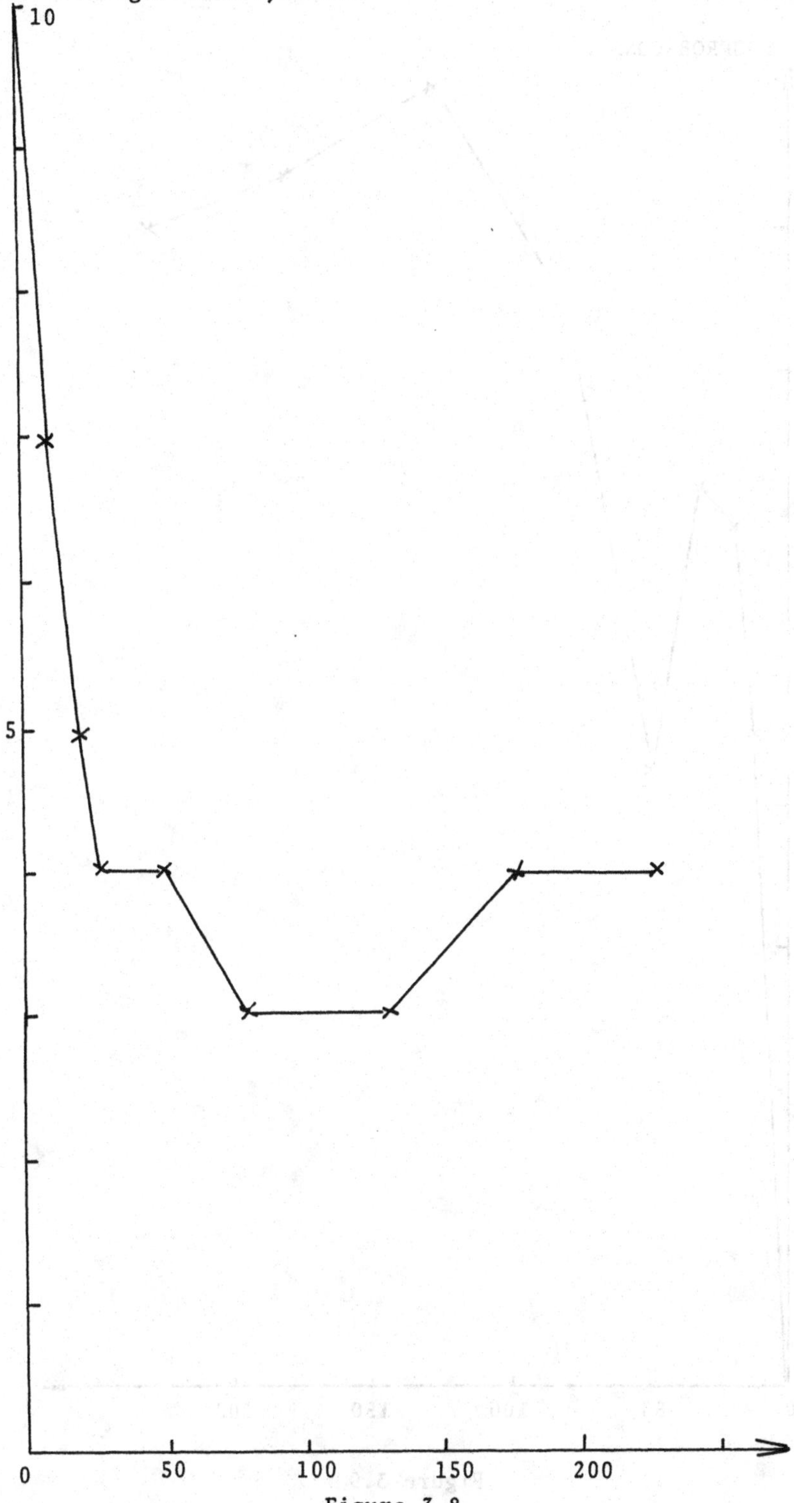

Figure 3.8

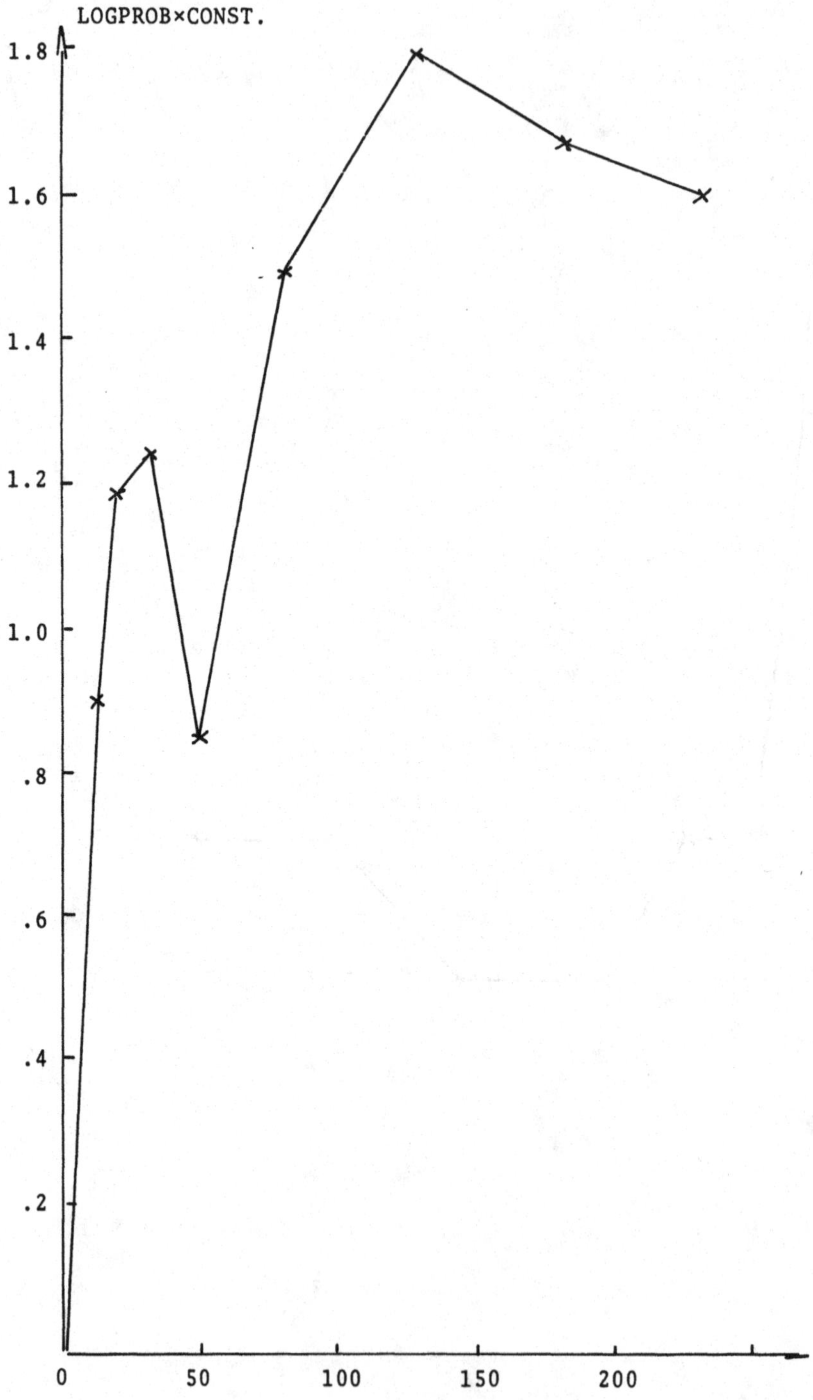

Figure 3.9

7.4. System in equilibrium

The limiting distribution P seems to be approached quickly, at least in the cases we have studied so far. To be able to proceed we must learn more of its analytical properties which will be needed in the later sections.

Theorem 1. *For a regular configuration* $c = \sigma(g_1,g_2,\ldots g_n)$ *the equilibrium distribution has the frequency function* p *over* $\mathscr{C}[g_1,g_2,\ldots g_n]$

$$p(c) = Z^{-1} \prod_{(k,\ell)\in\sigma} \frac{\lambda_{k\ell}}{\mu_{k\ell}} \tag{4.1}$$

where the product is extended over the bond couples (k,ℓ) *connected by* σ. *In* (4.1) Z *is a constant, the partition function, that should be chosen so that the sum of all* P(c) *equals one.*

Proof: We shall first show that

$$p(c) = Z^{-1} \prod_{(k,\ell)\in\sigma} \lambda_{k\ell} \prod_{(k,\ell)\in\sigma} \mu_{k\ell} \tag{7.2}$$

(with a different value of Z) is an equilibrium distribution. Recall that we know that the equilibrium solution of a dynamical system is uniquely determined. The equations of equilibrium are

$$p(c)q(c) = \sum_{c'} p(c')q(c' \to c) \tag{4.3}$$

where the sum is extended over all c' such that $c' \in N(c)$ and

$$q(c' \to c) = \begin{cases} \lambda_{k\ell} & \text{if } c = C_{k\ell}c' \\ \mu_{k\ell} & \text{if } c = O_{k\ell}c'. \end{cases} \tag{4.4}$$

We also have $q(c)$ expressed in terms of $\lambda_{k\ell}$ and $\mu_{k\ell}$ in Eq. (3.1), so that we can write

$$q(c) = \sum_{c'} q(c \to c') \tag{4.5}$$

with the sum extended over $N(c)$.

Say now that $c' = \sigma'(g_1, g_2, \ldots g_n)$ has a bond couple (ν,μ) connected but open in c; otherwise with the same connector so that $c' \in N_+(c)$. That corresponds to a term on the left side of (4.2) of the form

$$Z^{-1} \prod_{(k,\ell)\in\sigma}{}' \lambda_{k\ell} \prod_{(k,\ell)\in\sigma}{}' \mu_{k\ell} \cdot \mu_{\nu\mu} \cdot \lambda_{\nu\mu} \tag{4.6}$$

where the prime on the products indicates that they do not include $(k,\ell) = (\nu,\mu)$, where $\mu_{\nu\mu}$ is due to $(\nu,\mu) \notin \sigma$ and $\lambda_{\nu\mu}$ is due to a term in (4.5) since σ' has the bond (ν,μ) closed.

But on the right side of (4.3) we will have some $c' = \sigma'(g_1, g_2, \ldots g_n)$. This will give rise to a term

$$Z^{-1} \prod_{(k,\ell)\in\sigma}{}' \lambda_{k\ell} \prod_{(k,\ell)\in\sigma}{}' \mu_{k\ell} \cdot \lambda_{\nu\mu} \cdot \mu_{\nu\mu} \tag{4.7}$$

where the factor $\lambda_{\nu\mu}$ is due to the fact that (ν,μ) is closed in σ' while $\mu_{\nu\mu}$ comes from opening the bond to get c. But (4.7) equals (4.6).

We can argue in the same way when $c' \in N_-(c)$ so that to get to it from c we must close a bond. In either case we encounter the same terms on both sides of (4.3), so that equality holds.

Divide the right side of (4.2) by the product of *all* $\mu_{k\ell}$, and modify the constant Z by multiplying it by the same number. Then some of the $\mu_{k\ell}$ factors cancel and the result is the one stated in (4.1). Hence this p is the equilibrium frequency function. Q.E.D.

It will be convenient to write (4.1) in the form

$$p(c) = Z^{-1} \prod_{(k,\ell)\in\sigma} \kappa_{k\ell} \tag{4.8}$$

where $\kappa_{k\ell} = \lambda_{k\ell}/\mu_{k\ell}$. This ratio between the intensities for closing and opening bonds is what really matters, not the λ and μ values themselves. The absolute values of λ and μ influence the speed of convergence to the equilibrium but not the equilibrium itself. It can also be expressed in terms of the *bond affinity* $a_{k\ell} = \ell n\ \kappa_{k\ell}$.

Remark. Introducing the *interaction energy*

$$H_k = -a_k \tag{4.9}$$

the frequency function p can be expressed as

$$p(c) = Z^{-1} \exp[-H(c)] \tag{4.10}$$

where the total interaction energy can be expressed in the *total affinity* $a(c)$ of the configuration

$$H(c) = \sum_{(k,\ell)\in\sigma} H_{k\ell} = -\sum_{(k,\ell)\in\sigma} a_{k\ell} = -a(c). \tag{4.11}$$

This is exactly the form we have postulated repeatedly for *regularity controlled probabilities*, and it seems remarkable that we have arrived at exactly this model from what seems to be a quite different starting point.

Returning to (4.8), the $\kappa_{k\ell}$ can take only certain values that we shall enumerate as κ_ν and we shall use the corresponding bond frequencies

$$n_\nu(\sigma) = n_\nu(c) = \#(\text{bond couples in } \sigma \text{ for which } \kappa_{k\ell}=\kappa_\nu). \quad (4.12)$$

Then

$$p(c) = Z^{-1} \prod_\nu \kappa_\nu^{n_\nu} \quad (4.13)$$

a form that will be very useful in Section 6.

Consider now the time average of $\ell n\ p[c(t)]$, the quantity graphed in Figure 3.9 for the simulation experiment,

$$\frac{1}{T}\int_0^T \ell n\ p[c(t)]dt. \quad (4.14)$$

Since $c(t)$ is (at least asymptotically) a stationary ergodic process the individual ergodic theorem tells us that (4.14) converges a.c. to the limit

$$\sum_{c\in \mathscr{C}[g_1,\dots g_n]} \ell n[p(c)]p(c) = E[a(c)] - \ell n\ Z. \quad (4.15)$$

Except for an additive constant this is *the expected total affinity of the random configuration.* The additive constant only amounts to shifting the affinity level by choosing a new zero point.

It should be noted that the limit in (4.15) is also the entropy of the dynamical system.

Let us now *relate the algebraic properties of the configurations to the probabilistic ones* that we have just established. Introduce the three configuration spaces, over the same regularity $\mathscr{R}$ as before,

$$\begin{cases} \mathscr{C}_1 = \mathscr{C}[g_1,g_2,\dots g_{n_1}] \\ \mathscr{C}_2 = \mathscr{C}[g_1',g_2',\dots g_{n_2}'] \\ \mathscr{C} = \mathscr{C}[g_1,\dots g_{n_1},g_1',\dots g_{n_2}']. \end{cases} \quad (4.16)$$

Any $c \in \mathscr{C}$ can be written as $\sigma(c_1,c_2)$; $c_1 \in \mathscr{C}_1$, $c_2 \in \mathscr{C}_2$, in a way that determines σ uniquely. Let the same symbol σ stand for the bonds closed by σ. We then have

Theorem 2. *The measures* P_1, P_2, *and* P *over* $\mathcal{C}_1$, $\mathcal{C}_2$, *and* $\mathcal{C}$ *respectively have frequency functions* p_1, p_2, *and* p, *satisfying*

$$p(c) = \text{constant } p_1(c_1)p_2(c_2) \prod_{(k,\ell)\in\sigma} \kappa_{k\ell}. \tag{4.17}$$

Proof: The expression (4.1) gives us immediately

$$\begin{cases} p_1(c_1) = Z_1^{-1} \prod_{(k,\ell)\in\sigma_1} \kappa_{k\ell} \\ p_2(c_2) = Z_2^{-1} \prod_{(k,\ell)\in\sigma_2} \kappa_{k\ell} \end{cases} \tag{4.18}$$

where σ_1 and σ_2 stand for the closed bond couples of c_1 and c_2. Together with (4.1) it gives, since

connector of c =
{closed inner bonds of c_1}∪{closed inner bonds of c_2}
∪{closed bonds between c_1 and c_2}; (4.19)

the relation between the probabilities

$$p(c) = Z^{-1} \prod_{\substack{\text{inner bonds}\\ \text{of } c_1}} \kappa_{k\ell} \prod_{\substack{\text{inner bonds}\\ \text{of } c_2}} \kappa_{k\ell} \prod_{\substack{\text{bonds between}\\ c_1 \text{ and } c_2}} \kappa_{k\ell}$$

$$= Z^{-1}Z_1Z_2\, p(c_1)p(c_2) \prod_{(k,\ell)\in\sigma} \kappa_{k\ell} \tag{4.20}$$

where Z_1 and Z_2 are the partition functions belonging to P_1 and P_2. Choosing the constant in (4.17) as $Z^{-1}Z_1Z_2$ we get the stated result. Q.E.D.

This means that the probability of obtaining in $\mathcal{C}$ two configurations connected to one another is proportional to the product of their respective probabilities *times* a factor depending upon the way c_1 is coupled to c_2. If the total affinity of σ, the coupling connector, is fixed to constant we have conditional independence between marginal $(c)_{1,2,\ldots n_1}$ and marginal $(c)_{n_1+1,\ldots n_1+n_2}$; otherwise not.

This leads to a serious analytical complication. When we are going to study large configurations, $n \to \infty$, we cannot just decompose them into stochastically independent subconfigurations and appeal to classical limit theorems from probability theory. Such limit theorems are for the most part stated for independence, or for, in some sense, limited dependence. It is not obvious how to deal with this, but in the next section we shall show how to overcome this difficulty.

Lemma 4.1. *Similar configurations are equally likely.*

Proof: If $c = \sigma(g_1, g_2, \ldots g_n)$ and $c' = \sigma(g_1', g_2', \ldots g_n')$ are similar there exists a permutation $G \to G'$ such that $g_\nu \to g_\nu'$ preserving bonds: $B(g_\nu) = B(g_\nu')$, see Section 2. Since c and c' must have the same connector σ they have the same bond structure and we can use a single system of configuration coordinates to describe both. If σ connects the bond k to the bond ℓ then

$$\begin{cases} \lambda_{k\ell} = \lambda(\beta_k, \beta_\ell) = \lambda(\beta_k', \beta_\ell') \\ \mu_{k\ell} = \mu(\beta_k, \beta_\ell) = \mu(\beta_k', \beta_\ell') \end{cases} \tag{4.21}$$

since $\beta_k = \beta_k'$, $\beta_\ell = \beta_\ell'$. Hence $\kappa_{k\ell} = \kappa_{k\ell}'$, the affinities are the same for the two configurations, and (4.10) and (4.11) imply that $p(c) = p(c')$. Q.E.D.

Once $c(t)$ has reached a particular configuration c one can ask how quickly does it move on to other configurations, how unstable is c? This depends upon $q(c \to c')$, $c' \in c$ in the neighborhood.

<u>Lemma 2</u>. *The instability of* c *is given by*

$$\sum_{C_{k\ell} c \in N_+(c)} \lambda_{k\ell} + \sum_{O_{k\ell} c \in N_-(c)} \mu_{k\ell}. \tag{4.22}$$

<u>Proof</u>: Follows directly from (3.2). Q.E.D.

Note that the probability of a configuration can be expressed in terms of the sum of the affinities of all its closed bonds, while its instability is the sum of the opening and closing bond intensities associated with operators leading to the neighborhood configurations.

<u>Lemma 3</u>. *Given disjoint sets* $\{K_\nu;\ \nu = 1,2,\ldots r\}$ *of out-bonds and of in-bonds* $\{L_\nu;\ \nu = 1,2,\ldots r\}$ *consider the event* E *that any out-bond in* K_ν *is either open or connected to an in-bond in* L_ν *and any in-bond in* L_ν *is either open or connected to one in* K_ν. *Conditioned by* E *the connectivities of the* r *groups of out-in-bonds* $K_\nu \to L_\nu$ *are stochastically independent.*

<u>Proof</u>: For any $c = \sigma(g_1, g_2, \ldots g_n) \in E$ we can write its probability as

$$p(c) = Z^{-1}\left[\prod_{\nu=1}^{r} \prod_{\substack{k \in K_\nu \\ \ell \in L_\nu}} \kappa_{k\ell}^{e_{k\ell}}\right] \times \prod_{\sigma'} \kappa_{k\ell} = \left[\prod_{\nu=1}^{r} P_\nu\right] \times P \tag{4.23}$$

where $e_{k\ell} = 1$ if k connects to ℓ and zero otherwise. The connector σ' contains all connections in σ that are not taken care of in the earlier products. To see that this is true consider an arbitrary bond couple (r,s), open or not in σ. If $r \in K_\nu$ then $\kappa_{k\ell}$ can appear as a factor in (4.8) only if $s \in L_\nu$. The factor then belongs to P_ν in (4.23). On the other hand, if r belongs to no K_ν then (r,s) is either open, so that κ_{rs} does not appear in (4.8) or s is not an element in any L_ν. Hence we get the factoring in (4.23). But then the connectivity of $K_\nu \to L_\nu$ bonds is described by the factor P_ν, and the multiplicative form of (4.23) establishes the conditional independence.

Q.E.D.

Remark. Bonds are certainly not stochastically independent under (4.8), this is easily demonstrated by small examples. Lemma 3 gives us instead a weaker form of (conditional) independence. It is not known if the conditional independence described in the lemma is also a sufficient condition for (4.8) to hold.

Remark. It is perhaps tempting to believe that the marginal probability of a bond couple (k,ℓ) to be closed would be

$$\frac{\kappa_{k\ell}}{1+\kappa_{k\ell}} = \frac{\lambda_{k\ell}}{\mu_{k\ell}+\lambda_{k\ell}} . \tag{4.24}$$

This is not true. However, we believe, without proof, (4.24) holds asymptotically for small κ's.

We now come to the crucial concept of a *typical configuration* in P over $\mathcal{L}[g_1, g_2, \ldots g_n]$. Since the configuration space does not possess linear structure we do not have access to integrals (expected values) with respect to P. Instead one would be tempted to select the configuration for which $p(c)$ attains its maximum, or one of them if the maximum is attained for more than one configuration.

In other words we would choose the *mode* of P, c_{mode}, as a typical representative. Because of (4.10) and (4.11) we should solve

$$\sum_{(k,\ell)\in\sigma} a_{k\ell} = \max_{\sigma} . \tag{4.24}$$

In the graph where the set of out-bonds are connected (legally) to the set of in-bonds we should select a subgraph consisting of a subset of the edges indicated and where no vertex appears more than once. But this is the problem of *maximum matching* in graph theory for which fast algorithms exist.

There is however another, more intrinsic, difficulty associated with using the mode as a representor. To bring this out clearly let us look at the simple case shown in Figure 4.1. Here we have 7 out-bonds, grouped in two subsets, B_{out1} and B_{out2}, and 6 in-bonds, grouped in two subsets B_{in1}, B_{in2}. Assume that the affinities from any $k \in B_{out1}$ to any $\ell \in B_{in1}$ are the same, a_{11}, the affinities from any $k \in B_{out1}$ to any $\ell \in B_{in2}$, are the same, a_{12}, and so on as indicated in the figure.

Say that the maximum in (4.24) is attained by choosing 2 bonds from B_{out1} to B_{in1} and one to B_{in1}; also 2 bonds from B_{out2} to B_{in1} and 2 bonds to B_{in2}. Since $p(c)$ depends only upon the total affinity it is clear that we can attain the mode in many ways, namely

$$\frac{3!}{2!1!}\,\frac{4!}{2!1!1!}\,\frac{4!}{2!2!}\,\frac{2!}{1!1!}\,2!2!1!1! = 1706. \tag{4.25}$$

The lack of uniqueness is not serious in itself - it is all right to have a set of representors rather than a single one - but it is clear that the number will increase extremely fast as n increases. This means that when we compare the total probability contained in the mode set and compare it to

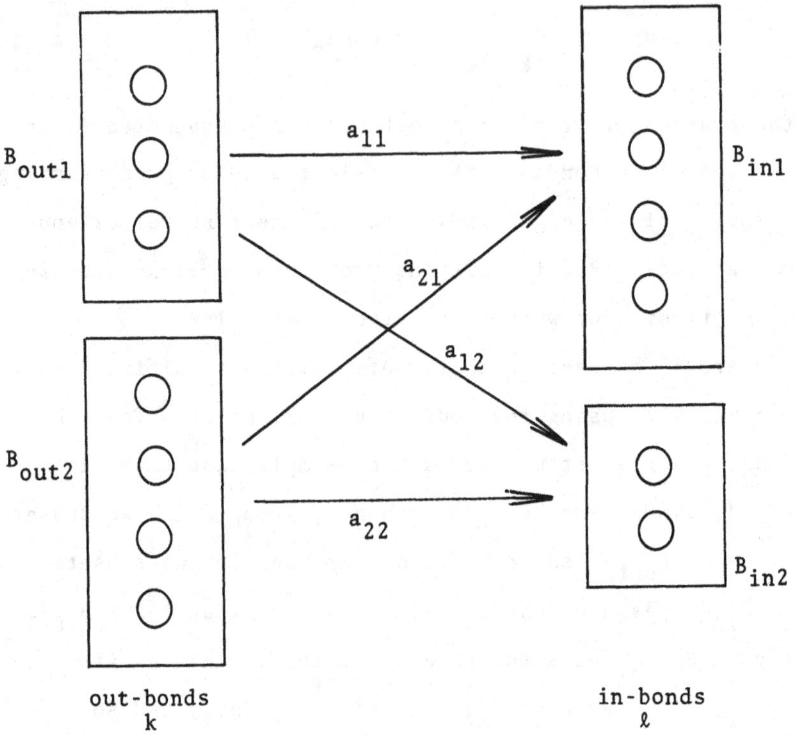

Figure 4.1

alternative choices of distributing bonds between the groups (only the sum of affinities matter!) it may be that alternative sets of constant total affinity may contain larger mass although not situated at the mode set. This would be due to the larger number of elements in the set, and in spite of the fact that the probability of individual configurations is smaller.

This was just an example but the same holds in general. Let us split up the out-bonds in groups with the same, or almost the same, out-bond value. Do the same for the in-bonds. Then the total affinity of a configuration depends only (exactly or approximately) on the number of bonds from group to group. We should therefore *choose as our representor a configuration whose group-to-group frequencies maximize* p(c)·#(elements in the set); the set being all elements with probability equal to p(c).

Note that all configurations in the set have the same probability so that the conditional distribution over the *representative set* is uniform. How to calculate the representative set is another question, that we shall return to in Section 6.

A reader familiar with statistical mechanics, especially Bose-Einstein and Fermi-Dirac models, will recognize their resemblance to this approach to selecting representors.

7.5. Large configurations - simulation results

So far we have kept the configuration size $n = \#(c)$ constant and let time t increase in order to get convergence to the equilibrium distribution. We now approach the more difficult question for the equilibrium distributions as n tends to infinity: *as the configuration size is made large, can we assert that the configuration in some sense tends to a typical one with large probability?* Are there any laws of large numbers?

We have already pointed out that the classical limit theorems in the calculus of probability assume stochastic independence, or some variation on this theme, and therefore do not provide an answer to our question: here we are dealing with *interacting individuals*; their social coupling is what is of interest.

But in statistical mechanics we also have mass phenomena with interacting particles and can still claim that *macroscopic limits* exist. This is encouraging and we shall carry out simulation experiments to guide us further. When we do this, executing the programs in the Appendix, we must make sure that the number of iterations is enough to bring the structure close to statistical equilibrium. For $n = 30$ it seems that about 500 iterations suffice, but we have occasionally used more to be safe. This seemed to change the results little if at all.

We found already in the first few experiments that the *herds* - the connected components of c - grow fast in size. This is of course compensated by a slow growth of the number NCS of components. The number of bonds NBOND is also growing fast.

This is not surprising. What we did not expect, however, is that *the configuration tended to be saturated,* few bond couples that could be legally closed under $\mathcal{R}$ remained open. More precisely, almost all in-bonds that could be legally connected to any out-bond were indeed connected. What is the reason for this peculiar behavior?

Suppose several in-bond values belonging to unconnected in-bonds are small enough to encourage several unconnected out-bonds to try to connect with them. Even if these two numbers of bonds are moderate the number of combinations (the product of the two) is large which will tend to make them connect rapidly. The same reasoning does not apply to the opening of already connected bonds: each one has some probability of disconnecting but with no multiplying factor due to the combinatorial effect.

This explanation also leads us to an important modification of the model in the DYNAMICS of Section 3. When n increases the possibility of an out-bond to connect will grow in relation to the number of unconnected in-bonds. This is unnatural; it means that the influence of a generator increases with n. To compensate for this undesirable effect we shall assume *a modified closing bond intensity*

$$\lambda_{mod}(\beta,\beta') = \frac{1}{n}\lambda(\beta,\beta'). \tag{5.1}$$

Note that the above reasoning leads to *no modification of the opening bond intensity* $\mu(\beta,\beta')$. Hence we shall have

$$\kappa_{mod}(\beta,\beta') = \frac{1}{n}\kappa(\beta,\beta'). \tag{5.2}$$

The modification can also be given the following interpretation: *a generator has mainly a local influence, it connects willingly to some close generators but less so with more distant one,* where "close" can mean geographically close, socially close, etc. Although we believe that the modified dynamics is the more natural one, the previous version deserves more attention than we will give it here. If the configuration size n is constant the two versions are of course equivalent except for a scale change.

Executing the program DYNAMICS with the appropriate change in line [2] of LAM, see Appendix, we have obtained the following experimental data.

Plotting the relative number of closed bonds NBOND N we get the remarkably stable curve in Figure 5.1. It points to the existence of a limit around the value 1/2 for the present choice of parameters.

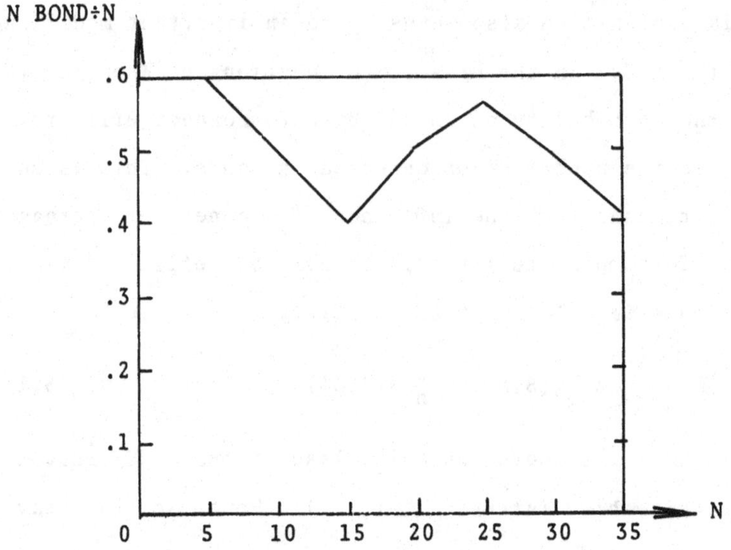

Figure 5.1

In Figure 5.2 we have plotted the relative number of components NCS÷N. It also seems to settle down around the value 1/2.

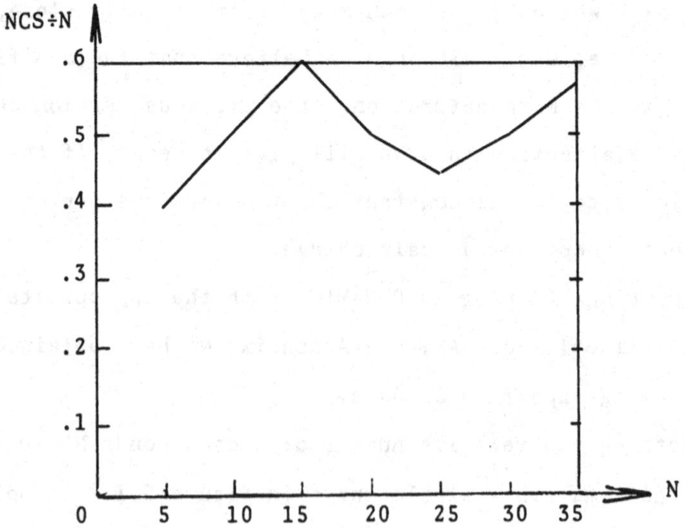

Figure 5.2

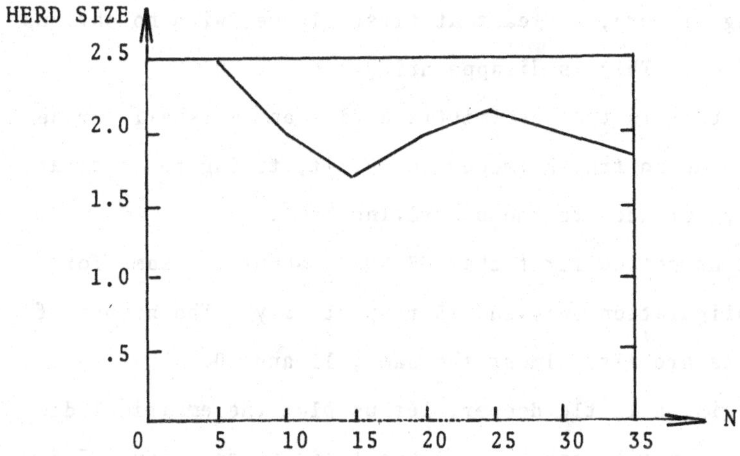

Figure 5.3

The average size of each herd is calculated in the ANALYSIS program and we display the number (the absolute one, not relative in Figure 5.3. There is a clear tendency towards a limit around the value 2.

The unmodified model did not show the limiting behavior that we have inferred from these three graphs.

The graph in Figure 5.1 varies in the opposite sense to that of Figure 5.2. This is reasonable since many connected bonds leads to few components. It will also make the average herd size bigger, see Figure 5.3.

We have also studied the standard deviation of the component sizes for a given configuration. They are consistently a good deal lower than the square root of the average, which indicates that the size distribution is far from Poisson. We have not formed any conjecture on the possible limit of the distribution as n tends to infinity.

Now let us look at two configurations over the same $\mathscr{L}[g_1,g_2,\ldots g_{25}]$ shown in Figures 5.4-5.5. They present a confusing picture, at least at first glance, with no obvious resemblance. This is disappointing.

But this is too pessimistic a view and we shall pursue the question in true hermeneutic spirit, trying to penetrate under the surface to the underlying laws.

Let us notice first that NBOND is about the same for both configurations, 14 and 15 respectively. The number of components are also almost the same, 11 and 10.

Digging a little deeper, let us plot the empirical distribution function for the component sizes, see Figure 5.6. The two graphs are fairly close indicating that *the statistical topologies of the two configurations are also close.* The roles of the individual generators vary a good deal, but the two connectivities are qualitatively similar to each other.

In a slightly larger simulation, $n = 40$, we have repeated the generation of P several times, keeping content(c) fixed, and display the result in Figures 5.7-5.8. Here we have chosen MAXIN=MAXOUT=2, so that Σ is no longer FOREST, but has full POSET structure.

The complete configuration diagrams would present an even more confusing picture than the ones in Figures 5.4-5.5. Therefore we have displayed the components of the configurations together and left out the unconnected bonds.

Comparing the two diagrams we see that, again, individual generators appear in quite different roles. At the same time it is striking how the two topologies resemble each other statistically.

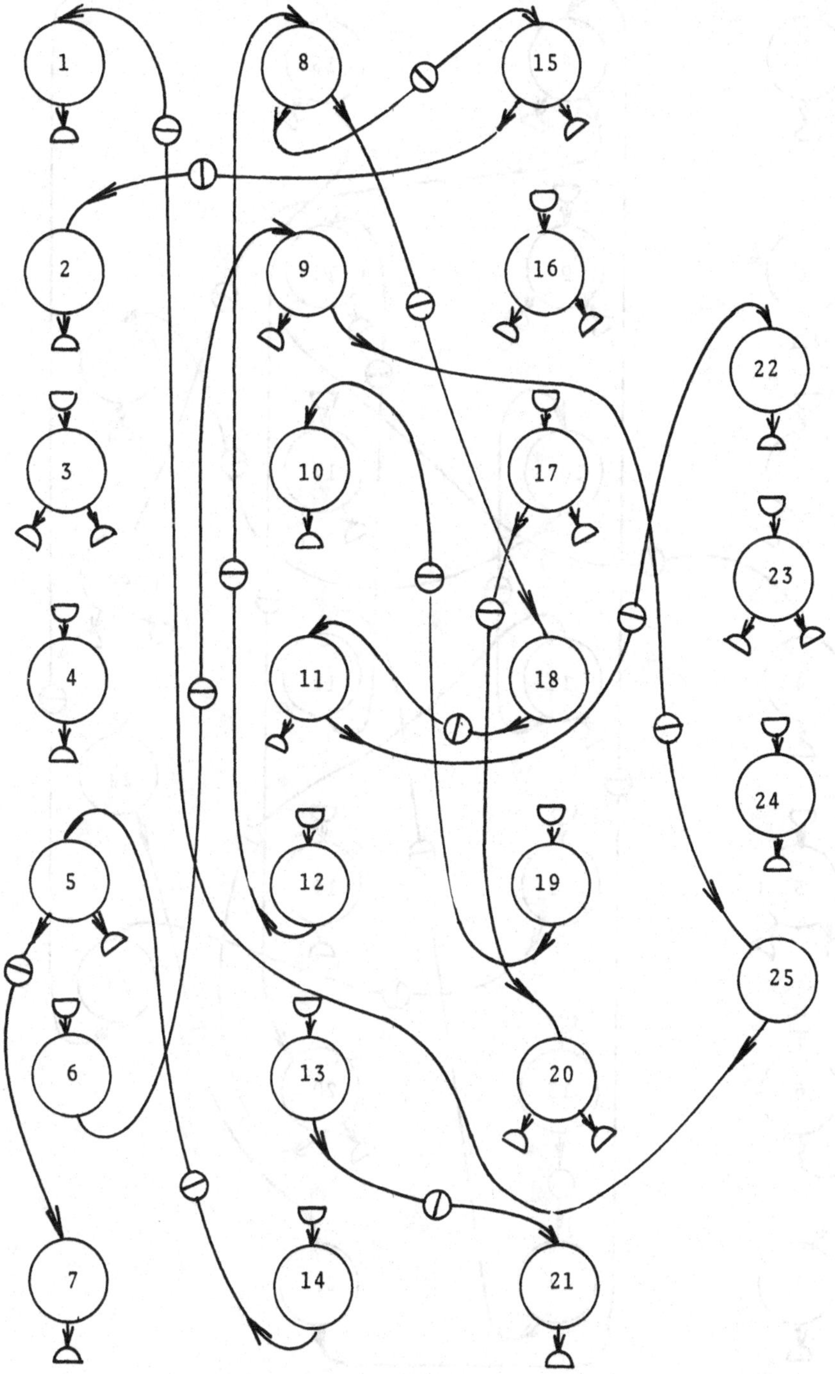

Figure 5.4

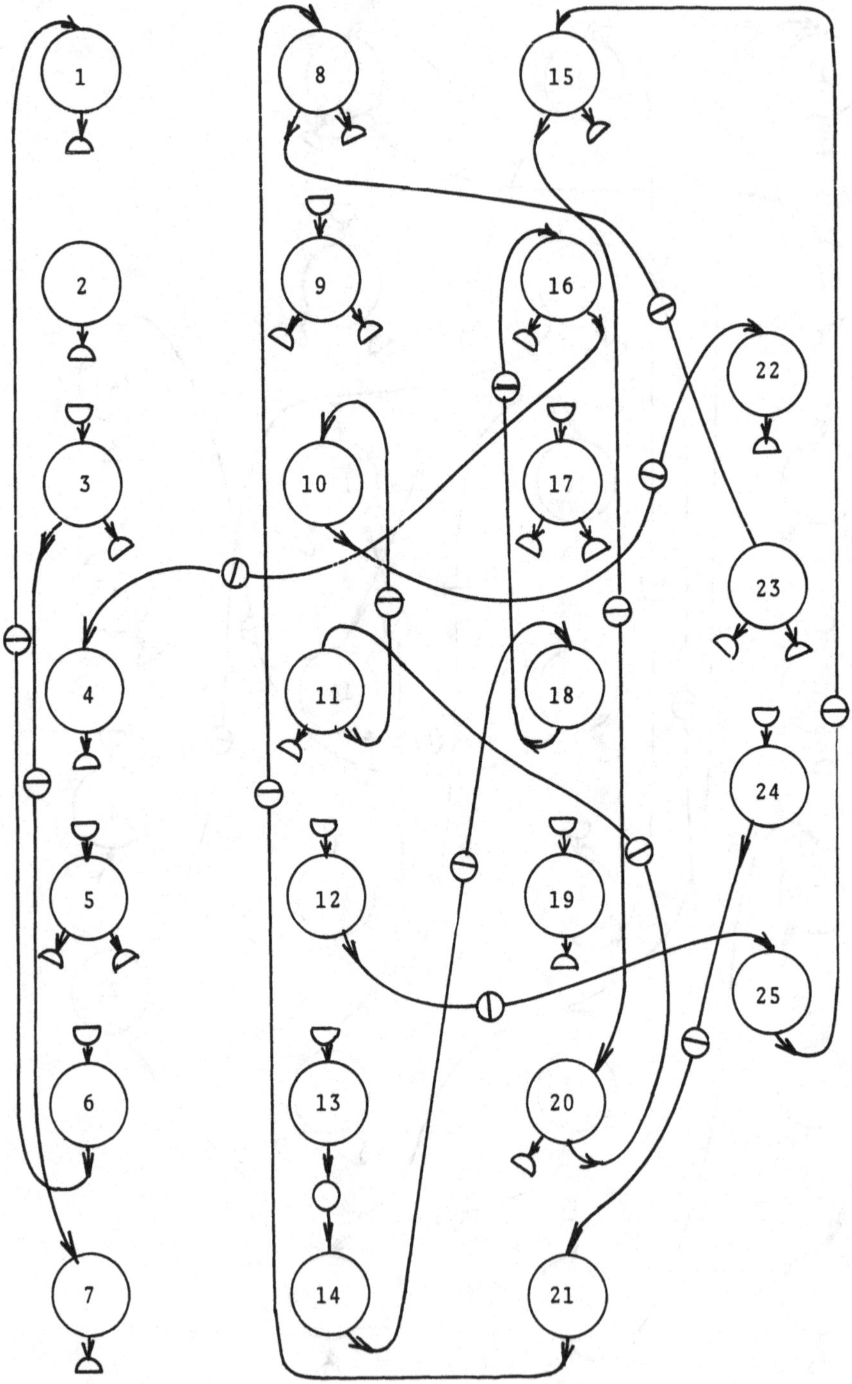

Figure 5.5

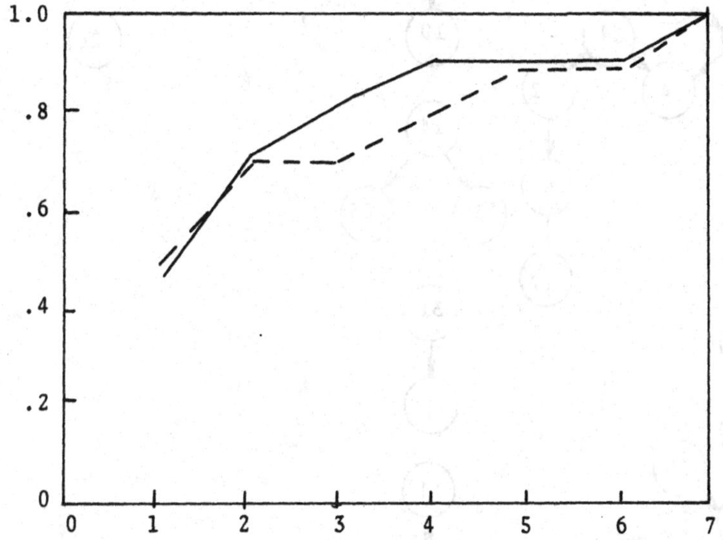

Figure 5.6

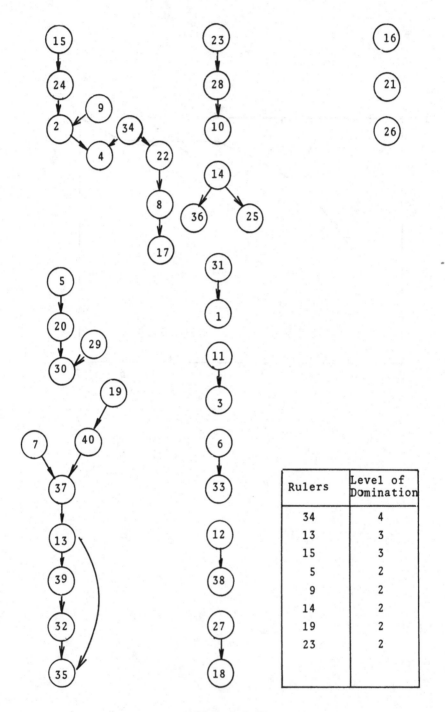

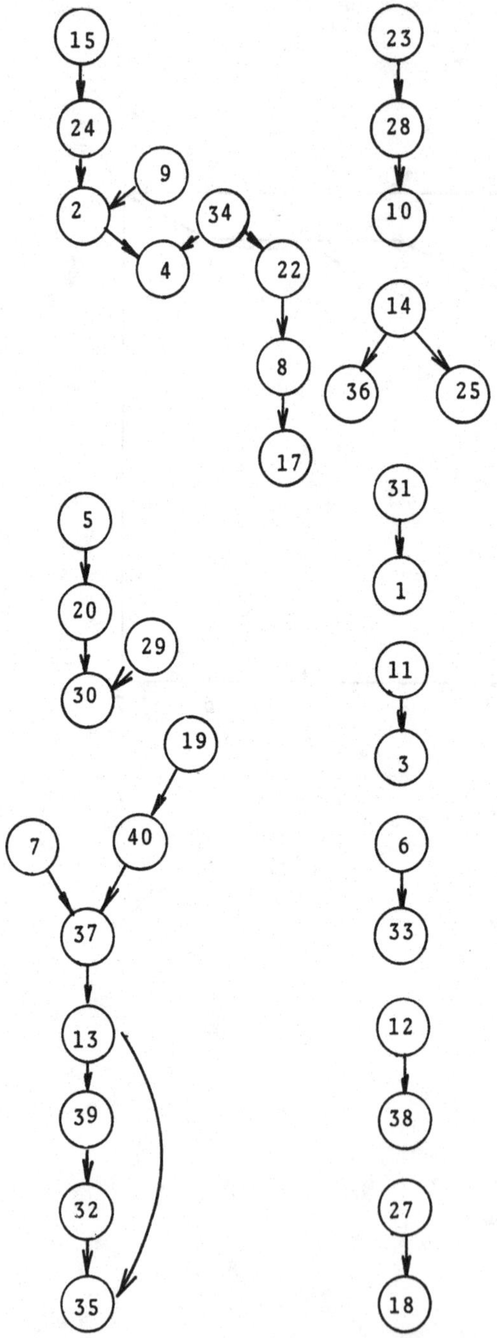

Rulers	Level of Domination
34	4
13	3
15	3
5	2
9	2
14	2
19	2
23	2

Figure 5.7

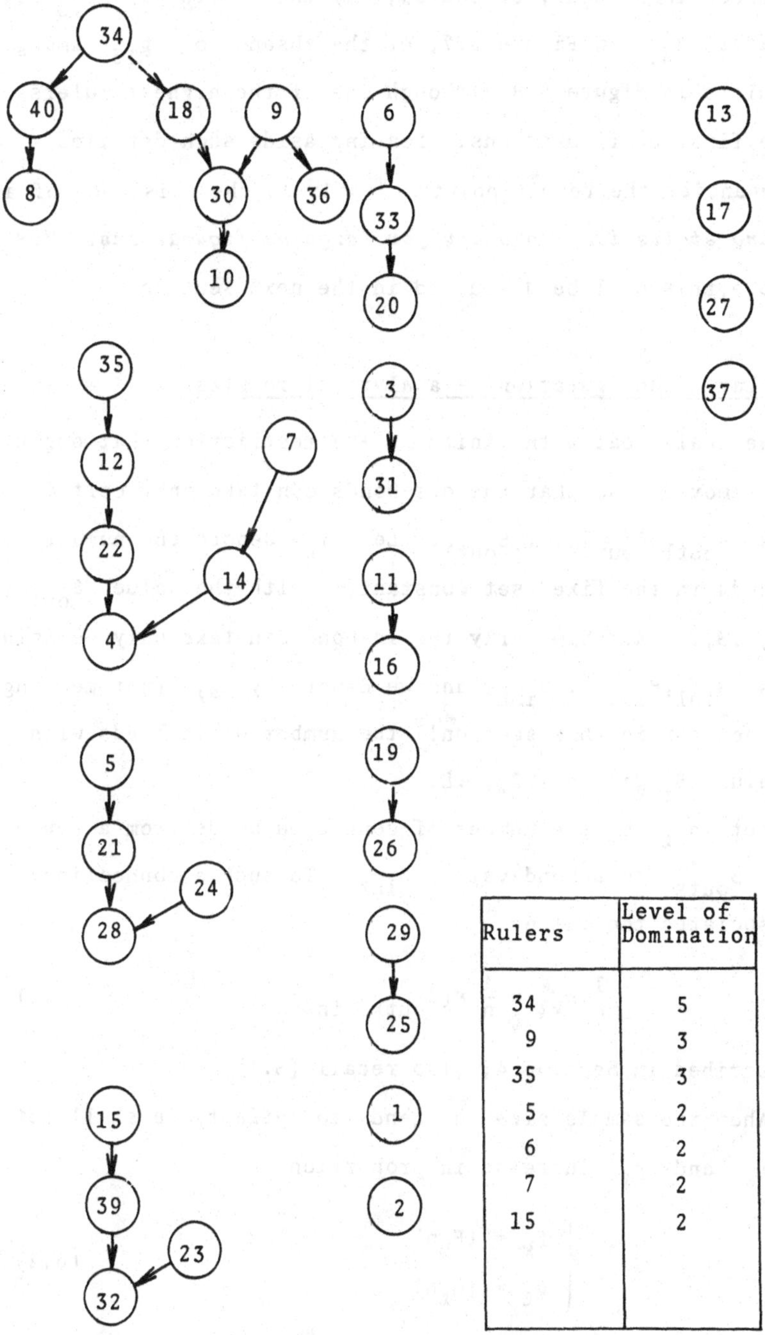

Rulers	Level of Domination
34	5
9	3
35	3
5	2
6	2
7	2
15	2

Figure 5.8

Exceptions occur, of course, say the double way g_{13} dominates g_{35} in Figure 5.7, or the absence of g_{13} among the rulers in Figure 5.8 although one of the highest rulers in the first configurations. Leaving aside such detailed discrepancies the result points clearly to the existence of a *limiting statistical topology for large configurations*. How to prove this will be discussed in the next section.

7.6. Large configurations - analytical results

We shall deal with finite G - a restriction that ought to be removed - so that the out-bonds can take only certain values $\beta_{out1},\beta_{out2},\ldots,\beta_{outK}$. Let f_k denote the number of out-bonds in the fixed set constant(c) with the value β_{outk}; $k = 1,2,3,\ldots,K$. Similarly the in-bond can take only certain values $\beta_{in1},\beta_{in2},\ldots\beta_{inL}$, and we denote by g_ℓ (not meaning any generator in this section!) the number of in-bonds with the value $\beta_{in\ell}$; $\ell = 1,2,\ldots L$.

Let $n_{k\ell}$ be the number of connected bonds from a bond-value β_{outk} to a bond-value $\beta_{in\ell}$. To such a connection is associated the value

$$\frac{1}{n}\kappa_{k\ell} = \frac{1}{n}\kappa(\beta_{outk},\beta_{in\ell}) \tag{6.1}$$

as described in Section 4; also recall (5.1).

When the sample size n tends to infinity we shall let the f_k and g_ℓ increase in proportion

$$\begin{cases} f_k = [F_k n] \\ g_\ell = [G_\ell n] \end{cases} \tag{6.2}$$

where F_k, G_ℓ are positive constants. This corresponds to sampling the generators i.i.d. according to the measure Q

over G. Indeed, this would result asymptotically in the proportionality (6.2) where

$$\begin{cases} F_k = \sum_r Q[\omega_{out}(g)=r] \sum_{j=1}^{r} Q[\beta_{outj}(g)=\beta_{outk}|\omega_{out}(g)=r] \\ G_\ell = \sum_s Q[\omega_{in}(g)=s] \sum_{j=1}^{r} Q[\beta_{inj}(g)=\beta_{in\ell}|\omega_{in}(g)=s] \end{cases} \tag{6.3}$$

In addition to the (absolute) bond frequencies we shall use the *relative bond (value) frequencies*

$$x_{k\ell} = \frac{n_{k\ell}}{n} \tag{6.4}$$

and the *marginal bond (value) frequencies*

$$\begin{cases} n_{k.} = \sum_\ell n_{k\ell} = \#\text{connections with out-bond values } \beta_{outk} \\ n_{.\ell} = \sum_k n_{k\ell} = \#\text{connections with in-bond value } \beta_{in\ell} \\ x_{k.} = \sum_\ell x_{k\ell} \\ x_{.\ell} = \sum_k x_{k\ell}. \end{cases} \tag{6.5}$$

Note that since $n_{k.} \le f_k$, $n_{.\ell} \le g$ and relation (6.2) we must have the $\{x_{k\ell}\}$-array representing a point in the simplex

$$S: \begin{cases} 0 \le x_{k.} \le F_k \\ 0 \le x_{.\ell} \le G \\ x_{k\ell} \ge 0 \end{cases} \tag{6.6}$$

The relative bond (value) frequencies describe an important aspect of the *statistical topology* of the configurations, a crucial concept in our approach.

We shall now show that the topology of large configurations in $\mathscr{L}[g_1,g_2,\ldots g_n]$, where the g's should satisfy (6.2), will converge statistically. The main result is the following surprising and non-intuitive limit theorem. The proof is complicated; we have not been able to simplify it.

Theorem 1. *The relative bond frequencies* $x_{k\ell}$ *tend to* $\xi_{k\ell}$ *in probability as* n *tends to infinity, where* $\xi = \{\xi_{k\ell}\}$ *is the unique solution of the system of equations*

$$\xi_{k\ell} = \kappa_{k\ell}(F_k - \sum_{\ell} \xi_{k\ell})(G_\ell - \sum_{k} \xi_{\ell k}); \; k=1,2,\ldots K, \ell=1,2,\ldots L. \quad (6.7)$$

Proof: Let us first show that (6.7) has at least one solution; to begin with let all $\kappa_{k\ell}$ be strictly positive which is possible for some regularities. Note that S is a convex and compact set which will be used later.

It is clear that all the functions

$$H_{k\ell}(x) = \frac{x_{k\ell}}{(F_k - x_{k.})(G_\ell - x_{.\ell})} \quad (6.8)$$

are continuous and strictly increasing in every variable x_k. On the other hand $H_{k\ell}(0) = 0$. Also, when the (relative) bond frequency vector x approaches the "outer" boundary of S (with the $x_{k\ell} > 0$) the function blows up.

If $\xi = \{\xi_{k\ell}\}$ is a solution we can write it in "multiplicative form"

$$\begin{cases} \xi_{k\ell} = \kappa_{k\ell}\alpha_k\beta_\ell \\ \alpha_k = F_k - \sum_{\ell} \xi_{k\ell}, \; \beta_\ell = G_\ell - \sum_{k} \xi_{k\ell} \end{cases} \quad (6.9)$$

so that

$$\alpha_k = F_k - \alpha_k \sum_{\ell} \kappa_{k\ell}\beta_\ell. \quad (6.10)$$

Note that $\alpha_k \geq 0$, $\beta_\ell \geq 0$ because of (6.6). Hence

$$\alpha_k(1 + \sum_{\ell} \kappa_{k\ell}\beta_\ell) = F_k \tag{6.11}$$

which determines $\{\alpha_k\}$ uniquely if $\{\beta_\ell\}$ is specified and vice versa for the dual equation

$$\beta_\ell(1 + \sum_{k} \kappa_{k\ell}\ell_k) = G_\ell. \tag{6.12}$$

We can now establish the existence of a solution; we know of no simpler way then the following which may appear somewhat contrived.

Consider the function $\mathbb{R}^{K+L} \to \mathbb{R}^{K+L}$ with argument $(\alpha_1,\ldots\alpha_K,\beta_1,\ldots\beta_L)$ and as values the vector with $K+L$ components

$$\phi_k = \alpha_k(1 + \sum_{\ell} \kappa_{k\ell}\beta_\ell) \tag{6.13}$$

for $k = 1,2,\ldots K$ and with

$$\psi_k = \beta_\ell(1 + \sum_{k} \kappa_{k\ell}\alpha_k) \tag{6.14}$$

for the next L arguments. Form the Jacobian matrix of derivatives in block form

$$J = \begin{Bmatrix} A & B \\ C & D \end{Bmatrix} \tag{6.15}$$

so that $A = \{a_{ij}\}$ where

$$a_{ij} = (1 + \sum \kappa_{i\ell}\beta_\ell)\delta_{ij} \tag{6.16}$$

and $B = \{b_{ij}\}$ with

$$b_{ij} = \alpha_i\kappa_{ij}. \tag{6.17}$$

Similarly $D = \{d_{ij}\}$ with

$$d_{ij} = (1 + \sum \kappa_{kj}\alpha_k)\delta_{ij} \tag{6.18}$$

and $C = \{c_{ij}\}$ with

$$c_{ij} = \beta_i \kappa_{ji}. \tag{6.19}$$

We shall now prove that J is nonsingular. With $J = I+M$ it is seen that all the diagonal elements m_{ii} are positive and that summing along columns but leaving out the diagonal element

$$\sum_{i \neq j} m_{ij} = m_{jj}. \tag{6.20}$$

Then any eigenvalue λ of M must have $\text{Re}(\lambda) \geq 0$. This follows by a simple variation of a classical argument: if $Mx = \lambda x$ and x_i is the component of x with largest absolute value, then

$$(\lambda - m_{ii})x_i = \sum_{j \neq i} m_{ij} x_j. \tag{6.21}$$

Hence

$$|\lambda - m_{ii}| \cdot |x_i| \leq |x_i| \sum_{j \neq i} |m_{ii}| = |x_i|\, m_{ii} \tag{6.22}$$

so that λ is contained in a circle of radius m_{ii} in the complex plane, and with the center at m_{ii}; $\text{Re}(\lambda) \geq 0$. Now just transpose M and the intermediate result follows.

Thus all eigenvalues of J have real parts at least equal to one: J is nonsingular. Consider the differential equation

$$J(\alpha,\beta)\begin{pmatrix} d\alpha \\ d\beta \end{pmatrix} = \begin{pmatrix} d\phi \\ d\psi \end{pmatrix} = \begin{pmatrix} F-\phi \\ G-\psi \end{pmatrix} dt. \tag{6.23}$$

Here $F = \text{col}(F_1, F_2, \ldots F_K)$, $G = \text{col}(G_1, G_2, \ldots G_L), \ldots$ $\phi = \text{col}(\phi_1, \phi_2, \ldots \phi_K)$, $\psi = \text{col}(\psi_1, \psi_2, \ldots \psi_L)$. Start with some value $\alpha(0)$, $\beta(0)$ in the region mentioned and follow the trajectory of (6.23). Along the trajectory we have, putting

$$Q(t) = \|F-\phi\|^2 + \|G-\psi\|^2, \tag{6.24}$$

the differential equation

$$-\frac{1}{2}\,dQ = (F-\phi,d\phi)+(G-\psi,d\psi) = Q(t)dt > 0 \tag{6.25}$$

or

$$\frac{d\,\ell n\,Q(t)}{dt} = -2 \tag{6.26}$$

so that

$$Q(t) = Q(0)e^{-2t}. \tag{6.27}$$

Following the trajectory $[\alpha(t),\beta(t)]$ we will never leave the region. Indeed if $\alpha_k = 0$ then (6.13) tells us that $\phi_k = 0$, and that making $d\alpha_k < 0$ will make $\phi_k < 0$. It is clear that this will not move (ϕ,ψ) toward the target point (F,G), which is in the positive orthant. Similarly for $\beta_\ell = 0$. On the other hand if $\alpha_k = F_k$ then (6.9) tells us that $\phi_k = F_k$ and that all $\beta_\ell = 0$. To make $d\alpha_k > 0$ would force some $d\beta_\ell$ to be negative and hence not move (ϕ,ψ) toward the target point. Thus the trajectory cannot cross the boundary of the region. Hence, selecting a subsequence $t \to \infty$ if necessary, it will tend to a limit point α,β belonging to the interior of the region. But then $\phi(\alpha,\beta) = F$, $\psi(\alpha,\beta) = G$ so that if we define $\{\xi_{k\ell}\}$ as in (6.9) we have

$$\begin{cases} F_k-\xi_{k.} = F_k-\alpha_k \sum_\ell \kappa_{k\ell}\beta_\ell = \alpha_k \\ G_\ell-\xi_{.\ell} = G_\ell-\beta_\ell \sum_k \kappa_{k\ell}\alpha_k = \beta_\ell \end{cases} \tag{6.28}$$

implying

$$\kappa_{k\ell}(F_k-\xi_{k.})(G_\ell-\xi_{.\ell}) = \xi_{k\ell}\alpha_k\alpha_\ell = \xi_{k\ell}$$

and Eq. (6.7) has been shown to have a solution. Later on we shall show that is unique.

If the configuration c has the (absolute) bond frequencies $\{n_{k\ell}\}$ its probability is

$$P(c) = Z^{-1} \prod_{k,\ell} \kappa_{k\ell}^{n_{k\ell}} \tag{6.29}$$

where Z, the partition function, should be adjusted to make $P(\mathscr{L}[g_1,g_2,\ldots g_n]) = 1$. We now calculate the number $N(\{n_{k\ell}\})$ of configurations with a given set $\{n_{k\ell}\}$ of bond (value) frequencies. Since we select $n_{k\ell}$ out-bonds from A_k, the set with bond values $\beta_{\text{out}k}$, $\#(A_k) = f_k$, and the same number of in-bonds from B_ℓ, the set with bond values $\beta_{\text{in}\ell}$, we get the number of combinations

$$\prod_k \frac{f_k!}{n_{k1}!n_{k2}!\ldots(f_k-n_{k1}-n_{k2}\ldots)!} \prod_\ell \frac{g_\ell!}{n_{1\ell}!n_{2\ell}!\ldots(g_\ell-n_{1\ell}-n_{2\ell}\ldots)!} . \tag{6.30}$$

In (6.30) the first multinomial coefficient is due to relating subsets of out-bonds from each A_k to each B_ℓ, the second one to select subsets of in-bonds from each B_ℓ to be connected to each A_k. But to each of these combinations we can find $n_{k\ell}!$ permutations of the respective connections. Hence the probability of getting the set $\{n_{k\ell}\}$ is

$$P(\{n_{k\ell}\}) = Z^{-1} \prod_k \frac{1}{(f_k-n_{k.})!} \prod_\ell \frac{1}{(g_\ell-n_{.\ell})!} \prod_{k,\ell} \frac{1}{n_{k\ell}!} \prod\left(\frac{\kappa_{k\ell}}{n}\right)^{n_{k\ell}} \tag{6.31}$$

where Z has been changed to include the factorials in the numerators of (6.30).

Let us compare this probability with that of the set $\{\nu_{k\ell}\}$, $n_{k\ell} = \nu_{k\ell} + \delta_{k\ell}$, where $\delta_{k\ell}$ is not the Kronecker delta but an arbitrary integer. Using (6.31) and observing the cancellations of factorials that take place we get

$$R_n = \frac{P(\{\nu_{k\ell}\})}{P(\{n_{k\ell}\})} = \Pi_1 \cdot \Pi_2 \cdot \Pi_3 \cdot \Pi_4 \cdot \Pi_5 \tag{6.32}$$

where, $\delta_{k\ell} = nd_{k\ell}$ and with obvious notation for summed subscripts, $\Pi_1 = \Pi_1^+ \cdot \Pi_1^-$ with (multiplying over k only)

$$\Pi_1^+ = \prod_{d_{k.}>0} \frac{1}{(f_k - {}_{k.} - \delta_{k.}+1)(f_k-\nu_{k.}-\delta_{k.}+2)\cdots(f_k-\nu_{k.})} \tag{6.33}$$

and

$$\Pi_1^- = \prod_{d_{k.}<0} (f_k-\nu_{k.}+1)(f_k-\nu_{k.}+2)\cdots(f_k-\nu_{k.}-\delta_{k.}). \tag{6.34}$$

Similarly we have $\Pi_2 = \Pi_2^+ \cdot \Pi_2^-$ with (multiplying over only)

$$\Pi_2^+ = \prod_{d_{.\ell}>0} \frac{1}{(g_\ell-\nu_{.\ell}-\delta_{.\ell}+1)(g_\ell-\nu_{.\ell}-\delta_{.\ell}+2)\cdots(g_\ell-\nu_{.\ell})} \tag{6.35}$$

and

$$\Pi_2^- = \prod_{d_{.\ell}<0} (g_\ell-\nu_{.\ell}+1)(g_\ell-\nu_{.\ell}+2)\cdots(g_\ell-\nu_{.\ell}-\delta_{.\ell}). \tag{6.36}$$

Also $\Pi_3 = \Pi_3^+ \cdot \Pi_3^-$ with (multiplying over both k and ℓ)

$$\Pi_3^+ = \prod_{d_{k\ell}>0} (\nu_{k\ell}+1)(\nu_{k\ell}+2)\cdots(\nu_{k\ell}+\delta_{k\ell}) \tag{6.37}$$

and

$$\Pi_3^- = \prod_{d_{k\ell}<0} \frac{1}{(\nu_{k\ell}+\delta_{k\ell}+1)(\nu_{k\ell}+\delta_{k\ell}+2)\cdots\nu_{k\ell}}. \tag{6.38}$$

Furthermore,

$$\begin{cases} \Pi_4 = \prod_{k,\ell} \kappa_{k\ell}^{-\delta_{k\ell}} \\ \Pi_5 = n^{\prod_{k,\ell} \delta_{k\ell}} \end{cases} \tag{6.39}$$

Now choose

$$\begin{cases} f_k = n\,F_k, \quad g_\ell = nG_\ell \\ \nu_{k\ell} = n\,\xi_{k\ell},\ n_{k\ell} = nx_{k\ell},\ \delta_{k\ell} = nd_{k\ell} \end{cases} \tag{6.40}$$

where $\{F_k\}$, $\{G_\ell\}$, $\{\xi_{k\ell}\}$, $\{x_{k\ell}\}$, $\{d_{k\ell}\}$ are kept fixed when n tends to infinity.

We then get, by extracting a number of n-factors from each Π_i

$$\begin{cases} n^{\sum\limits_{d_{k.}>0} \delta_{k.}} & \text{from } \Pi_1^+ \\ n^{\sum\limits_{d_{k.}<0} \delta_{k.}} & \text{from } \Pi_1^- \end{cases} \tag{6.41}$$

and similarly from Π_2^+ and Π_2^{-1}, etc. Combining all these factors together we get for the factor in R_n containing a power to n to the power of

$$-\sum_k \delta_{k.} - \sum_\ell \delta_{.\ell} + \sum_{k,\ell} \delta_{k\ell} + \sum_{k,\ell} \delta_{k\ell} = 0 \tag{6.42}$$

so that the n-factors just cancel each other.

We now consider

$$\frac{1}{n}\,\ell n\, R_n = S_1+S_2+S_3+S_4+S_5 \tag{6.43}$$

referring to the five products on the right side of (6.32). We have seen in (6.42) that $S_5 = 0$. For example $S_1 = S_1^+ + S_1^-$ with

$$S_1^+ = -\sum_{d_{k.}>0} \frac{1}{n} \sum_{p=0}^{d_{k.}n} \ell n(F_k - \xi_{k.} - d_{k.} + \tfrac{p}{n}). \tag{6.44}$$

The inner sum tends to

$$\int_0^{d_{k.}} \ln(F_k-\xi_{k.}-d_{k.}+x)\,dx = \int_{F_k-\xi_{k.}-d_{k.}}^{F_k-\xi_{k.}} \ln x \, dx \tag{6.45}$$

$$= (F_k-\xi_{k.})\ln(F_k-\xi_{k.})-(F_k-\xi_{k.}-d_{k.})\ln(F_k-\xi_{k.}-d_{k.})-d_{k.}.$$

Similarly for S_1^- we get an inner sum

$$\frac{1}{n}\sum_{1}^{-\delta_{k.}} \ln(F_k-\xi_{k.}+\frac{p}{n}) \to \int_0^{-d_{k.}} \ln(F_k-\xi_{k.}+x)\,dx$$

$$= \int_{F_k-\xi_{k.}}^{F_k-\xi_{k.}-d_{k.}} \ln x \, dx = (F_k-\xi_{k.}-d_{k.})\ln(F_k-\xi_{k.}-d_{k.}) \tag{6.46}$$

$$- (F_k-\xi_{k.})\ln(F_k-\xi_{k.}) + d_{k.}.$$

For S_1^- we do not have the minus sign that appears for S_1^+, compare (6.33) with (6.34), so that the two resulting expressions have the same analytic form for both signs of d_k.

Proceeding in this manner we get the awesome expression

$$\begin{aligned}\lim_{n\to\infty} \frac{1}{n} \ln R_n = &\sum_k [(F_k-\xi_{k.}-d_{k.})\ln(F_k-\xi_{k.}-d_{k.}) \\ &- (F_k-\xi_{k.})\ln(F_k-\xi_{k.})+d_{k.}] \\ &+ \sum_\ell [(G_\ell-\xi_{.\ell}-d_{.\ell})\ln(G_\ell-\xi_{.\ell}-d_{.\ell}) \\ &- (G_\ell-\xi_{.\ell})\ln(G_\ell-\xi_{.\ell})+d_{.\ell}] \\ &+ \sum_{k,\ell} [(\xi_{k\ell}+d_{k\ell})\ln(\xi_{k\ell}+d_{k\ell}) \\ &- \xi_{k\ell}\ln \xi_{k\ell} - d_{k\ell}] \\ &- \sum_{k,\ell} d_{k\ell} \ln \kappa_{k\ell}.\end{aligned} \tag{6.47}$$

With the notation $h(x) = x \ln x$ this can be simplified slightly to

$$
\begin{aligned}
T = \sum_k & [h(F_k - \xi_{k.} - d_{k.}) - h(F_k - \xi_{k.})] \\
& + \sum_\ell [h(G_\ell - \xi_{.\ell} - d_{.\ell}) - h(G_\xi - \xi_{.\ell})] \\
& + \sum_{k,\ell} [h(\xi_{k\ell} + d_{k\ell}) - h(\xi_{k\ell})] \\
& + \sum_{k,\ell} d_{k\ell}(-\ell n\ \kappa_{k\ell} + 1).
\end{aligned} \tag{6.48}
$$

Put $d_{k\ell} = te_{k\ell}$, $0 \leq t \leq 1$, for $e_{k\ell}$ such that $\xi_{k\ell} + e_{k\ell}$ belongs to the domain we are working in and consider the function $T(t)$. It is clear that $T(0) = 0$. Differentiating with respect to t, observing that

$$
\begin{cases} h'(x) = 1 + \ell n\ x \\ h''(x) = \dfrac{1}{x} \end{cases} \tag{6.49}
$$

we get

$$
\begin{aligned}
\frac{\partial T}{\partial t} = & - \sum_{k,\ell} e_{k\ell} - \sum_{k,\ell} e_{k\ell}\ \ell n(F_k - \xi_{k.} - d_{k.}) \\
& - \sum_{k,\ell} e_{k\ell} - \sum_{k,\ell} e_{k\ell}\ \ell n(G_\ell - \xi_{.\ell} - d_{.\ell}) \\
& + \sum_{k,\ell} e_{k\ell} + \sum_{k,\ell} e_{k\ell}\ \ell n(\xi_{k\ell} + d_{k\ell}) \\
& + \sum_{k,\ell} e_{k\ell}(-\ell n\ \kappa_{k\ell} + 1).
\end{aligned} \tag{6.50}
$$

Putting $t = 0$ in (6.50) we get

$$
\begin{aligned}
\sum_{k,\ell} e_{k\ell} & [-\ell n(F_k - \xi_{k.}) - \ell_n(G_\ell - \xi_{.\ell}) \\
& + \ell n\ \xi_{k\ell} - \ell n\ \kappa_{k\ell}].
\end{aligned} \tag{6.51}
$$

Now choose $\{\xi_{k\ell}\}$ such that it satisfies the system of equations in (6.7) which we know is possible. Then the bracket in the above expression becomes just

$$\ln\left(\frac{\xi_{k\ell}}{(F_k-\xi_{k.})(G_\ell-\xi_{.\ell})}\frac{1}{\kappa_{k\ell}}\right) = \ln 1 = 0 \tag{6.52}$$

so that

$$\left(\frac{\partial T(t)}{\partial t}\right)_{t=0} = 0. \tag{6.53}$$

On the other hand, differentiating once more and using the second relation in (6.49) we get

$$\begin{aligned}\frac{\partial^2 T(t)}{\partial t^2} &= \sum_{k,\ell,m} \frac{e_{k\ell}e_{km}}{F_k-\xi_{k.}-d_{k.}} \\ &+ \sum_{k,\ell,m} \frac{e_{k\ell}e_{m\ell}}{G_\ell-\xi_{.\ell}-d_{.\ell}} \\ &+ \sum_{k,\ell} \frac{e_{k\ell}^2}{\xi_{k\ell}+d_{k\ell}} .\end{aligned} \tag{6.54}$$

Some simplification reduces this to

$$\begin{aligned}\frac{\partial^2 T(t)}{\partial t^2} &= \sum_k \frac{1}{F_k-\xi_k-d_{k.}}\left(\sum_\ell e_{k\ell}\right)^2 \\ &+ \sum_\ell \frac{1}{G_\ell-\xi_{.\ell}-d_{.\ell}}\left(\sum_k e_{k\ell}\right)^2 \\ &+ \sum_{k,\ell} \frac{e_{k\ell}^2}{\xi_{k\ell}+d_{k\ell}} \geq 0\end{aligned} \tag{6.55}$$

with equality only for $e_{k\ell} \equiv 0$.

This means that $T(t) \geq 0$ with strict inequality for $t \neq 0$. Hence, with the same assumptions and notation as before,

$$\lim_{n\to\infty} \frac{1}{n} \ln \frac{p(\{\nu_{k\ell}\})}{p(\{n_{k\ell}\})} = T(1) > 0 \tag{6.56}$$

so that

$$\frac{p(\{\nu_{k\ell}\})}{p(\{n_{k\ell}\})} \to +\infty. \tag{6.57}$$

Here $\{\nu_{k\ell}\}$ corresponds to a solution ξ and $\{n_{k\ell}\}$ to some point $\{x_{k\ell}\}$. If both ξ' and ξ'' are solutions to (6.7) it is clear that we can choose ξ as ξ' and x and

ξ'', or ξ as ξ'' and x as ξ'. Then (6.57) gives a contradiction unless $\xi' = \xi''$; the solution to the system (6.7) must be unique.

Now we can complete the proof of Theorem 2 and let ε be an arbitrarily small positive number and consider the event

$$E : \left|\frac{n_{k\ell}}{n} - \xi_{k\ell}\right| \leq \varepsilon, \text{ all } k,\ell. \tag{6.58}$$

The toal number of $\{n_{k\ell}\}$-points is bounded by

$$\prod_{\ell=1}^{L} (g_\ell n) \prod_{k=1}^{K} (f_k n) = 0(n^{K+L}). \tag{6.59}$$

On the other hand if $\{n_{k\ell}\} \in E$, so that for some r,s

$$\left|\frac{n_{rs}}{n} - \xi_{rs}\right| > \varepsilon \tag{6.60}$$

then the limit relation (6.57) shows that

$$p(\{n_{k\ell}\}) \leq (1-\delta)^n p(\{\nu_{k\ell}\}) \leq (1-\delta)^n \tag{6.61}$$

and using the bound (6.59)

$$P(E^c) = 0[n^{K+L}(1-\delta)^n] \tag{6.62}$$

which tends to zero as n tends to infinity. Since this is true for any $\varepsilon > 0$ the assertion made in the theorem holds.

This was done, however, under the assumption that all $\kappa_{k\ell} > 0$. If some of them are zero it is clear that the corresponding values of $n_{k\ell}$ and $\xi_{k\ell}$ should be made zero too in order to get configurations with positive probability, and kept zero all through the derivation. With the corresponding modifications the proof goes through as before. Q.E.D.

Remark 1. To actually solve the system numerically we could use the procedure obtained by solving the differential equation (6.23) by, for example, the Runge-Kutta algorithm. This was actually our original motivation for introducing the equation (6.23). To avoid computing the inverse J^{-1} repeatedly we have instead used the iterative scheme

$$\begin{cases} \alpha_k^{(\nu+1)} = \dfrac{F_k}{1+\sum\limits_{\ell} \kappa_{k\ell}\beta_\ell^{(\nu)}} \\ \beta_\ell^{(\nu+1)} = \dfrac{G_\ell}{1+\sum\limits_{k} \kappa_{k\ell}\beta_k^{(\nu)}} \end{cases} \tag{6.63}$$

In each numerical case that we have tried this procedure converged fast to a solution. We have no proof of convergence for the numerical scheme in (6.63), however, so that this suggestion should be treated with caution.

Remark 2. Is it possible to prove a law of large numbers for the statistical topology in a more detailed sense? More precisely, if ¢ is a given, fixed, subconfiguration skeleton, and if N(¢) is the number of subconfigurations of c with skeleton ¢, does N(¢)/n converge in probability? We do not know if this is true.

Remark 3. The proof establishes the truth of the assertion but does not shed any light on what is really the reason for the limiting behavior. One needs a better intuitive understanding of this peculiar limit theorem.

7.7. Further problems and extensions

The results obtained show clearly that the social patterns synthesized in Section 2 can be mathematically understood. We have shown that the social dynamics studied leads to a probability measure over configuration space that is of the regularity controlled type. We have also derived a notion of typical set in configuration space, and that a law of large number is valid for large configurations. Without the notion of typical set that we have used we would not have arrived at our law of large numbers. Several questions should be answered before one goes ahead to more general patterns, and we suggest that one start by the following ones.

A. Can one prove that the relative number NCS÷N of components tends to a limit as N tends to infinity, and if so, what is the limit?

B. If G is not finite so that the number of possible κ-values is infinite, how can Theorem 6.1 be extended to deal with this infinite (perhaps continuous) case? Does the solution of the associated non-linear integral equation exist, is it unique, what smoothness assumptions on F and G and κ are needed for this?

C. Minimum instability probably leads, for large configurations to solving

$$\sum_{k,\ell} x_{k\ell}\mu_{k\ell} + \Sigma(F_k - x_{k.})(G_\ell - x_{.\ell}) = \min. \qquad (7.1)$$

Show that this is true, and study (7.1).

D. Is there a central limit theorem for large configurations corresponding to Theorem 6.2, and if so, what is the asymptotic covariance operator?

The success of this pattern synthesis also suggests several promising extensions of the model.

A. Let the strategies (and arities) be time-dependent, adjustable to fit the existing social environment of any generator. Strategies may be chosen so that they are directed toward dominating a particular individual or group of individuals.

B. Allow generators to carry more information in its attribute vector, for example sex and age, and let these attributes influence the behavior of g.

C. There could be several types of bonds acting in parallel in addition to domination (active) and submission (passive), for example

$$\begin{cases} \text{cooperative, for finding food or repelling enemies} \\ \text{sexual attraction} \\ \text{exchange of information.} \end{cases}$$

D. Introduce an interaction matrix on $\mathscr{L}[g_1,\ldots g_n]$, not necessarily in terms of physical distance, so that

$$\begin{cases} \lambda = \lambda(\beta',\beta'', \text{distance}) \\ \mu = \mu(\beta',\beta'', \text{distance}). \end{cases} \tag{7.2}$$

If λ is a decreasing function of distance this can probably be shown to correspond to the modified model (with the factor $1/n$).

E. The big problem of pattern *analysis* for social regular structures is: having observed social systems empirically, how can we *make inferences about the underlying mathematical structure*? This is an open question in the present context.

7.8. Appendix

The following APL-code was used with small modifications for the mathematical experiment described above. All variables are kept global for ease of modification/debugging.

```
      ∇ GEN N
[1]   NUVECTOR←,0
[2]   NUVALUE←0
[3]   NGEN←N
[4]   I←1
[5]   ARITY←(N,2)⍴0
[6]   X←XDIST N
[7]   INDEPF←1+1E¯6×?N⍴1000000
[8]   IN←(N,MAXIN)⍴1000
[9]   OUT←(N,MAXOUT)⍴0
[10]  NBOND←0
[11]  BOND←(0 4)⍴0
[12] ARITYLOOP:ARITY[I;1]←?MAXIN
[13]  ARITY[I;2]←?MAXOUT
[14] BONDLOOP:OUT[I;⍳ARITY[I;2]]←X[I]÷ARITY[I;2]+INDEPF[I]×ARITY[I
[15]  IN[I;⍳ARITY[I;1]]←INDEPF[I]×OUT[I;1]
[16]  I←I+1
[17]  →(I≤N)/ARITYLOOP
```

In line [7] a factor IND may be introduced multiplying 10^{-6}x?10^{6} to express the independence strategy. The value 1000 in line [8] represents an in-bond that cannot be connected. The value could be replaced by any other large number. We also use the auxiliary programs

```
      ∇ Z←XDIST N
[1]   Z←10×1E¯6×+/?(N,2)⍴1000000
      ∇
      ∇ Z←OUTV LAM INV
[1]   DIFF←OUTV-INV
[2]   Z←(DIFF>0)×1-*-(DIFF÷A2)*P2
      ∇
      ∇MU[⎕]∇
      ∇ Z←OUTV MU INV
[1]   DIFF←OUTV-INV
[2]   Z←(DIFF<0)+(DIFF≥0)×*-(DIFF÷A1)*P1
      ∇
      ∇NU[⎕]∇
      ∇ Z←OUTV NU INV
[1]   Z←(⍟OUTV LAM INV)-⍟OUTV MU INV
      ∇
```

In simulations with variable n line [2] in LAM should be multiplied by constant ÷n.

```
      ∇DYNAMICS[⎕]∇
      ∇ DYNAMICS ITER
[1]     N←NGEN
[2]     T←1
[3]   LOOP:N1←?N
[4]     N2←?N-1
[5]     N2←((⍳N1-1),N1+⍳N-N1)[N2]
[6]     K1←?ARITY[N1;2]
[7]     K2←?ARITY[N2;1]
[8]     J←(∧/BOND=(NBONDρ1)∘.×(N1,K1,N2,K2))/⍳NBOND
[9]     →(0=ρJ)/CLOSE
[10]    PROB←OUT[N1;K2] MU IN[N2;K2]
[11]    E←PROB≥1E¯6×?1000000
[12]    →(1-E)/CONT
[13]    BOND←BOND[(⍳J-1),J+⍳NBOND-J;]
[14]    NBOND←NBOND-1
[15]    NUVALUE←NUVALUE-OUT[N1;K1] NU IN[N2;K2]
[16]    NUVECTOR←NUVECTOR,NUVALUE
[17]    →CONT
[18]  CLOSE:J←(∨/∧/BOND[; 1 2]=(NBONDρ1)∘.×(N1,K1))∨(∨/∧/BOND[; 3 4]=(NBONDρ1)∘.×(N2,K2))
[19]    →J/CONT
[20]    PROB←OUT[N1;K1] LAM IN[N2;K2]
[21]    E←PROB≥1E¯6×?1000000
[22]    →(1-E)/CONT
[23]    NBOND←NBOND+1
[24]    BOND←BOND,[1] N1,K1,N2,K2
[25]    NUVALUE←NUVALUE+OUT[N1;K1] NU IN[N2;K2]
[26]    NUVECTOR←NUVECTOR,NUVALUE
[27]  CONT:T←T+1
[28]    →(T≤ITER)/LOOP
      ∇
```

In DYNAMICS the bonds are selected at random, favoring bonds of low arity. If one wishes to modify this line [3]-[7] should be changed. A simple way of doing this is by inserting a rejection rule that rejects the choice with a probability that depends upon the arity.

To analyze the patterns we execute ANALYSIS with the code

```
      ∇ANALYSIS[⎕]∇
      ∇ ANALYSIS
[1]   Z←CONNECT ELIST BOND
[2]   NCS←⌈/Z
[3]   Z←Z,NCS+⍳NGEN-⍴Z
[4]   NCS←⌈/Z
[5]   'NUMBER OF COMPS, REL NUMBER'
[6]   NCS,NCS÷NGEN
[7]   'COMPS ARE'
[8]   J←1
[9]   SIZEVECTOR←⍳0
[10] LOOP:⎕←V←(J=Z)/⍳NGEN
[11]  SIZEVECTOR←SIZEVECTOR,⍴V
[12]  'SIZE'
[13]  ⍴V
[14]  ⍳0
[15]  J←J+1
[16]  →(J≤NCS)/LOOP
[17]  ⍳0
[18]  'AVERAGE SIZE'
[19]  ⎕←AVE←(+/SIZEVECTOR)÷NCS
[20]  'STAND DEV'
[21]  ((+/(SIZEVECTOR-AVE)*2)÷NCS)*0.5
[22]  'RELATIVE NUVALUE'
[23]  NUVALUE÷NGEN
      ∇
```

To find the herds we use CONNECT and ELIST

```
      ∇CONNECT[⎕]∇
      ∇ SRCHD←CONNECT E;LST;V;C
[1]   LST←0⍴0
[2]   SRCHD←(⌈/,E)⍴0
[3]   C←0
[4]   E←E,[1]⌽E
[5]   E←E[⍋E[;1];]
[6]  NXTGRP:→((⍴SRCHD)<V←SRCHD⍳0)/0
[7]   C←C+1
[8]  SRCHED:SRCHD[V]←C
[9]   LST←LST,(E[;1]=V)/E[;2]
[10] NXTV:→(0=⍴LST)/NXTGRP
[11]  V←LST[⍴LST]
[12]  LST←LST[⍳-1-⍴LST]
[13]  →(SRCHD[V]=0)/SRCHED
[14]  →NXTV
      ∇
```

```
∇ EDGES←ELIST BOND
  EDGES←(1 2)ρBOND[1; 1 3]
  →(0=NBOND)/0
  I←2
LOOP:→(∨/∧/(((I-1)ρ1)∘.×BOND[I; 1 3])=EDGES)/CONT
  EDGES←EDGES,[1] BOND[I; 1 3]
CONT:I←I+1
  →(I≤NBOND)/LOOP
∇
```

When repeating simulations on the *same* content(c), so that no new Q-randomness should be introduced, one can initialize the bond structure by executing INIT.

```
∇INIT
BOND←(0 4)ρ0
NBOND←0
NUVALUE←0
NUVECTOR←ι0∇
```

If this will be done several times, execute REPEAT TIMES, where the right argument is the number of repetitions

```
∇REPEAT TIMES
S←1
RL:INIT
DYNAMICS ITER
'REPETITION NO. ',⍕S
ANALYSIS
ι0
S←S+1
→(S≤TIMES)/RL∇
```

The program SOLVE is not needed for the simulation, it is used to solve the system of non-linear equations in Section 6. Since the underlying algorithm has not been analyzed for convergence it should be used with caution.

```
  ∇ SOLVE
LOOP:AOLD←A
  BOLD←B
  A←F÷1+KAPPA+.×B
  B←G÷1+A+.×KAPPA
  →(TOL≤(+/|A-AOLD)++/|B-BOLD)/LOOP
  KSI←KAPPA×A∘.×B
  ∇
```

The following variables should be set before the simulation: A1,A2,ITER,IND,INDEPF,MAXIN,MAXOUT,P1,P2.

CHAPTER 8
TAXONOMIC PATTERNS

8.1. A logic for taxonomic patterns

When *Carolus Linnaeus* writes that the major activities of the taxonomist are *divisio et denominatio* - separation into classes, families, orders, and naming them - he emphasizes distinctness or dissimilarity of specimens. In abstract terms if the specimens are represented as elements x of some space X, $x \in X$, we look for a way of decomposing X into subsets X_ν; $\nu = 1,2,\ldots$; such that the elements in X_ν are quite different from those of X_μ when $\nu \neq \mu$, see Notes A.

The question of how to construct such partitions $\{X_\nu\}$ of the background space X becomes meaningful only when *we specify a logic in terms of which the partitions can be defined and analyzed.* One way of doing this is to specify a set $\mathcal{F}$ of binary features f, $f \in \mathcal{F}$, and allow only partitions that can be described by certain Boolean functions $B \in \mathcal{B}$ of such features.

Naturally one tries to make the descriptions, i.e., B, simple. For example we can ask that B be made up only of conjunctions of some f's and their negations, so that when B is written in the normal conjunctive form it would appear as

$$B(f_1,f_2,\ldots) = \bigwedge_i f_i^{e_i}; \quad e_i = 0,-1,1 \tag{1.1}$$

with the interpretation that $f^1 = f$, $f^0 \equiv \text{TRUE}$, $f^{-1} = {\sim}f$. Or we could allow the logical factors in (1.1) to be conjunctions of at most two features and their conjunctions and so on.

This is the *feature logic* introduced by this author; see e.g. Grenander (1976), p. 329-359. It enables us to give a mathematical basis for analyzing such notions as clustering, size, separation of clusters.

This makes it possible to give a precise meaning to notions such as "most efficient clustering for a given probability measure": "what is the set E of given cluster size of maximum probability?" This is the *isoperimetric clustering problem* which deserves a careful study. We shall not do this here, only mention in passing a conjecture. In the example mentioned above, and with $X = \mathbb{R}^2$, we believe that the isoperimetric sets are characterized by the property that the curvature at the boundary is proportional to the density of the probability measure.

We now turn to a complementary view: instead of dividing X into very dissimilar subsets, try to establish connections between very similar elements x and y. To do this we need of course some measure of similarity between x and y but it is not easy to suggest global similarity measures in a general situation.

To get a handle on this problem consider the situation in Figure 1.1 where, for simplicity, we have chosen $X = \mathbb{R}^2$.

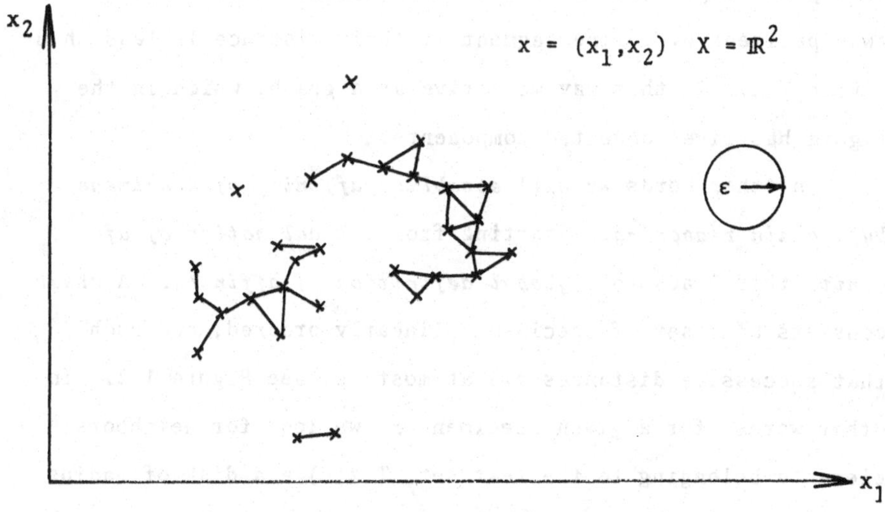

Figure 1.1

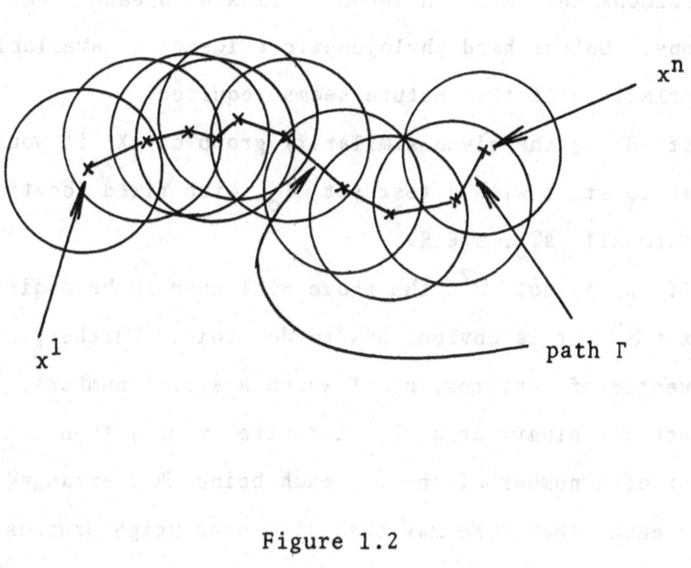

Figure 1.2

Each point represents an observed specimen and we connect two specimens by a line segment if their distance is less than a given ε. In this way we arrive at a graph, which in the figure has five connected components.

In other words we will *establish affinity of specimens by a chain reasoning*. Starting from a *local notion of affinity* this leads to a *global definition of affinity*. A chain consists of a set of specimens, linearly ordered, and such that successive distances are at most ε, see Figure 1.2. In other words, for a given specimen x we look for neighbors: elements belonging to the test set $T(z;\varepsilon)$ = a disk of radius ε centered at z. Of course we need not use disks, based on a given local metric, but can use more general sorts of neighborhoods. This is to some extent a matter of computational convenience.

This sort of chain reasoning amounts basically to the tenet that "*natura non facit saltus*", a species is made up of specimens that are continuously linked to each other with no jumps. Unless hard phylogenetic evidence is available some principle of this nature seems required.

If S is the given similarity group on X it would be natural to start with a test set T_0 with fixed location and then form all sT_0, $s \in S$.

If X is not $\mathbb{R}^2$ the above will have to be modified. For $x = \mathbb{R}^n$ it is obvious how to do this. Further, if x is a vector of features, n of which are real numbers, and the rest are binary or at least finite valued, then X will consist of a number of sheets, each being $\mathbb{R}^n$, arranged on top of each other. We may then also need neighborhoods relating some of these sheets.

It should be noted that we may start with a very highly dimensional feature space and replace it by a projection to a lower dimensional one keeping only the important features that may be certain linear functions of the original features. Such reduction can sometimes be achieved by principal component analysis. In any way we shall assume in the following that this reduction has already been done.

Suppose now that in X, again say $\mathbb{R}^2$ for simplicity, we are given a density function f (not to be confused with the features). This function could be interpreted as the intensity of an inhomogeneous Poisson process or as a frequency function. Can we join $u = x^1$ and $v = x^n$ in Figure 1.2 by a chain with "small" steps? In order that this be possible there must be no gap of size greater than ε among the observed elements in a band around the path Γ. We shall take *the negative value of the logarithm of the probability of this event suitably normalized, as a measure of non-affinity between* u *and* v. The analytic reason for this choice will become apparent later.

Considering the different paths that join u and v we can then define the (global) distance from u to v as the minimum of the above functional. Of course it must be demonstrated that this is indeed a distance, which is rather obvious once it is realized that the functional is additive over paths due to the stochastic independence along two distinct bands.

8.2. Logic of taxonomic affinity patterns

Given two points z_1 and z_2 in the plane, connect them by a C_2-curve Γ, of course without double points and with arc length s (not denoting similarities here!) measured from z_1, with $s = 0$, to z_2, with $s = L$ = length of Γ.

We divide Γ into L equal arcs by points $\xi_0 = z_1, \xi_1, \xi_2, \ldots \xi_n = z_2$ so that $|\xi_{i+1} - \xi_i| \sim d = L/n$. Introduce a local coordinate system as in Figure 2.1 with a local y-axis in the direction of the normal and form the sets

$$A_i = \{(s,y) \mid \xi_i < s < \xi_{i+1}, |y| < w/2\}; \quad i = 0,1,2,\ldots n-1 \tag{2.1}$$

bounded by normals to Γ and by the two parallel curves at distance $w/2$ from Γ; w is the width of the band formed in this way. One of the test sets A_i is shown shaded in the figure.

With a *continuous density* (this will be assumed throughout the chapter) $\nu f = \nu f(z)$ for the inhomogeneous Poisson process we should calculate the probability that each *test set* A_i contains at least one specimen

$$P_n = P\{\text{no } A_i \text{ empty}\} = \prod_{i=0}^{n-1} \left[1 - e^{-\nu F(A_i)}\right] \tag{2.2}$$

with the set function F given as the integral of f over A and ν is a sampling intensity parameter.

To study the limiting behavior of (2.2), when the parameters ν and n tend to infinity, we must normalize it appropriately to get a meaningful and non-degenerate limit. In order to understand the asymptotic behavior of (2.2), let us first remark that $F(A)$ behaves asymptotically as w times the

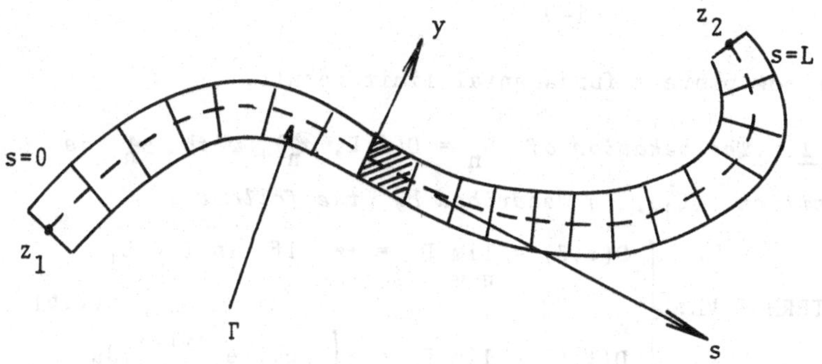

Figure 2.1

integral of $f(z)$ along the arc from ξ_i to ξ_{i+1}. This arc will be denoted by E_i, and with arc length as the measure $m(E_i) = L/n$. The form of (2.2) is reminiscent of product integrals so that it is natural to take logarithms and normalize these. The appropriate form is

$$D(f;\Gamma,\mathscr{P}) = -\sum_{i=0}^{n-1} n\left[1-e^{\nu w F(E_i)}\right]m(E_i) \tag{2.3}$$

where $\mathscr{P}$ stands for the partition of Γ into the arcs (ξ_i,ξ_{i+1}) and, with a different meaning for the symbol F,

$$F(E_i) = \int_{E_i} f(s)ds. \tag{2.4}$$

The quantity D_n is automatically non-negative. If it happens that one $F(E_i) = 0$ the natural interpretation of D_n is to give it the value $+\infty$. We shall let $\nu \to \infty$ with $w = L/n = d$ such that $\nu w \to \lambda$ as $\nu \to \infty$, in other words

$$\nu \sim \lambda\left(\frac{n}{L}\right)^2 = \frac{\lambda}{d^2}\,.$$

Note that we can write our functional as

$$D(f;\Gamma,\mathscr{P}) = -\sum_{i=0}^{n-1} \ln\left[1-e^{-\lambda \frac{F(E_i)}{m(E_i)}}\right] m(E_i). \qquad (2.5)$$

We shall now prove a fundamental limit result.

Theorem 1. *The behavior of* $D_n = D(f;\Gamma,\mathscr{P}_n)$, *with* $\mathscr{P}_n$ *as the partition* $\{\xi_i\}$, *is described by the following*

$$\text{ALTERNATIVE:}\begin{cases} D(f;\Gamma) = \lim_{n\to\infty} D_n = +\infty \quad \text{if} \quad \ln f \notin L_1 \\ D(f;\Gamma) = \lim_{n\to\infty} D_n = -\int_\Gamma \ln[1-e^{-\lambda f(s)}]ds \\ \qquad\qquad\qquad\qquad \text{if} \quad \ln f \in L_1 \end{cases} \qquad (2.6)$$

Remark. In the important special case when λf does not take small values we can approximate the functional in (2.6) by the expression

$$-\int_\Gamma \ln[\lambda f(s)]ds \qquad (2.7)$$

which approximation may suffice for some purposes.

Proof: To bring out the essential ideas in the proof as clearly as possible we shall organize it into a series of lemmas. Note that (2.5) makes sense for more general partitions $\mathscr{P}$ not necessarily consisting of intervals in terms of arc length s, but arbitrary Borel sets E_i of positive Lebesgue measure, again with the natural interpretation if any $F(E_i)$ happen to be zero. This trivial case will be excluded for the moment.

Allowing such general partitions we shall first show monotonicity of D_n as the partitions $\mathscr{P}$ are made finer. This is similar to how the Hellinger integral behaves.

Lemma 1. *If* E *is the disjoint union of the two Borel sets* A *and* B *then*

$$-m(E)\ln\left[1-e^{-\frac{\lambda F(E)}{m(E)}}\right] \leq -m(A)\ln\left[1-e^{-\frac{\lambda F(A)}{m(A)}}\right] - m(B)\ln\left[1-e^{-\frac{\lambda F(B)}{m(B)}}\right] \tag{2.8}$$

<u>Proof</u>: Consider the function $h(x) = -\ln(1-e^{-x})$ with

$$h''(x) = \frac{e^{-x}}{(1-e^{-x})^2} \geq 0. \tag{2.9}$$

Since h is convex we have for $0 < c < 1$,

$$ch(x) + (1-c)h(y) \geq h(cx+(1-c)y). \tag{2.10}$$

Putting

$$\begin{cases} x = \dfrac{\lambda F(A)}{m(A)} \\ y = \dfrac{\lambda F(B)}{m(B)} \\ c = \dfrac{m(A)}{m(E)} \end{cases} \tag{2.11}$$

we get

$$h\left(\frac{\lambda F(E)}{m(E)}\right) \leq \frac{m(A)}{m(E)}\, h(x) + \frac{m(B)}{m(E)}\, h(y) \tag{2.12}$$

which proves (2.8). Q.E.D.

A consequence of this lemma is that if $\mathscr{P}'$ is a finer partition than $\mathscr{P}$ then

$$D(f;\Gamma,\mathscr{P}) \leq D(f;\Gamma,\mathscr{P}') \tag{2.13}$$

and this is the monotonicity property announced earlier.

The next lemma deals with the influence of E-sets with small values of f and shows that they are asymptotically negligible. More precisely

<u>Lemma 2</u>. *If* $\ln f \in L_1$ *we have for* $E_\varepsilon = \{s|f(s) < \varepsilon\}$

$$\lim_{\varepsilon\downarrow 0} m(E_\varepsilon)\ln\left[1-e^{-\frac{\lambda F(E_\varepsilon)}{m(E_\varepsilon)}}\right] = 0. \tag{2.14}$$

Proof: The arithmetic-geometric mean inequality gives

$$\frac{F(E_\varepsilon)}{m(E_\varepsilon)} \geq \exp\left\{\frac{1}{m(E_\varepsilon)} \int_{E_\varepsilon} \ln f(s)ds\right\}. \tag{2.15}$$

But for small positive u we have $1-e^{-u} \geq u/2$ so that

$$1-e^{-\frac{\lambda F(E_\varepsilon)}{m(E_\varepsilon)}} \geq \frac{1}{2}\frac{\lambda F(E_\varepsilon)}{m(E_\varepsilon)}. \tag{2.16}$$

Combining this with (2.15) we get

$$-m(E_\varepsilon)\ln\left[1-e^{-\frac{\lambda F(E_\varepsilon)}{m(E_\varepsilon)}}\right] \leq m(E_\varepsilon)\left[\ln 2 - \frac{\lambda}{m(E_\varepsilon)}\int_{E_\varepsilon} \ln f(s)ds\right]. \tag{2.17}$$

But since $\ln f$ is integrable

$$\int_{E_\varepsilon} \ln f(s)ds \to 0 \quad \text{as} \quad \varepsilon \to 0 \tag{2.18}$$

and this proves the lemma. Q.E.D.

This shows that we do not have to worry about the behavior for f close to zero, as long as $\ln f$ is integrable. If it is not integrable the situation is quite different but we postpone the discussion of this to Lemma 4. First we deal with what could be called *the well behaved density case:* the second alternative in Theorem 1.

Lemma 3. *If* f *is bounded away from zero,* $f(s) \geq \varepsilon > 0$, *all* s, *we have the limit*

$$-\sum_i m(E_i)\ln\left[1-e^{-\frac{F(E_i)}{m(E_i)}}\right] \to -\int_\Gamma \ln\left[1-e^{-\lambda f(s)}\right]ds \tag{2.19}$$

as the partition is made finer indefinitely and with the maximum diameter of its subsets tending to zero.

Proof: Introduce the function for a given partition

$$g(s;\mathscr{P}) = -\ln\left[1-e^{-\frac{\lambda F(E_i)}{m(E_i)}}\right], \text{ when } s \in E_i \tag{2.20}$$

so that our criterion can be written

$$D(f;\Gamma,\mathscr{P}) = \int g(s;\mathscr{P})ds. \tag{2.21}$$

But the non-negative functions $g(s;\mathscr{P})$ are uniformly bounded by the constant

$$g(s;\mathscr{P}) \leq -\ln[1-e^{-\lambda\varepsilon}] \tag{2.22}$$

since, according to the mean value theorem,

$$\frac{F(E_i)}{m(E_i)} = \frac{1}{m(E_i)}\int_{E_i} f(s)ds \geq \varepsilon. \tag{2.23}$$

When the partition is made finer we get the limit

$$\frac{F(E_i)}{m(E_i)} \to f(s), \quad s \in E_i, \tag{2.24}$$

so that

$$-\ln\left[1-e^{-\frac{\nu F(E_i)}{m(E_i)}}\right] \to -\ln[1-e^{-\nu f(s)}] \tag{2.25}$$

for any fixed value s and E_i is selected as the arc containing s. The bounded convergence theorem then proves (2.19). Q.E.D.

Note that it does not matter what partition we use as long as the maximum diameter of its subsets tends to zero. In particular the partition by the points ξ_i leads to the limit in (2.19). Now we shall study the *badly behaved density case.*

<u>Lemma 4</u>. *If* $\ln f \notin L_1$, *which is only possible by divergence of its integral to minus infinity, so that*

$$\int_\Gamma \ln f(s)ds = -\infty, \tag{2.26}$$

then

$$\lim_{\mathscr{P}} D(f;\Gamma,\mathscr{P}) = +\infty. \tag{2.27}$$

Proof: Consider the functions $g(s;\mathscr{P})$ introduced in the proof of Lemma 3. We still have (2.24) but can no longer proceed in the same way. Instead we appeal to Fatou's lemma which implies that (2.25) holds. Indeed we have

$$\lim D(f;\Gamma,\mathscr{P}) \geq -\int_\Gamma \ln[1-e^{-\lambda f(s)}]ds \tag{2.28}$$

with the usual interpretation in case the integral diverges. It does indeed diverge since we have

$$-\int_\Gamma \ln(1-e^{-\lambda f})ds \geq -\lambda \int_\Gamma \ln f(s)ds = +\infty \tag{2.29}$$

and this proves the assertion. Q.E.D.

The inequality (2.29) was used to prove that $\ln f \notin L_1$ implies $\ln(1-e^{-\lambda f}) \notin L_1$, which can be extended to show that the implication also goes the other way.

Having proved these lemmas we go ahead with the main part of the proof of Theorem 1. Because of the last lemma we can exclude the case $\ln f \notin L_1$, so that we now assume that $\ln f \in L_1$, and introduce, for a small positive and fixed ε, the set

$$F = \{s|f(s) \geq \varepsilon\}. \tag{2.30}$$

If $\mathscr{P}_n$ represents the partition by $\{\xi_i\}$ into sets E_i let $\mathscr{P}'_n$ be the partition into sets $E'_i = E_i \cap F$, $i = 0,1,\ldots n-1$ and $E'' = F^c$. Then, according to (2.13)

$$D(f;\Gamma,\mathscr{P}_n) \le D_n(f;\Gamma,\mathscr{P}_n') = -\sum_{i=0}^{n-1} \ln\left[1-e^{-\lambda\frac{F(E_i')}{m(E_i')}}\right]m(E_i')$$

$$-\ln\left[1-e^{-\lambda\frac{F(E'')}{m(E'')}}\right]m(E''). \qquad (2.31)$$

But Lemma 2 shows that the last term can be made arbitrarily small if ε is small enough. The other term, the sum in (2.31), can be handled just as in Lemma 3 and converges to

$$-\int_\Gamma \ln\left[1-e^{-\lambda f_\varepsilon(s)}\right]ds \qquad (2.32)$$

where

$$f_\varepsilon(s) = \begin{cases} f(s) & \text{if } s \in F \\ +\infty & \text{else} \end{cases} \qquad (2.33)$$

The $+\infty$ value of f_ε can be handled by a slight modification of the proof of Lemma 3. Hence

$$\overline{\lim_{n\to\infty}}\, D(f;\Gamma,\mathscr{P}_n) \le \overline{\lim_{\varepsilon\downarrow 0}} - \int_\Gamma \ln\left[1-e^{-\lambda f_\varepsilon(s)ds}\right]ds$$

$$= -\int_\Gamma \ln[1-e^{-\lambda f(s)}]ds \qquad (2.34)$$

using the remark made after Lemma 4 about the integrability of $-\ln[1-e^{-\lambda f}]$ together with the theorem of dominated convergence. Using Fatou's lemma again we have

$$\underline{\lim_{n\to\infty}}\, D(f;\Gamma,\mathscr{P}_n) \ge -\int_\Gamma \ln[1-e^{-\lambda f(s)}]ds \qquad (2.35)$$

which together with (2.34) completes the proof of the theorem. Q.E.D.

We can now go ahead and ask what happens if we adopt a *more cautious inductive logic for forming taxa.* If we ask not only that no E_i be empty, but that each one contain at least $k-1$ observed specimens, then the Poisson probability

$$\sum_{\nu=0}^{k-1} e^{-\lambda \frac{F(E_i)}{m(E_i)}} \left[\lambda \frac{F(E_i)}{m(E_i)}\right]^{\nu} \frac{1}{\nu!} \tag{2.36}$$

should be used in (2.5) instead of just the exponential expression of form $1-e^{-x}$. This leads us to the following limit theorem.

<u>Theorem 2</u>. *The behavior of* $\mathscr{D}_n^k = D_n^k(f;\Gamma,\mathscr{P}_n)$, *asking for at least* $k-1$ *observations in each test set, is governed by the*

$$\text{ALTERNATIVE:} \begin{cases} D^k(f,\Gamma) = \lim_{n\to\infty} D_n^k = +\infty & \text{if } \ln f \notin L_1 \\ D^k(f,\Gamma) = \lim_{n\to\infty} D_n^k = -\int_\Gamma \ln \frac{\Gamma(\lambda f(s),k)}{\Gamma(k)}\, ds & \text{if } \ln f \in L_1 \end{cases} \tag{2.37}$$

where $\Gamma(x,k)$ *is the incomplete* Γ*-function.*

The proof is practically the same except for the form of the integrand that replaces $1-e^{-x}$ which can be expressed in terms of the incomplete Γ-function

$$1-e^{-x} - \frac{x}{1!}e^{-x} - \frac{x^2}{2!}e^{-x} - \dots - \frac{x^{k-1}}{(k-1)!}e^{-x} = \frac{\Gamma(x,k)}{\Gamma(k)} . \tag{2.38}$$

Note that the singularity condition for the alternative remains the same: $\ln f \notin L_1$, whatever is the value of k. On the other hand, the resulting metric will certainly depend upon what k is, since $D^k(f,\Gamma)$ is an increasing function of k. This can be seen from its definition but is also intuitively clear.

Now we can take a more extreme conservative attitude toward the inductive logic underlying this taxonomic procedure. Instead of asking that $\#(E_i)$ be at least equal to some fixed integer we can demand that it be of the same order of magnitude as the number of test sets in $\mathscr{P}_n$:

$$\#(E_i) \geq \alpha n. \tag{2.39}$$

In order that this lead to a non-degenerate limit we must let λ increase, say proportionately to n, so that ν should behave as

$$\nu \sim \frac{n^3}{L^2} = \frac{n}{d^2}\ . \tag{2.40}$$

We then get a drastically different result

Theorem 3. *Under the given conditions the functional*

$$D^{\alpha}(f;\Gamma,\mathscr{P}_n) = -\sum_{i=1}^{n-1} m(E_i)\ln[P(E_i^{\alpha})] \tag{2.41}$$

where E_i^{α} *stands for the event in* (2.39) *behaves asymptotically according to the*

$$\text{ALTERNATIVE:}\begin{cases} D^{\alpha}(f;\Gamma) = \lim\limits_{n\to\infty} D^{\alpha}(f;\Gamma,\mathscr{P}_n) = +\infty \\ \qquad\qquad \text{if}\ \ f(s) < \alpha\ \ \text{for some}\ \ s \\ D^{\alpha}(f;\Gamma) = 0\ \ \text{if}\ \ f(s) > \alpha\ \ \text{for all}\ \ s. \end{cases} \tag{2.42}$$

Proof: Our $\mathscr{D}_n^{\alpha}$ functional now takes the form

$$D^{\alpha}(f;\Gamma,\mathscr{P}) = -\ \Sigma\ m(E_i)\ln P(E_i,\alpha) \tag{2.43}$$

where $P(E_i,\alpha)$ denotes the probability of finding at least αn specimens in the test set E_i belonging to the partition $\mathscr{P}$. As before we have started with a band of width w, asked that each test area A_i contain at least αn specimens, and then let w tend to zero as ν tends to infinity, see relation (2.40).

Let us study the asymptotic behavior of $P(E_i,\alpha)$ in the neighborhood $N(s)$ of a point s on the path Γ for which $f(t) < \alpha$, $t \in N(z)$. We should evaluate the Poisson probability

$$P(E_i,\alpha) = \sum_{k\geq\alpha n} \frac{m^k}{k!} e^{-m}, \quad m = \frac{n^2}{L} \int_{N(s)} f(f)dt. \qquad (2.44)$$

Note that as $N(s)$ shrinks to the point s then $m \sim nf(s)$. Now we use a time honored bound, N = smallest integer $\geq \alpha n$,

$$P(E_i,\alpha) = \frac{m^N e^{-m}}{N!}\left[1 + \frac{m}{N+1} + \frac{m^2}{(N+1)(N+2)} + \cdots\right] \leq \frac{m^N e^{-m}}{N!(1-a)} \qquad (2.45)$$

with $a = m/N$ which tends to $\rho = f(s)/\alpha < 1$ as $N(s)$ shrinks to $\{s\}$. Hence with Stirling's approximation we can bound $P(E_i,\alpha)$ by, asymptotically, the expression

$$\frac{\text{constant}}{\sqrt{n}} u^{n\alpha}; \quad u = \rho e^{(1-\rho)} < 1 \text{ for } \rho < 1. \qquad (2.46)$$

Hence $-\ln P(E_i,\alpha)$ can be bounded from below, asymptotically, by

$$\text{constant} + \frac{1}{2} \ln n - n\alpha \ln u \qquad (2.47)$$

which tends to $+\infty$ as n tends to infinity. This implies one part of the alternative: if there is any point s on the part Γ for which $f(s) < \alpha$, then $D^\alpha = +\infty$.

Now assume instead that $f(s) > \alpha$ for all s on the path. Then for any s and corresponding E_i covering s, we have the opposite inequality

$$\begin{aligned} 1-P(E_i,\alpha) &= \sum_{k<\alpha n} \frac{m^k}{k!} e^{-m} \\ &= \frac{m^N e^{-m}}{N!}\left[1 + \frac{N+1}{m} + \frac{(N+1)(N+2)}{m} + \cdots\right] \\ &\leq \frac{m^N e^{-m}}{N!} \frac{1}{1-b} \end{aligned} \qquad (2.48)$$

where N is the largest integer $< \alpha n$ and $b = \frac{N}{m} \sim \frac{\alpha}{f(s)} < 1$. Hence, again with Stirling's approximation, we get the upper bound for $1-P(E_i,\alpha)$ with another constant but otherwise as in (2.45)-(2.46)

$$\frac{\text{const}}{\sqrt{n}} u^{n\alpha}, \quad u = \rho e^{1-\rho} < 1 \quad \text{for} \quad \rho > 1. \tag{2.49}$$

Hence

$$D^{\alpha}(f;\Gamma, {}_{n}) \leq -\sum \frac{L}{n} \ln\left[1 - \frac{\text{const}}{\sqrt{n}} u_o^{n\alpha}\right] \tag{2.50}$$

with $u = \rho_o(1-e^{-\rho_o})$, $\rho = \min_s f(s)/\alpha$. But the right hand side of (2.50) tends to zero as n tends to infinity which proves the second part of the alternative. Q.E.D.

8.3. Synthesis of taxonomic affinity patterns

For two points z_1 and z_2 connected by a path Γ the expression (2.6) gives us the possibility of evaluating our functional $D(f;\Gamma)$. If z_1 is a point on Γ dividing this path into the union $\Gamma_1 \cup \Gamma_2$ we have because D is now given as an integral

$$D(f;\Gamma_1 \cup \Gamma_2) = D(f;\Gamma_1) + D(f;\Gamma_2). \tag{3.1}$$

It is then natural to introduce the metric

$$d(z_1,z_2) = \inf_{\Gamma} D(f;\Gamma), \quad \Gamma \text{ connecting } z_1 \text{ and } z_2. \tag{3.2}$$

In other words we use the geodesic paths and measure distance along them. It is clear that $d(\cdot,\cdot)$ is indeed a distance. It is non-negative, and it can equal zero only if the two points coincide. Finally the triangle inequality is satisfied since if Γ_1 connects z_1 and z_2, and Γ_2 connects z_2 and z_3 then

$$d(z_1,z_2) \leq \left[\inf_{\Gamma_1,\Gamma_2} D(f;\Gamma_1)+D(f;\Gamma_2)\right] = d(z_1,z_2)+d(z_2,z_3).$$

It should be noted that this is a *very special case of a Riemannian metric* since it is isotropic locally at each z.

The specimens related to a given z on the level d are now given by the *affinity set*

$$A(z;d) = \{z' | d(z,z') \leq d\} \tag{3.4}$$

in other words by the *geodesic disks centered at* z. In an area where f varies slowly, so that grad f is small, the geodesic disk is almost an ordinary disk with Euclidean distance. The distance d^k defined as in (2.37) but with the functional D^k instead of k is more sensitive to variations in f. The same holds a fortiori for the distance d^α, corresponding to the functional D^α.

If f is given in analytical form we may be able to find the geodesics, and hence the affinity sets, explicitly. Indeed, the Euler equation for the calculus of variation problem in (3.2) is known classically as

$$y'' = (\phi_y' - \phi_x' y')[1+(y')^2] \tag{3.5}$$

where

$$\phi = -\ln[1-e^{-\lambda f(x)}]. \tag{3.6}$$

With the knowledge of the similarity group S we may use Lie's method of reducing a second order equation to an equation of the first order and may be able to solve the geodesics in terms of a quadrature.

An example may be illuminating. Say that λf does not take large values so that we can use the approximation in (2.7). This means that ϕ is approximated by $-\ln(\lambda f)$. Consider $\lambda f(x,y) = e^{-(x^2+y^2)}$ so that we should put ϕ equal to x^2+y^2 in (3.6). Now this ϕ has rotational symmetry around the origin, so that it is natural to introduce polar coordinates (r,ϕ). We can then write down directly a first integral of the second order differential equation which in polar coordinates takes the form

$$\frac{r^4}{\sqrt{\left(\frac{dr}{d\phi}\right)^2+r^2}} = \text{constant} \tag{3.7}$$

or, with another c,

$$\frac{dr}{d\phi} = \pm\sqrt{cr^8-r^2}\ . \tag{3.8}$$

Solving this by quadrature we get after some elementary calculations

$$r = A\sin^{-1/3}[3(\phi-\phi_o)]. \tag{3.9}$$

These geodesics have the qualitative shape indicated in Figure 3.1 and rotations of the curves shown. The trajectories tend to curve toward $x = y = 0$ where the density f is largest. This example is of more general scope than may be thought at first glance: the reduction to a first order equation is possible as soon as S is known.

We can now synthesize the taxonomic patterns from the affinity sets $A(z;d)$ as generators. The connection type Σ connects generators with the same *central specimen* z with an arrow from $A(z;d_1)$ to $A(z;d_2)$ if $d_1 < d_2$ and with the bond relation ρ = INCLUSION. Σ does *not* connect generators with different central specimens. Therefore Σ can be seen as several LINEAR connection types in parallel. The bond values are the sets themselves. The pure image is a set of such generators, satisfying the regularity requirement $= \langle\Sigma,\rho\rangle$.

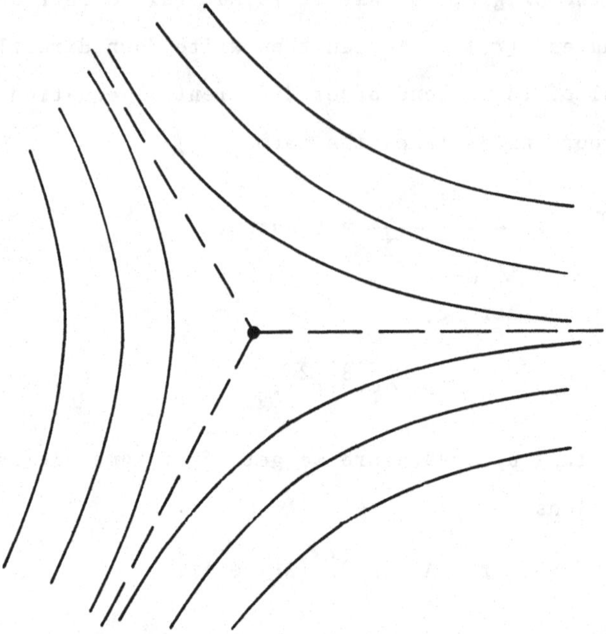

Figure 3.1

Note that the taxa, in this case the affinity sets, can very well overlap as in Figure 3.2, where three central specimens z_1, z_2, z_3 are shown with four of their affinity sets.

Hence the relation between a central specimen and the other points in $A(z;d)$ is not an equivalence relation and *transitivity will usually not hold.* These taxa express the phenetic classification achieved by local similarity in z-space extended to global similarity by the affinity logic described.

The same holds for any D^k functional, although of course the precise shape of the taxa will change. When we go to D^α, however, we meet a drastic difference. Here the

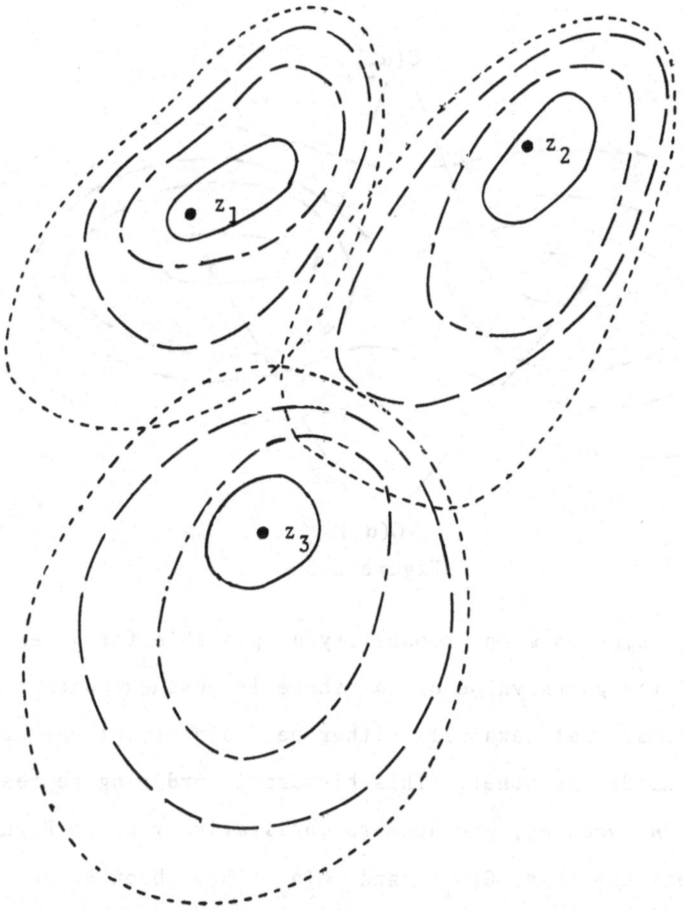

Figure 3.2

taxa will be the connected components of the *concentration sets*

$$C(\alpha) = \{z \mid f(z) > \alpha\} \tag{3.10}$$

well known from Lebesgue integration. The notion of concentration ellips (-oid) was used by Cramér to denote the ellips containing maximum probability mass for given area. In exactly the same way the Neyman-Pearson lemma guarantees that

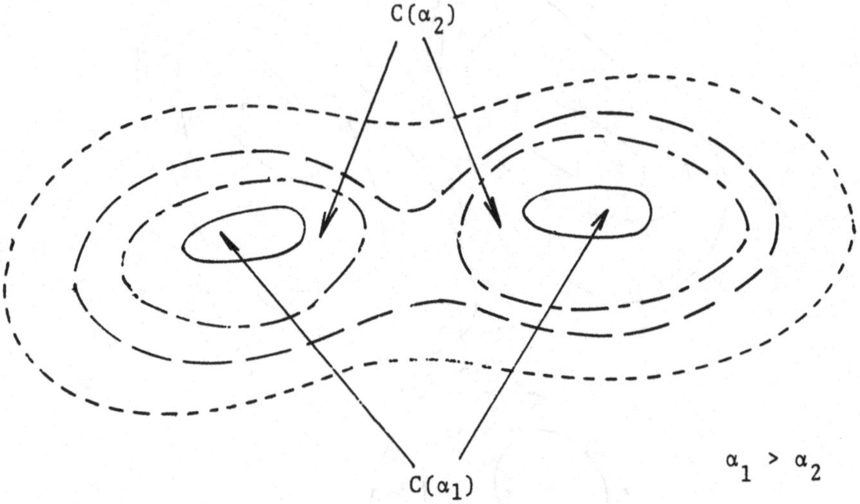

Figure 3.3

$C(\alpha)$ contains as much probability as possible for given area. For a given value of α there is just one concentration set, so that taxa will either be disjoint, or one contained inside the other. This hierarchic ordering represents a *Linnaean taxonomy*, and appears qualitatively as in Figure 3.3, where the taxa $C(\alpha_1)$ and $C(\alpha_2)$ have been shown with two connected components each, while the lower values of α have only a single connected component each.

We therefore get another connection type, namely TREE, and the reader is referred to Section 2.6 in Volume I for a brief discussion of this regular structure.

Remark 1. When we pass from the taxonomic logic based on the metric D^k to the one based on D^α the connection type changes from many LINEAR in parallel to TREE. Intuitively one can think of this as a bunching together of LINEAR, joining some of them at the apex of a tree.

8.4. Analysis of affinity patterns

To relate these ideas to empirically observable patterns we shall study the inverse problem of pattern inference. Having observed the sample $g^{\mathcal{D}}$ of points over a bounded background space X, how can we restore the true but unknown generator g that has been deformed to $g^{\mathcal{D}}$? Let us deal with the case when the generators are connected concentration sets, $g = C(\alpha)$. A partial answer is given by the following result. We denote by $D(z,\nu)$ a disk of some radius r_ν, centered at z, $r_\nu \downarrow 0$, and by $\#\{D(z,\nu)\}$ the number of observed sample points inside $D(z,\nu)$.

Theorem 1. *To restore the generator* $g = C(\alpha)$, $\alpha > 0$, *from the deformed image* $I^{\mathcal{D}}$, *where* $\mathcal{D} = \mathcal{D}_3'$ *with Poisson intensity* νf *and* $C(\alpha)$ *is a bounded set, form the restoration*

$$g^* = \{z \,|\, \#\{D(z;\nu)\} \geq \alpha a(\nu)\nu\} \qquad (4.1)$$

where $a(\nu) = m[D(z,\nu)]$ *is such that* $\lim_{\nu\to\infty} \nu a(\nu) = +\infty$. *If* $m(\partial C(\alpha)) = 0$ *we have consistent restoration in expected area*

$$\lim_{\nu\to\infty} E[m(g^* \Delta g)] = 0. \qquad (4.2)$$

Remark. This $\mathcal{D}$ is a generalized version of $\mathcal{D}_3$ in Volume II, p. 281.

Proof: Consider a point $z \in \text{interior}[C(\alpha)]$ so that $f(z) > \alpha$. We have for the indicator functions 1_g and 1_{g^*} respectively when ν is big enough so that r_ν has become small

$$E[|1_{g^*}(z)-1_g(z)|] = P\{1_{g^*}(z) = 0\} \tag{4.3}$$

which can be written as

$$P\{\#\{D(z;\nu)\} < \alpha a(\nu)\nu\} = P\{N_\nu(z) < \alpha a(\nu)\nu\}. \tag{4.4}$$

Here $N_\nu(z)$ is a Poisson variable with mean

$$m_\nu = \nu\int_{D(z,\nu)} f(\xi)d\xi \sim \nu f(z)a(\nu) \quad \text{as} \quad \nu \to \infty. \tag{4.5}$$

Since $f(z) > \alpha$ this implies that (4.4) tends to zero as can be shown by an elementary argument.

On the other hand if $z \notin \text{closure}[C(\alpha)]$, so that $f(z) < \alpha$ we get for ν big enough

$$E[|1_{g^*}(z)-1_g(z)|] = P\{1_{g^*}(z) = 1\} = P\{N_\nu(z) > \alpha a(\nu)\nu\} \tag{4.6}$$

which tends to zero as $\nu \to \infty$ since $N_\nu(z)$ is Poisson with mean $\sim\nu f(z)a(\nu)$.

Finally, for z on the boundary of $C(\alpha)$, the contribution to the expected area error is zero since $m[\partial C(\alpha)] = 0$. Together with

$$m(g^*\Delta g) = \int_X |1_{g^*}(z)-1_g(z)|dz \tag{4.7}$$

and the bounded convergence theorem this proves (4.2). Q.E.D.

Remark. The test set, say T, has here been chosen as a disk $D(z;\nu)$ but other shapes can be dealt with by the same method.

The numerical implementation of Theorem 1 can be carried out in several ways. For a given shape of the test set T, for example a square with side σ, we could attempt to construct g^* literally as in (4.1). To do this efficiently would require a fast geometric algorithm, and while we believe that this could be achieved, we shall not attempt this here.

A modification of some interest would be to consider only test sets centered at the observed specimens z_ν and form the set of those points z_ν for which the inequality in (4.1) holds. Compute the connectivity graph that connects that subset of these points whose distance is at most constant × σ. The graphs obtained by a transitive closure computation would then approximate the topology of the connected components of the concentration set $C(\alpha)$.

An even simpler algorithm can be constructed as follows. Divide the background space X, say the unit square, into squares of side σ

$$SQ_{\nu\mu} = \{(x,y) \mid \nu d \leq x < (\nu+1)d,\ \mu d \leq y < (\mu+1)d\} \quad (4.8)$$

for $\nu,\mu = 0,1,2,\ldots 1/d-1$, where $1/d$ should be an integer. In each such test set $SQ_{\nu\mu}$ find out whether the inequality in (4.1) is satisfied or not. Connect those of the resulting $SQ_{\nu\mu}$ that are nearest neighbors. (We could use an 8-neighbor definition for example) and compute the transitive closures. Again one would believe that the resulting graphs would approximate the topology of the connected components of $C(\alpha)$.

To make this precise some care is needed, since for two given specimens z_1 and z_2 there are "many" paths Γ connecting them and the probabilistic statement about convergence of the empirically established topology toward the correct one is not trivial. We shall prove a partial result on this which will be formulated in terms of conditions that are probably much more stringent than needed.

Theorem 2. *Assume that* $C(\alpha)$ *has* n_c *connected components separated by positive distances. Further let each component have a piecewise analytic boundary. If* $d \downarrow 0$ *as* $\nu \to \infty$ *in such a way that* $d^2\nu/\ln \nu \to \infty$ *then the probability that no chain will connect two components will tend to one.*

Remark 1. Our problem can be categorized as the question of *statistical estimation of the topology of the concentration set.*

Remark 2. Note that the requirement $d^2\nu/\ln \nu \to +\infty$ is slightly stronger than the condition in Theorem 1.

Proof: See Figure 4.1 where $d = 1/10$, n_c = number of connected components of $C(\alpha) = 3$, $C(\alpha) = C_1 \cup C_2 \cup C_3$, and where the band B separates C_1 and C_3. What is the probability P_ν that none of the test sets $SQ_{\nu\mu}$ in B satisfy the inequality in 1)? This will tell us how likely it is that our estimation of the topology of $C(\alpha)$ will actually separate C_1 from C_3.

We have (distinguish between the Poisson parameter ν and the ν,μ coordinates)

$$P_\nu = \prod_{SQ_{\nu\mu}\subset B} P\{\#(SQ_{\nu\mu}) < \alpha a(\nu)\nu\}. \tag{4.9}$$

But the $n_{\nu\mu} = \#(SQ_{\nu\mu})$ are independent Poisson variables with the means

$$m_{\nu\mu} = \nu \iint_{SQ_{\nu\mu}} f(z)dz \sim \nu a_{\nu\mu} f(z), \quad z \in SQ_{\nu\mu}. \tag{4.10}$$

Hence using the reasoning that led to (2.46) gives us asymptotically the lower bound for $\ln P_\nu$

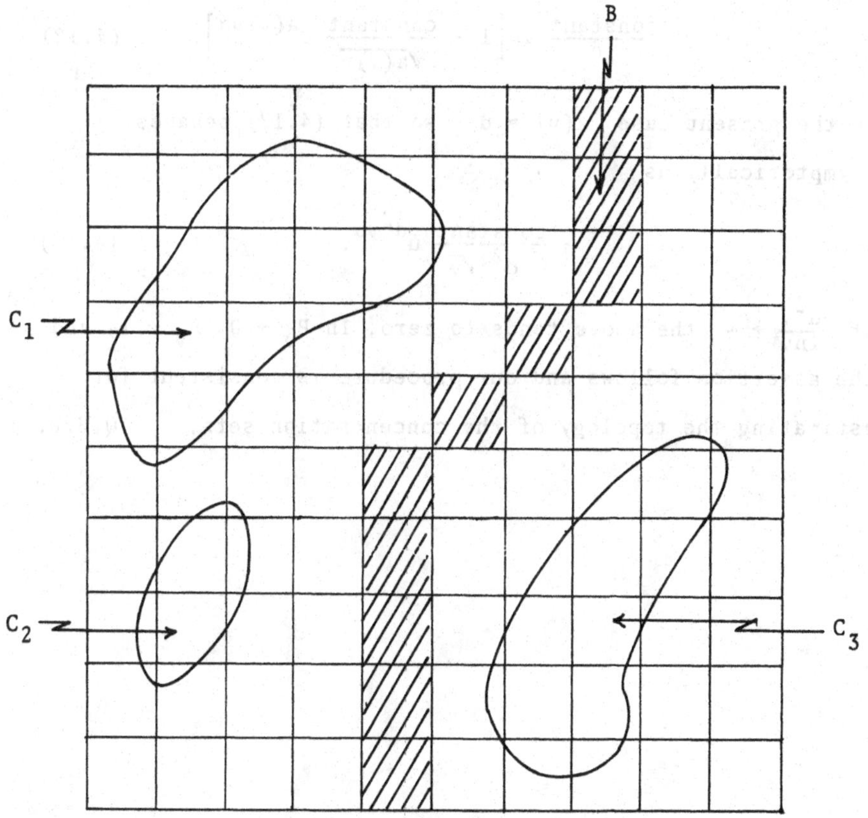

Figure 4.1

$$\sum_{SQ_{\nu\mu}\subset B} \ln\left[1 - \frac{\text{constant}}{\sqrt{a(\nu)\nu}} u_{\nu\mu}^{a(\nu)\nu\alpha}\right] \tag{4.11}$$

where $u_{\nu\mu} = \rho_{\nu\mu} e^{(1-\rho_{\nu\mu})}$; $\rho_{\nu\mu} = f(z)/\alpha$ in $SQ_{\nu\mu}$. Hence we can bound u away from the value one uniformly by some $u < 1$ since $f(z)/\alpha$ is uniformly bounded away from one in a continuous path separating C_1 from C_3.

But when we make d smaller the number of test sets in B will be asymptotically proportional to $1/d$ so that we get the lower bound for $\ell n\ P_\nu$

$$\frac{\text{constant}}{d} \ln\left[1 - \frac{\text{constant}}{\sqrt{a(\nu)\nu}} u^{a(\nu)\nu\alpha}\right]. \tag{4.12}$$

In the present case $a(\nu) = d^2$ so that (4.12) behaves asymptotically as

$$-\frac{\text{constant}}{d^2\sqrt{\nu}} u^{d^2\nu\alpha}. \tag{4.13}$$

If $\frac{d^2\nu}{\ln\nu} \to \infty$ the above tends to zero, $\ln P_\nu \to 0$, $P_\nu \to 1$, and the assertion follows and our procedure is consistent for estimating the topology of the concentration sets. Q.E.D.

CHAPTER 9
PATTERNS IN MATHEMATICAL SEMANTICS

9.1. Introduction

In this chapter we shall introduce mathematical semantics as the pattern theoretic study of mappings between image algebras and formal languages.

The image algebra will be synthesized using generators that represent relations. This will serve as the semantic counterpart of a formal language. Then the image algebra is studied in terms of similarities, local and global regularity.

The semantic map will be seen to form a category, in the algebraic sense of the term, and we shall examine its morphisms.

We shall also present strategies for constructing semantic maps with special properties related to memory requirements.

Some examples and computer experiments will be given to supplement the analytical treatment.

9.2. Introducing mathematical semantics

2.1.1. Can pattern theory contribute anything to the study of semantics and to the study of how semantics is learned (should be learned) by man (machines)? The word semantics is of fairly recent origin, dating back to the XIX century, but the subject itself goes back to the beginnings of philosophy. Most of the major figures in the history of philosophy devoted some of their thinking to the relation between words, sentences, grammar, and language, on the one hand, with phenomena in the real world on the other.

Such studies have traditionally been carried out by informal means and involved no explicit use of mathematics.

2.1.2. More recently attempts have been made to formalize semantic ideas, which can be seen especially in two disciplines: linguistics and computer science. In *formal* linguistics this seems to have been started at about the same time as when the study of syntax was formalized during the 1960's. The earliest reference that we are aware of is Katz-Fodor (1963), where syntactic structures were transformed into what has become known as K-F trees. The K-F trees are formal constructs attributing meaning to linguistic utterances.

Linguists have continued along this avenue of approach, which has resulted in a large literature. An important idea in this literature is the semantic net which has been applied many times. One has typically taken a subset of a natural language, usually English, and tried to formalize its semantics by a computer program. In this way one would hope that the logical discipline and precision required when writing the program would bring out the basic difficulties clearly.

An important contribution can be found in Woods (1970). The interested reader will find an interesting presentation of this approach in Simmons (1973).

2.1.3. These endeavors overlap to a considerable extent with work done in artificial intelligence, although the emphasis differs. In the latter the goal is often to build a question-answer program for some sufficiently narrow domain of discourse. The well-known work by Winograd (1972) belongs in this group.

The many attempts that have been made in this direction aim at, not just a computer program, sometimes possibly of utilitarian value, but insight and understanding of semantic structures. In spite of skeptical comments to the contrary we believe that these efforts, some of which were mentioned above, have indeed led to an increased understanding.

2.1.4. As far as we know, mathematical formalization has not been employed except in a few publications. One is in Sandewall (1971), where the mathematical tool is predicate calculus.

In 1977 the author together with P. Wegner organized a seminar series in formal semantics at Brown University. During this series the voluminous literature was surveyed, most of it from the linguistic and computer science journals. Formalization in mathematical terms seems to have been attempted only sporadically, and we came across little of mathematical content.

One reason why mathematics has been used so little is probably that no mathematical theory has appeared suitable for the analysis of semantic structures. We believe that pattern theory offers a tool suitable for this purpose. The present section is a continuation of work begun in Volume II, Section 2.4. It was reported in Grenander (1978b).

In particular we shall attempt to show that mathematical semantics can be expressed in terms of mappings of configuration spaces and image algebras. Such mappings are fundamental to pattern theory, just as morphisms are fundamental in algebra in general.

2.2.1. Our perspective is conformal to that of the early Wittgenstein in his *Tractatus Logico-Philosophicus*, except, of course, that we shall proceed in a mathematically formalized manner. In the next sections we shall remind the reader of Wittgenstein's view of the issues that will concern us here. Some of his aphorisms have been reproduced in an Appendix.

Wittgenstein is often as obscure as he is thought provoking, perhaps intentionally so. When he speaks of "things" for example, it is not clear if these are material objects or, say, sensory data. See Notes A.

2.2.2. The world consists of facts, T1.1-1.12 (this refers to the numbered sections of *Tractatus*). A fact is a collection of things related to each other, T2.0272, 2.031. The things make up the substance of the world, T2.021.

Some facts can be seen to be made up from other facts, others cannot be split up. The latter are the atomic facts.

2.2.3. Let us denote the set of things by T and consider a set O of operations. The operations act upon things and produce simple facts. An operation can operate on just one thing, or two things, and so on. It is a function with, say, n places (or arguments).

When we apply all operators to all combinations of things we get the set S of atomic facts. Wittgenstein probably does not assume that an operator with n places can be applied to any combination of n things. If this is so the operations are partial functions.

Another set U of operations acts upon atomic facts, from S, and results in composite facts. The set F of all such facts is the ontological base for understanding the world.

2.2.4. Of course Wittgenstein did not formalize his thinking in this way, perhaps he would be opposed to *any* formalization attempt. It would be *too* precise, losing the "multi-dimensional" ambiguity.

2.2.5. A picture in *Tractatus* is a model of the world, grouping elements that correspond to things (T.2.13) into structures. A picture is also a fact, T.2.141.

A proposition is made up of names. It is a fact, its elements are related to each other, T.3.14, and it is a picture of a possible grouping of things.

In some sense the structure of the picture should be "congruent" to the real situation it represents. "Congruent" does not mean identical, the correspondence can be more complicated.

This correspondence, if it could be articulated exactly, would associate meaning to propositions. It is likely that Wittgenstein did not have ordinary natural language in mind when he discusses propositions. Perhaps he meant "scientific language", or language as it *ought* to be.

2.2.6. A reader familiar with pattern theory will recognize the similarity between some of its basic concepts with the thinking in *Tractatus*. The generators correspond to things and operators, $T \cup O$. The operators in O have arities, the number of places. Configurations correspond to facts and the connectors allowed in the configuration space correspond to the operators in U. The totality F is the configuration space.

2.2.7. In Sections 3-7 a mathematical formalization of semantics will be given expressed as mappings between two image algebras. The philosophical view of *Tractatus* has influenced this formalization.

In his later years Wittgenstein renounced Tractatus, the work of his youth. We shall have something to learn also from the later Wittgenstein, however, namely about learning semantics.

2.3.1. Our speaker/listener will be immersed in a world of sensory impressions. Based on these sensory data and with the aid of a priori knowledge he, the observer, makes statements or receives statements about the state of the world expressed, we assume, in some formal language L. Since our approach will be abstract, we need not specify whether these statements are just declarative, affirmative, or whether they can be questions, expressing doubt, containing judgments, or be imperative, and so on.

The fact that we shall use examples where the statements look like simple English sentences should not be taken to mean that we are modelling the semantics of English, not even a subset of it. Our goal is to understand certain mathematical phenomena, not linguistic ones. If this can be achieved we hope that the results will in due time have applications to linguistics, but this would be too early to claim at present.

2.3.2. The observer's statements should be correlated to his view of the world. His view will be expressed formally as an image algebra to be examined in Section 3. The image algebra should be mathematically consistent, as will be proved for the one we propose, but it need not be a "true" description of the world.

We are therefore operating on three levels. The "true" world, the formal description of the way the observer views the world, and the linguistic utterances prompted by the view. It is only the relation between the two latter levels that we shall study here.

2.4.1. All natural languages can be ambiguous. This has been pointed out so many times that we need not elaborate this trite fact any further. In context, and with access to linguistic deep structure, ambiguity may perhaps be removed. Whether this is so or not, we shall simply *require* that the grammatical utterances have a unique semantic content.

Most of our attention will then be paid to the study of such semantic maps, their mathematical construction and analysis of their properties, especially of their memory requirements and limitations. This will be done in Sections 6 and 7. In the last sections of this chapter we shall study the learning of semantic maps.

2.4.2. When mathematics is applied to any subject matter one is forced to simplifications, sometimes drastic ones. This is certainly true here; a narrow range of situations will be analyzed in some depth at the cost of introducing specializing assumptions. The abstract treatment is hoped to bring out the logical essence of the problem as clearly as possible. This will avoid vague generalities and bring into the open hidden assumptions, albeit at the price of restricting the scope of the results.

2.4.3. In order to pinpoint the concepts needed for the mathematical analysis our reasoning, we shall be dialectic, arguing for and against adopting certain notions and assumptions. In this way we have arrived at a formalization that we hope will be useful for our later work.

2.5.1. The abduction machine analyzed in Grenander (1978), Chapter 7, *creates syntactic hypotheses* sequentially, tests them and accepts or rejects them. In a certain well-defined linguistic situation it was proved to yield, ultimately, a set of correct hypotheses.

In an early theorem (see Notes B) the author showed how syntactic abduction can be achieved for languages of a very general type. This theorem is, however, only of theoretical interest since the algorithm would be very slow due to the fact that it is too general; it does not exploit any underlying structure. Another drawback is that the learning would not be incremental. The syntactic abduction machine mentioned seems better suited to the problem.

2.5.2. Is it possible to build an *abduction machine for semantic hypotheses?* We shall show that mathematically this amounts to estimating a relation from a finite set (consisting of productions for L) to the morphisms of a category. As far as we know this mathematical problem has never been studied up till now; it will be done in 9.8.

9.3. Formalization through regular structures

3.1. Any coherent view of the world must be based on some *notion of regularity*. Otherwise it would be without laws and constancies, with nothing permanent to learn, no structure to discover.

The regularity need not be deterministic. On the contrary, many of the phenomena that we encounter in every day life are ruled by *statistical laws* only. Statistical regularity should therefore be allowed. A mathematical consequence of this is that the state space becomes more sophisticated.

3.2. To formalize a view of the world we need a precise notion of regularity. We shall show in the following that *combinatory regularity* (pattern theory) is logically conformal to the ideas of Section 2.2.

3.3. Let us remind the reader that pattern theory is of algebraic nature and based on the idea of an *image algebra*

$$\mathcal{I} = \langle G, S, \mathcal{R}, R \rangle. \tag{3.1}$$

An image algebra is made up of a set G of *generators*, from which *configurations* are formed following the *rule of regularity*, $\mathcal{R}$. The group S of transformations of G onto G, the *similarities*, expresses which generators are similar to each other. The set of regular configurations $\mathcal{C}(\mathcal{R})$, formed according to $\mathcal{R}$, is divided into equivalence classes, the *images*, by means of the equivalence relation R: the *identification rule*. The images form a partial universal algebra $\mathcal{I}$ with respect to certain connection operations.

We now discuss the choice of each component in (3.1) for the purpose of this study.

3.4.1. The generators $g \in G$ shall be thought of as *relations* in a general sense that will become clearer as we go along.

In Section 6 we shall relate the image algebra to language. Formal linguistics is dominated by the finitistic attitude so that it would seem natural to assume that G is finite.

On the other hand, we would like to let the generators carry attributes such as location, orientation, frequency, time, etc. These are usually thought to be continuous in nature so that we would be led to allow G to be infinite.

For the time being we shall choose the first alternative, $\#(G) < \infty$, reserving the possibility of extending the results to infinite generator spaces.

3.4.2. Generators shall carry two sorts of *bonds*, in-bonds and out-bonds, leading us to directed regularity. The *out-arity* shall be finite and, since G is finite, bounded over G

$$\omega_{out}(g) \leq \omega_{max} < +\infty. \tag{3.2}$$

We are less certain about the *in-arities* $\omega_{in}(g)$. After having examined a large number of cases it became clear that generators should be allowed to accept many in-bonds. Whether this number should be bounded or not is less clear. We choose for the moment to make it unbounded

$$\omega_{in}(g) = +\infty, \quad \forall g \in G. \tag{3.3}$$

Note that all generators have in-bonds but not necessarily out-bonds.

The arities as well as the values "in", "out", associated with every bond belong to the bond structure. Sometimes the different out-bonds have different functions so that it will be necessary to indicate this by other *bond structure parameters*, see Chapter 3. This will be done by markers "1", "2", etc. We then rule out the possibility that some markers are equal, at least for now. For the in-bonds no such markers will be used at present; again this may have to be modified when we have learnt more about the use of these regular structures.

3.4.3. To each bond is associated a *bond value* v, taking values in some set B. We suspect that it would be convenient to make these values subsets of G

$$V = 2^G, \tag{3.4}$$

but in the present discussion this will not be done.

For a given generator the bond values associated with out-bonds may differ, expressing their difference in function. The in-bond values, on the other hand, will be assumed to be the same. The rationale behind this assumption is that out-bonds shall express active properties of a generator (relation) that may vary from bond to bond. The in-bonds express passive properties that are constant for all in-bonds of the generator.

We are aware of examples where this assumption will lead to logical inconsistencies. A generator may accept two in-bonds belonging to two generators, that, viewed as unary relations, expresses properties that are not compatible with each other. Recalling the discussion in Section 2, however, this will be allowed: the observer's view of the world need not be consistent with the "true" state of the world.

3.4.4. To be able to refer to the bonds of a given generator we need *bond coordinates*. Therefore we shall enumerate the out-bonds by $1,2,3,\ldots\omega_{out}(g)$, with the convention that if some of them have already been marked by the bond structure parameters $1,2,\ldots r$, then this numbering will be adhered to for the bond coordinates. In configuration diagrams bond coordinates will sometimes be put inside parentheses when needed for clarity.

Since all the in-bonds carry the same bond value, at least for now, we need not distinguish between them and shall not use any bond coordinates for them.

3.4.5. Consider a generator g with $\omega_{out}(g) = \omega$, so that its (out-) bond coordinates are $1,2,\ldots\omega$. Let

$$\pi : (1,2,\ldots\omega) \to (i_1,i_2,\ldots i_\omega) \tag{3.5}$$

be a permutation of the ω first natural numbers. To each ν, $1 \le \nu \le \omega$, correspond bond structure parameters $B^s_\nu(g)$ and bond values $B^v_\nu(g)$. If

$$\begin{cases} B^s_\nu(g) = B^s_{i_\nu}(g) \\ B^v_\nu(g) = B^v_{i_\nu}(g), \end{cases} \tag{3.6}$$

for all ν, the renumbering π does not affect the connectivity properties of g. The set of all such permutations π forms a subgroup $\pi(g)$ of the symmetric group over ω objects: the *symmetry group* of g.

In the special case when all out-bonds of g carry distinct markers $1,2,\ldots\omega$ the symmetry group consists of the identity element.

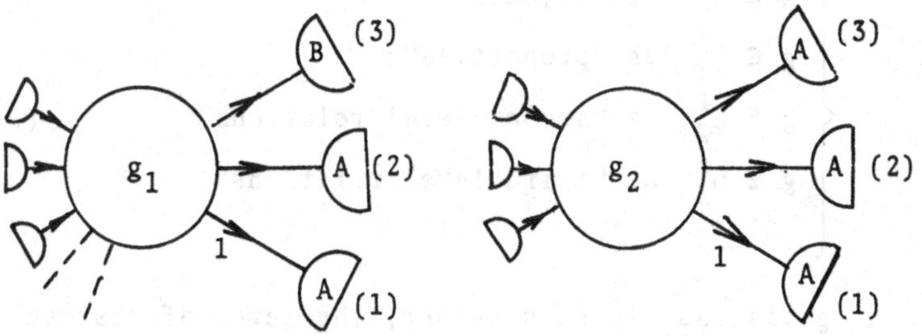

Figure 3.1

In Figure 3.1 the symmetry group $\pi(g_1)$ consists of the identity if $A \neq B$, but $\pi(g_2)$ is of order 2 allowing (2) and (3) to be exchanged without changing the connectivity properties of g_2. Note the bond structure marker "1" at the bottom out-bond in the diagram.

3.4.6. Bonds shall take values in sets $\mathscr{B}_\nu \subset \mathscr{B}$, $\nu \geq 0$. Any generator shall have one and the same in-bond value from some $\mathscr{B}_\nu$ and then its out-bonds, if there are any, shall be from $\mathscr{B}_{\nu-1}$. If $\nu \geq 1$ the generator shall have at least one out-bond. The value of ν expresses the *level of abstraction* of the generator, $g, \ell(g) = \nu$ in a way that will become clear as we go along.

We have one partition of G into sets G_k^ω where $k = \omega_{out}(g)$. Here ω is just a label for "arity", not a variable superscript. Another partition is induced by the level of abstraction into sets

$$G_\nu^\ell = \{g \mid \ell(g) = \nu\}; \quad \nu = 0,1,\ldots ; \tag{3.7}$$

with ℓ a label for "level". We shall refer to generators from these classes as follows:

$$\begin{cases} g \in G_0^{\ell} & \text{as "objects"} \\ g \in G_1^{\ell} & \text{as "properties"} \\ g \in G_2^{\ell} & \text{as "second level relations"} \\ g \in G_3^{\ell} & \text{as "third level relations"} \\ \cdots \end{cases} \tag{3.8}$$

To each g is associated a number, the level of abstraction, $\ell = \ell(g) = \nu$ denoting the number of the set family $\mathscr{B}_\nu$ to which the in-bonds belong.

Combining all the elements with the same out-arity we get, as mentioned above, a partition into the sets

$$G_\mu^\omega = \{g | \omega_{out}(g) = \mu\}. \tag{3.9}$$

Lemma 1. $G_0^\ell = G_0^\omega$; *objects, and only objects, have out-arity* 0.

Proof: If $g \in G_0^\ell$ its in-bond values are in $\mathscr{B}_0$. Since $\mathscr{B}_0$ has no predecessor to which the out-bond values should belong, g can have no out-bonds, so that $\omega_{out}(g) = 0$, $g \in G_0^\omega$, and $G_0^\ell \subseteq G_0^\omega$.

On the other hand, if $g \in G_0^\omega$, so that it has no out-bonds, then it cannot have in-bonds with values from any $\mathscr{B}_\nu$, $\nu \geq 1$, see above. Hence $g \in G_0^\ell$ which implies $G_0^\omega \subseteq G_0^\ell$.

Q.E.D.

3.5.1. The *similarities* will be chosen as the set S of all permutations $s: G \to G$ leaving bonds, i.e. bond structure and bond values, unchanged

$$B(sg) = B(g), \quad \forall g \in G. \tag{3.10}$$

It is immediately clear that the permutations s satisfying (3.10) form a group, the similarity group.

Since any s leaves the bond structure invariant, $B^s(sg) = B^s(g)$, it follows that our definition of S is correct, see Volume I, p. 9, except that (ii)(ibid) cannot yet be verified since the generator index has not been defined so far.

3.5.2. Since the present S leaves invariant, not only the bond structure as all similarities do, but also the bond values, it follows that the classification of any g in terms of the set families $\mathscr{B}_\nu$ is also S-invariant. A consequence is that *the level of abstraction is S-invariant*

$$\ell(g) = \ell(sg): \quad \forall g \in G, \quad \forall s \in S. \tag{3.11}$$

3.5.3. We now define a *generator index* class as the set of all g's with the same $B(g)$.

Lemma 2. *This partition is the finest partition by any generator index.*

Proof: If g_1 and g_2 both belong to the same α-class we have $B(g_1) = B(g_2)$. Appealing to (3.10) we see that $B(sg_1) = B(sg_2)$, $\forall s \in S$, which implies that the α-classes are invariant, $\alpha(sg_1) = \alpha(sg_2)$, and hence that α is a legitimate generator index corresponding to the similarity group, see Volume I, Chapter 1, Definition 1.1, (ii).

On the other hand, if α' is some other generator index and $\alpha(g_1) = \alpha(g_2)$ then by definition $B(g_1) = B(g_2)$. The permutation s_o of G that only permutes g_1 with g_2 is therefore a similarity; see (3.10). But all generator indices must be S-invariant so that $\alpha'(g_1) = \alpha'(s_o g_2) = \alpha'(g_2)$ and g_1, g_2 belong to the same α'-class. This shows that α-classes are contained in α'-classes. Q.E.D.

Note that generators with the same index are of the same level of abstraction, since if two generators have the same index α, then they have the same in-bond values. These values then belong to the same set family $\mathcal{B}_\nu$, which leads to the same level of abstraction.

3.5.4. Our choice of generator index could be criticized in that it is too narrow: in order that $\alpha(g_1) = \alpha(g_2)$ hold we must have exactly the same bond structure and bond values for g_1 and g_2. When we exemplify our construction by concrete image algebras this will lead to a classification of generators into very small classes, perhaps too small to be natural. Some modification may be needed as we go along.

3.6.1. We now come to the *rules* $\mathcal{R}$ *of combinatory regularity*

$$\mathcal{R} = \langle\rho,\Sigma\rangle \tag{3.12}$$

with some bond relation ρ, local regularity, and connection type Σ, global regularity. In accordance with the discussion in Section 2 we want our configurations to consist of relations combined together into a "formula". In order that the formula be "computable" we must choose ρ so that all the connections that are allowed by ρ make sense.

3.6.2. At first it seemed reasonable that the *bond relation* ρ ought to be chosen as INCLUSION. If we think of the generators as logical operators with domains and ranges we are led to operator configurations, see Volume I, Chapter 2, Case 7.1, where INCLUSION was the natural choice.

After examining a number of special cases we have concluded, however, that the more restrictive relation ρ = EQUAL suffices for the present purpose; we choose this definition for the rest of this chapter.

<u>3.6.3</u>. It is clear that EQUAL is a legitimate bond relation for the similarity group chosen. Indeed, if g_1 connects to g_2 via the bond-values β_1 and β_2, then β_1 must equal β_2. Applying the same similarity s to both g_1 and g_2 will not change the bond values. Hence sg_1 can connect to sg_2 via the same bonds, which shows that EQUAL is legitimate; see Volume I, Chapter 2, p. 27.

<u>3.6.4</u>. This choice of ρ has implications for the levels of abstractions of connected generators.

<u>Lemma 3</u>. *If a generator* g_1 *is connected by an out-bond to an in-bond of* g_2 *then*

$$\ell(g_1) = \ell(g_2) + 1. \tag{3.13}$$

<u>Proof</u>: See Figure 3.2 where the $(k)^{th}$ out-bond of g_1 is connected to an in-bond of g_2. The corresponding bond-values are denoted by β_{1k} and β_{in} respectively. If g_2 is of abstraction level $\ell = \ell(g_2)$ it follows that $\beta_{in} \in \mathcal{B}_\ell$. But ρ requires, in order that the connection be regular, that $\beta_{1k} = \beta_{in}$ so that β_{1k} is also in $\mathcal{B}_\ell$. Then the in-bond value of g_1 must be in the set family $\mathcal{B}_{\ell+1}$ so that $\ell(g_1) = \ell+1$. Q.E.D.

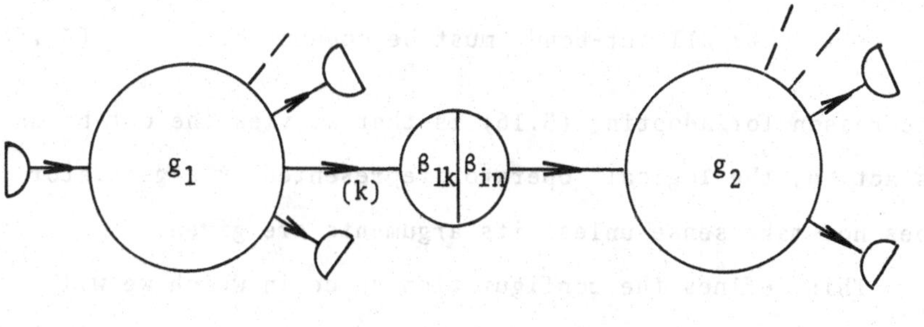

Figure 3.2

Lemma 4. *The generators in any regular configuration* c *have* POSET *structure.*

Proof: Consider a connected component of c with generators $g_1, g_2, \ldots g_n$. All connections go from some level ℓ to some level $\ell-1$. Defining $g_i > g_j$ if there is a connected chain

$$g_i \to g_{i_1} \to g_{i_2} \to \ldots \to g_j \tag{3.14}$$

it is clear that

$$\ell(g_i) = \ell(g_{i_1})+1 = \ell(g_{i_2})+2 = \ldots \tag{3.15}$$

so that loops cannot occur. It follows easily that ">" satisfies the postulates of a partial order. Q.E.D.

Generators belonging to two connected components that are not connected to each other, are not ordered with respect to each other. Generators belonging to a connected component are not ordered with respect to each other if they are of the same level of abstraction. Even if they are of different levels it can happen that they are not comparable via ">".

3.7.1. We are dealing with symmetric regularity: out-bonds can only connect to in-bonds. In this context only finite configurations will occur. The main restriction on Σ will be (in addition to POSET structure as shown in Lemma 4)

$$\Sigma\text{: all out-bonds must be connected.} \tag{3.16}$$

The reason for adopting (3.16) is that we view the out-bonds as active; the logical operator represented by a generator does not make sense unless its arguments are given.

This defines the configuration space in which we will be operating from now on

$$\mathscr{C}(\mathscr{R}) = <G,S,\mathscr{R}>. \tag{3.17}$$

3.7.2. It may be remarked that *this connection type is not monotonic:* if we open some of the bonds or delete some of the generators (and their bonds) from a regular configuration the resulting configuration is not always regular. The reason for this is that we may have opened up an out-bond belonging to the subconfiguration, and this violates (3.16).

Nevertheless we shall have occasion in what follows to deal with such $\mathscr{R}$-irregular subconfigurations. To get this configuration space we apply the functor $\mathscr{MTX}$ to our configuration space

$$\bar{\mathscr{C}}(\mathscr{R}) = \mathscr{MTX}\mathscr{C}(\mathscr{R}) \tag{3.18}$$

see Section 3.5. In $\bar{\mathscr{C}}(\mathscr{R})$ all closed bonds satisfy ρ but out-bonds may be left open.

3.7.3. Just as we need coordinates for a generator to be able to refer unambiguously to its bonds, it is convenient to have some way of numbering the generators in a configuration. A configuration will therefore be described here, as several times before, as an indexed set $\{g_i;\ i = 1,2,\ldots n\}$ of generators, each of which has out-bonds with absolute coordinates $(i,1),(i,2),\ldots(i,0_i)$, with $0_i = \omega_{out}(g_i)$; $i = 1,2,\ldots n$. The in-bonds of g_i will have the coordinates $(i,1),(i,2),(i,3),\ldots$. When referring to a bond (i,k) we must also specify whether it is an in- or out-bond.

Such *configuration coordinates* were discussed, but in a general setting, in Section 3.2.

Strictly speaking a configuration is not entirely specified unless expressed via a system of configuration coordinates, see Notes A. Therefore two configurations c, with generators g_i; $i = 1,2,\ldots,n$; and c', with generators g_i'; $i = 1,2,\ldots,n'$; and with bonds denoted as described, are identical from the functional point of view if and only if

$$\begin{cases} \text{(i)} & n = n' \\ \text{(ii)} & g_i = g_i';\ i = 1,2,\ldots n \\ \text{(iii)} & \text{bonds connected in } c \text{ should have their homo-} \\ & \text{logues in } c' \text{ connected, and vice versa.} \end{cases} \tag{3.19}$$

Note that (ii) implies that $B(g_i) = B(g_i')$ with homologue bonds given by the coordinate system.

More about this in Section 3.8 below when identification is introduced via R.

3.7.4. The cardinality of $\mathscr{L}(\mathscr{R})$ can never be more than denumerable, since we can enumerate $\mathscr{L}(\mathscr{R})$ by first a finite number of configurations in $\mathscr{L}_1(\mathscr{R})$, monatomic ones, then a finite number in $\mathscr{L}_2(\mathscr{R})$, biatomic ones, and so on.

If we exclude the trivial case when $G_0^\omega = \phi$, as will always be done, we can never have $\text{card}[\mathscr{L}(\mathscr{R})] < \infty$. Indeed if $g \in G_0^\omega$ then

$$c = \phi(\underbrace{g,g,\ldots g}_{n \text{ times}}) \tag{3.20}$$

is regular for any n. In (3.20) ϕ denotes the empty connector that does not close any bonds. That $c \in \mathscr{L}(\mathscr{R})$ follows from the fact that all out-bonds in c are connected (there are not any) and ρ holds trivially since no bonds are closed. Hence $\text{card}[\mathscr{L}(\mathscr{R})]$ = denumerably infinite.

3.7.5. The generators in $G_0^\omega = G_0^\ell$, the *objects* (see (3.8)), play a dominant role in regular configurations.

Lemma 5. *All regular non-empty configurations contain objects.*

Proof: Consider an arbitrary $g \in c$ and let ℓ be its level of abstraction. If $\ell = 0$ then g is an object and the assertion holds.

If $\ell \geq 1$ then it has out-bonds in $\mathscr{B}_{\ell-1}$ and Σ requires that they connect to some generator g' of level $\ell-1$. Either $\ell-1 = 0$ so that g' is an object, or we can repeat the argument; eventually we will arrive at some object in the configuration. Q.E.D.

Remark 1. In the monotonic extension $\overline{\mathscr{L}}(\mathscr{R})$ any monatomic configuration is allowed; the level of its generator can then be positive so that configurations consisting entirely of generators more abstract than objects can occur in $\overline{\mathscr{L}}(\mathscr{R})$.

Remark 2. A warning is motivated. "Object" need not represent an object in some material world. As usual, caution is required when mathematical entities are related to concepts used in common sense parlance.

A direct consequence of Lemma 5 is that the only monatomic configuration in $\mathscr{L}(\mathscr{R})$ consist of an object.

3.7.6. The *prime configurations* in $\mathscr{L}(\mathscr{R})$ are easy to characterize.

Lemma 6. *A configuration* $c \in \mathscr{L}(\mathscr{R})$ *is prime if and only if it is connected.*

Proof: If c is not connected it can be viewed as the ϕ-connection of two non-empty and regular configurations c' and $c'' \in \mathscr{L}(\mathscr{R})$. This follows immediately from the fact that the connected components of any c are regular: they satisfy Σ, since all out-bonds are connected, and ρ holds trivially in them. But if $c = \phi(c',c'')$, c' and $c'' \in \mathcal{L}(\mathcal{R})$ not empty, then c is *composite*, not prime.

On the other hand if c is connected it cannot be expressed as $\sigma(c',c'')$ with both c' and c'' non-empty and regular. Indeed, neither c' nor c'' can have any open out-bonds (this would violate the connection type Σ). But then $\sigma = \phi$ which implies that c' and c'' are not connected to each other, against the assumption. Q.E.D.

It is different for the notion *reducible/irreducible*, see Notes B. A configuration is called reducible if it has some regular proper subconfiguration; otherwise it is irreducible.

Lemma 7. *A regular configuration* c *is irreducible if and only if it is monatomic and consists of a single object.*

Proof: The "if" part is obvious. On the other hand if it is not just an isolated object we can select one of its objects, since Lemma 6 assures that it has at least one. Take the subconfiguration consisting of this object in isolation. It is regular, implying that c is reducible. Q.E.D.

Lemma 7 says that in the present $\mathcal{L}(\mathcal{R})$ we have only trivial irreducible configurations. This is an atypical situation in pattern theory.

3.7.7. What are the "simplest" regular configurations? Fixing a generator g, let us demand that the configuration contain g but has no (proper) regular subconfiguration also containing g. Such a configuration will be said to be a *simple* g-configuration.

Lemma 8. *A configuration* $c \in \mathcal{L}(\mathcal{R})$ *is a simple* g_1*-configuration if and only if* a) g_1 *is the unique solution in* c *of*

$$\ell(g_1) = \max_{g \in c} \ell(g) \triangleq (c) \tag{3.21}$$

and, b) *if all other* $g_i \in c$ *satisfy* $g_i \prec g_1$.

Proof: Assume that c is a simple g_1-configuration with $\ell(c) = \ell$ and that it has another generator g_2 with $\ell(g_2) = \ell$. No descending chain (see 3.6.2) can lead from g_1 to g_2, so that we can delete g_2 with its bonds from c without leaving any out-bonds open. Hence a) holds.

Now assume that c has some generator g_i which is not subordinated to g_1 as in b). Then we can remove g_i with its bonds without leaving any out-bonds open: b) holds.

On the other hand, if c is a regular configuration for which a) and b) hold it cannot be reduced, still keeping g in it. To see that this is so, say that the reduced configuration leaves out g_i from c. Since there is a descending chain from g_1 to g_i some out-bond will be left open, when we delete g_i, unless the entire chain is deleted. But then g_1 is not in the reduced configuration, so that the condition is both necessary and sufficient. Q.E.D.

Lemma 8 tells us that the simple g-configurations have the typical appearance of Figure 3.3, where $\ell(c) = \ell(g) = 3$ and c contains the two objects g_6 and g_7.

How many simple g-configurations are there? With $N = \#(G)$ we get the crude upper bound

$$N \times N^{\omega_{max}} \times N^{\omega_{max}^2} \times \dots \times N^{\omega_{max}^{\ell}} \tag{3.22}$$

If $\omega_{max} > 1$ this gives us the bound

$$N^{\frac{\omega_{max}^{\ell+1}-1}{\omega_{max}-1}} . \tag{3.23}$$

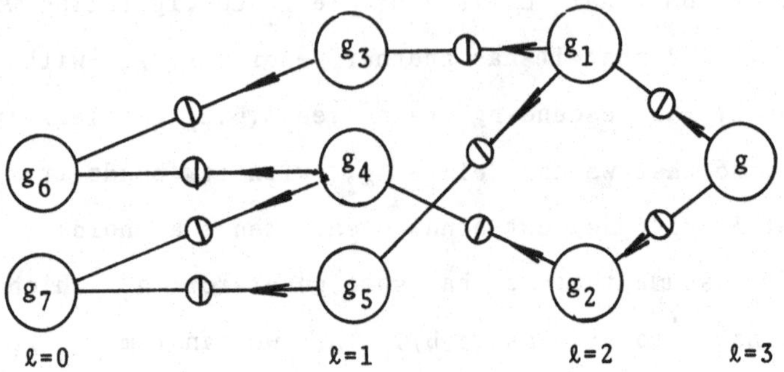

Figure 3.3

The number of simple g-configurations is certainly finite, but it can be extremely large if the abstraction level is big. This will have serious consequences for the ability to learn semantics later on.

3.7.8. Consider a configuration $c \in \mathscr{L}(\mathscr{R})$ with a sub-configuration $c_1 \in \mathscr{L}(\mathscr{R})$. Among the out-bonds of c_1 there may be some that are open. Close these bonds by adding the appropriate generators of c, close the new out-bonds left open, and continue like this. Since c is finite the process will end with some regular subconfiguration c_1^*. In extreme cases $c_1^* = c_1$ or c itself. We shall call c_1^* the *minimal extension* of c_1 in c. The process can be shown to lead to a unique result.

Lemma 9. *For a regular configuration containing the generator* g *the minimal extension* g^* *of the monatomic subconfiguration* $\{g\} \in \mathscr{L}(\mathscr{R})$ *in* c *is a simple* g-*configuration.*

Proof: The regular configuration g^* can have no (proper) regular subconfiguration containing g since if c' were one

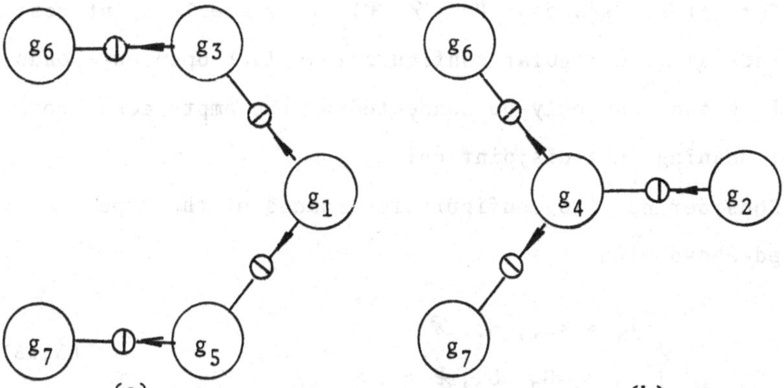

(a) Figure 3.4 (b)

such subconfiguration all the out-bonds of g must be closed, just as the out-bond from generators connected to the out-bonds of g, and so on. But then the above procedure gives $c' = g^*$, which proves the assertion. Q.E.D.

As an example, with c as in Figure 3.3, we get g_1^* as in Figure 3.4(a) and g_2^* as in (b).

Remark 1. It is clear that the minimal extension is a closure operation in the limited sense (for a fixed c)

$$\begin{cases} c_1 \subseteq c_1^* \\ (c_1^*)^* = c_1 \end{cases} \tag{3.24}$$

where the inclusion relation denotes the relation configuration-subconfiguration, not just inclusion of sets of generators. For fixed c the minimal extension maps a subset of $\mathscr{L}(\mathscr{R})$ into $\mathscr{L}(\mathscr{R})$.

Remark 2. The operation is also monotonic in the sense that, for fixed $c \in \mathscr{L}(\mathscr{R})$, $c_1 \subseteq c_2 \subseteq c$ implies $c_1^* \subseteq c_2^* \subseteq c$.

3.7.9. As in most pattern theory the *homomorphisms between configuration spaces* play an important role, see

Grenander (1977d). In the present case this is certainly true for $\overline{\mathscr{C}}(\mathscr{R})$ spaces. For $\mathscr{C}(\mathscr{R})$ they are less interesting since all its regular configurations lack open out-bonds. Therefore they can only be connected by the empty connector $\sigma = \phi$, meaning just disjoint union.

Consider now two configuration spaces of the type studied above with

$$\begin{cases} \mathscr{C}_1 = \langle G_1, S_1, \mathscr{R} \rangle \\ \mathscr{C}_2 = \ G_2, S_2, \mathscr{R} \rangle . \end{cases} \tag{3.25}$$

and with a surjective generator map $\mu: G_1 \to G_2$ preserving bonds $B(\mu g) = B(g)$ and where G_1 and G_2 have the "same" set families, in the sense that $\mu\mathscr{B}^1_\ell = \mathscr{B}^2$.

Extend the definition of μ to $h: \mathscr{C}_1 \to \mathscr{C}_2$ by putting hc, $c \in \mathscr{C}_1$, equal to the configuration with same connection but where each g_i in c is replaced by μg_i. We then have

Lemma 10. *The configuration map* $h: \mathscr{C}_1 \to \mathscr{C}_2$ *is a homomorphism in the sense of Grenander* (1977d).

Proof: First recall how μ sets up a correspondence between the two similarity groups S_1 and S_2. Each similarity group consists of permutations leaving bonds invariant. But μ preserves bonds, so that the two groups are isomorphic, $S_1 \cong S_2$, although μ need not be bijective. Actually, both of them are the (full) symmetric group of the generator index classes in G_1 and G_2 respectively and these are bijectively related since μ is surjective.

To prove Lemma 10 let $c \in \mathscr{C}_1$ and consider sc, $s \in S_1$. To calculate $h(sc)$ we first have to permute the generators appearing in c according to the similarity c, and then replace each generator g by μg. But this leads to the same result as if we first replaced each generator by its μ-map value and then permuted them by the isomorphic permutation in S_2: the two operations commute. Recall that the similarities just permute index classes that are defined in terms of equal bonds, and that bonds are preserved by our map.

Finally if $c = \sigma(c_1, c_2)$, with $c, c_1, c_2 \in \mathscr{C}_1$, calculate hc_1 and hc_2 and combine them by the *same* connector σ as before. This is possible since bond values are preserved and ρ is EQUAL; the connection type offers no restriction in the present case since σ is not changed. But $\sigma(hc_1, hc_2)$ will then have exactly the connection of $\sigma(c_1, c_2)$. Its generators have been exchanged according to the generator map μ. Hence $\sigma(hc_1, hc_2) = hc$ as required for h to be a homomorphism.

Q.E.D.

Remark 1. In order that the conclusions of Lemma 10 hold it is not necessary to require that $B(\mu g) = B(g)$. It suffices to ask that a) the bond structure remains the same, $B^s(\mu g) = B^s(g)$, and b) that if $\beta_i(g_1) = \beta_j(g_2)$ then $\beta_i(\mu g_1) = \beta_j(\mu g_2)$.

Remark 2. Also the bond values just mentioned need of course not be exactly the same in the general case when ρ is not EQUAL. It is enough if there exists a map $\beta \to \beta'$ between the two bond value sets for G_1 and G_2 respectively such that the relation $\beta_1 \rho \beta_2$ implies $\beta_1' \rho \beta_2'$. Here β_1 and β_2 refer to g_1 and g_2, while β_1' and β_2' refer to the homologue bonds of μg_1 and μg_2.

This simple fact will be useful later on.

One particular case of some interest later on is when in each index class of G_1 we single out one element g_α, a *prototype*, and define μ as $g \to g_\alpha$ if g belongs to G_α. It is easy to show that the conditions of Lemma 10 are satisfied and hence this generator map leads to a homomorphism between the two configuration spaces.

3.8.1. We are now ready to introduce the image algebra. The configurations play the role of "formulas," here satisfying the particular rules $\mathcal{R}$ of combinatory regularity that we have discussed in Section 3.6. The images, on the other hand, express the "function" of these formulas.

In the present case the identification rule R (see Volume I, Section 3.1) will be chosen to make the functions *coordinate free* - the choice of coordinate system for describing configurations should be irrelevant; see Notes C.

To formalize this, consider two regular configurations c and c' with n generators $g_1, g_2, \ldots g_n$ in c and n' generators $g_1', g_2', \ldots g_{n'}'$ in c' respectively. The bond coordinates will be denoted by

$$\begin{cases} (k,j), \quad j=1,2,\ldots n_k \text{ for } g_k;\ k=1,2,\ldots n \\ (k,j), \quad j=1,2,\ldots n_k' \text{ for } g_k';\ k=1,2,\ldots n' . \end{cases} \tag{3.26}$$

We shall let R identify c and c' if and only if (a) $n = n'$, (b) there exists a permutation $(1,2,\ldots n) \to (i_1, i_2, \ldots i_n)$ such that $g_k = g_{i_k}'$, $k = 1,2,\ldots n$, (c) for each k we have $n_k = n_{i_k}'$, (d) for each k there exists a permutation $(1,2,\ldots n_k) \to (j_1, j_2, \ldots j_{n_k})$ such that the ℓth bond of g_k equals in structure and value the j_ℓ^{th} bond of g_{i_k}, and (e) bonds corresponding to each other are connected/not connected in the same way.

Lemma 11. *This* R *is an identification rule for* $\mathcal{C}(\mathcal{R})$ *and for* $\overline{\mathcal{C}}(\mathcal{R})$.

Proof: That R is an equivalence is obvious. Also, if c and c' are regular, and if they are R-equivalent, cRc', then c and c' have the same unconnected bonds, related by a permutation. These are the external bonds, same for c and c'. We shall then in the following assume that the coordinates have been permuted, if necessary, so that the external bonds are the same for each coordinate. Further, if again cRc', then if we apply the same similarity s to c and to c' we can relate sc and its bonds to sc' and its bonds by the same permutation as for c and c'. Hence (sc)R(sc'). Finally, if c_1Rc_1' and c_2Rc_2', where all four configurations are regular, then c_1 and c_1' have the same external bonds, related by some permutation, and similarly for c_1 and c_2'. Connect c_1 and c_2 into some regular configuration $c = \sigma(c_1,c_2)$. We now connect c_1' and c_2' by the same σ expressed in the coordinate system mentioned above. It follows that $c' = \sigma(c_1',c_2')$ is also regular, since bonds connected in c correspond to bonds connected in c', and vice versa. The bond relation is therefore satisfied for all closed bonds in c'. The connection of c' is also in Σ. But the new c', now known to be regular, consists of the same generators as c, and with the same connections, related by permutations as described. Hence cRc' which shows that R has all the properties of an identification rule. Q.E.D.

Combining Lemma 11 with the properties shown for we have proved

Theorem 1. *With* $\mathscr{L}(\mathscr{R})$ *and* R *as given above* $\mathscr{T} = <\mathscr{L}(\mathscr{R}),R>$ *is an image algebra and so is* $\overline{\mathscr{T}} = <\overline{\mathscr{L}}(\mathscr{R}),R>$.

In the following all perceptions of the world will be expressed in image algebras of this form.

3.8.2. Any S-invariant class of images forms a *pattern*. Among these are those generated by a *template* I_0

$$P(I_0) = \{sI_0 | \forall s \in S\} \subseteq \mathcal{T}. \tag{3.27}$$

Two patterns $P(I_1)$ and $P(I_2)$ of the form in (3.27) are either identical or disjoint as subsets of $\mathcal{T}$.

Two distinct images in a pattern describe different perceptions of the world, since otherwise they would be identified by R, but they have the same *logical structure*. That this is so follows from the fact that they are made up of generators, see Notes D, say g for I_1, and then sg for I_2, that have the same generator index and play the same logical, but not substantial, role.

9.4. Two special image algebras

4.1. The construction of the image algebra in the last section is based on simple pattern theoretic ideas. Nevertheless, it takes some experience of manipulating configuration spaces and their images before one becomes familiar with such structures. To facilitate this for the reader we now present two fairly simple examples, the first completely abstract, the second with an intuitive interpretation.

They may appear as quite different from each other but, as we shall show in Section 4.4, they are closely related to each other.

4.2.1. Let G_1 consist of the 12 abstract generators in Table 4.1. As identifiers we have chosen simply the letters of the alphabet and as bond values Roman numerals.

TABLE 4.1

GENERATORS FOR $\mathcal{G}_1$

number	identifier	level	β_{in}	ω_{out}	β_{out1}	β_{out2}	β_{out3}	β_{out4}	β_{out5}
1	a	3	-	2	I	I	-	-	-
2	b	3	-	1	I	-	-	-	-
3	c	2	I	1	II	-	-	-	-
4	d	2	III	1	II	-	-	-	-
5	e	2	III	1	II	-	-	-	-
6	f	1	II	5	V	V	V	V	V
7	g	1	II	3	V	V	VI	-	-
8	h	1	II	1	VI	-	-	-	-
9	i	1	II	4	VI	VI	V	V	-
10	j	1	VII	1	VI	-	-	-	-
11	k	0	V	0	-	-	-	-	-
12	l	0	VI	0	-	-	-	-	-

The out-arities of these generators vary between 0 and 5, the objects with $\omega_{out} = \ell = 0$ consisting of the two generators k and ℓ.

The levels of abstraction vary between $\ell = 0$, for the two objects, to $\ell = 3$ for a and b. The two partitions $\{G_\mu^\omega\}$ and $\{G_\nu^\ell\}$ are given in Tables 4.3 and 4.4.

The bond values are from the sets $\mathcal{B}_\nu$ as described in Section 3.4.6 and tabulated in Table 4.2. One need not specify the in-bond values of the generators of maximum level of abstraction since no out-bond can connect to them. This is why the cells for β_{in} are empty for the two first rows of Table 4.1.

When the out-arity exceeds one we need generator coordinates to keep bonds apart, and this has been done by numbers 1,2,... marking the out-bonds. In-bonds are treated as identical, for any given generator, and therefore need no coordinates.

TABLE 4.2

SET FAMILY $\mathscr{B}$ FOR $\mathscr{T}_1$	
$\mathscr{B}_0$	{V, VI}
$\mathscr{B}_1$	{II, IV, VII}
$\mathscr{B}_2$	{I, III}

TABLE 4.3

ν	G_ν^ℓ
0	$\{k,\ell\}$
1	$\{f,g,h,i,j\}$
2	$\{c,d,e\}$
3	$\{a,b\}$

TABLE 4.4

μ	G_μ^ω
0	$\{k,\ell\}$
1	$\{b,c,d,e,h,j\}$
2	$\{a\}$
3	$\{g\}$
4	$\{i\}$
5	$\{f\}$

This is of course a quite small generator space, Actually, if we calculate the generator classes, putting the generator index equal to a constant, we see that each generator constitutes its own generator class, with the exception that d and e belong to the same class $\alpha(d) = \alpha(e)$.

For this reason we have a poor group of similarities, only allowing the permutation of d and e. This is not a typical situation, but occurred since we chose a very simple example.

4.2.2. We can now combine the generators a,b,c,... following the rules $\mathcal{R}$ of combinatory regularity. We get, for example the regular configuration (a) in Figure 4.1 consisting of the generators k, occurring twice, g,ℓ, and j. It contains the objects, of type k and ℓ, and is of abstraction level one due to the occurrence of the generators g and j. To prove that (a) satisfies $\mathcal{R} = \langle \rho, \Sigma \rangle$ we first note from Table 4.1 that $\omega_{out}(k) = \omega_{out}(\ell) = 0$, so that k and ℓ have no out-bonds. Also that g has three out-bonds, all indicated in the diagram, with coordinates shown and j has one out-bond. Since all these four out-bonds are closed in the diagram the connection type Σ is satisfied. Now look at the four closed bonds in the diagram, for example the one from g to the upper k. From 4.1 we find that the first out-bond of g has bond value V, and that the in-bond value of k is V, equality holds and ρ is satisfied for this bond. In the same way we check the other bonds and conclude that $\mathcal{R}$ holds: (a) is a regular configuration. It is not a simple g-configuration since j can be dropped without destroying the regularity. However, with j deleted, the resulting subconfiguration is g-simple.

Another example is given in (b) with $n(c) = 7$. It consists of the generators k, occurring twice, e, twice, and g, h, and ℓ. It is of abstraction level two, due to the

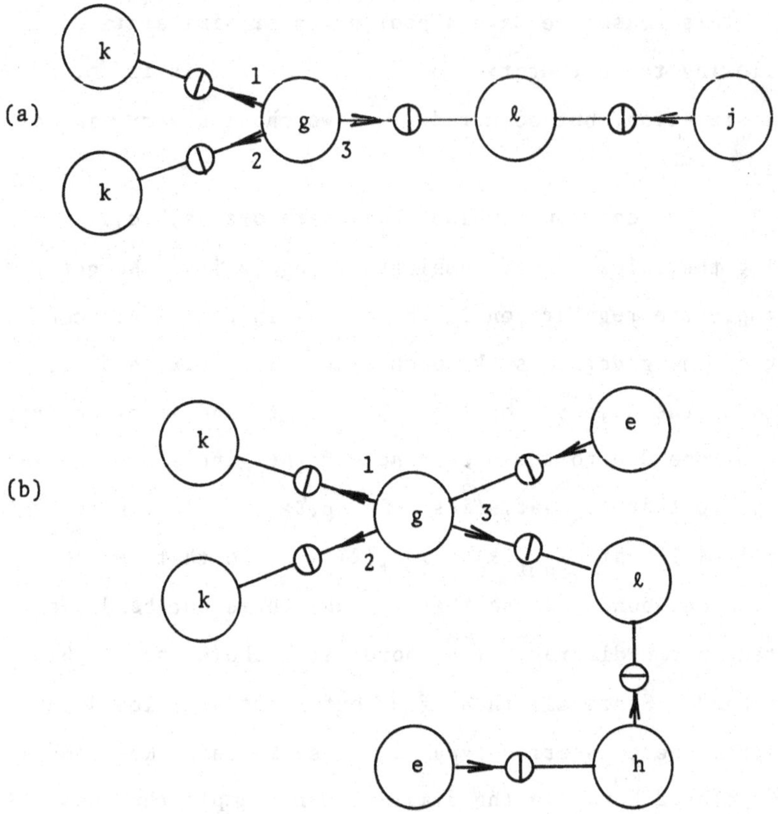

Figure 4.1

CONFIGURATION DIAGRAMS OVER G_1

occurrences of e. If we delete the bottom e and h we get an e-simple subconfiguration.

<u>4.3.1</u>. Let us now look at a more interesting example with the 34 generators listed in Table 4.5 together with their levels, out-arities, and bond values.

In addition the table contains identifiers, but now in the form of English word(s), intended to give the reader a concrete idea of the purpose of this regular structure. The

TABLE 4.5

GENERATORS FOR $\mathcal{T}_2$

number	identifier	level	β_{in}	ω_{out}	β_{out1}	β_{out2}	β_{out3}	β_{out4}	β_{out5}
1	individual 1	3	-	2	H	H	-	-	-
2	individual 2	3	-	2	H	H	-	-	-
3	left	3	-	1	H	-	-	-	-
4	right	3	-	1	H	-	-	-	-
5	horizontal	3	-	1	H	-	-	-	-
6	vertical	3	-	1	H	-	-	-	-
7	arm	2	H	1	G	-	-	-	-
8	idle	3	F'	1	G	-	-	-	-
9	strongly	2	F'	1	C	-	-	-	-
10	weakly	2	F'	1	C	-	-	-	-
11	clockwise	2	F'	1	D	-	-	-	-
12	counterclockwise	2	F'	1	D	-	-	-	-
13	left'	2	F'	1	G	-	-	-	-
14	right'	2	F'	1	G	-	-	-	-
15	horizontal'	2	F'	1	G	-	-	-	-
16	vertical'	2	F'	1	G	-	-	-	-
17	down	2	F'	1	E	-	-	-	-
18	up	2	F'	1	E	-	-	-	-
19	hand	1	G'	5	B	B	B	B	B
20	grasp	1	C	3	B	B	A	-	-
21	rotate	1	D	1	A	-	-	-	-
22	press	1	E	4	A	A	B	B	-
23	brass	1	F	1	A	-	-	-	-
24	steel	1	F	1	A	-	-	-	-
25	little	1	F	1	A	-	-	-	-
26	big	1	F	1	A	-	-	-	-
27	thumb	0	B	0	-	-	-	-	-
28	index finger	0	B	0	-	-	-	-	-
29	middle finger	0	B	0	-	-	-	-	-
30	ring finger	0	B	0	-	-	-	-	-
31	little finger	0	B	0	-	-	-	-	-
32	bolt	0	A	0	-	-	-	-	-
33	nut	0	A	0	-	-	-	-	-
34	cylinder	0	A	0	-	-	-	-	-

idea is to formalize hand motions of two individuals working with a few things made of metal. This should be compared to the discussion in Section 3.7 of Volume I on motion studies, and to Case 3.6.4 (anatomical patterns), ibid.

A few remarks will be in order. The in-bond values F and F' never occur among out-bond values. This implies that the generators 8-18 and 23-26 never accept out-bonds from other generators.

We have two generators *right* and *right'* that seem to play the same role and analogously for *left* and *left'*. This is not so, however. The generator g = *right* connects to g = *arm*, but not to *hand*. The generator *right'*, on the other hand, connects to *hand*, but not to *arm*. We have thought of an arm as being made up of various parts, one of them being a hand. It is then not appealing to allow the same unary relation (property) to be applicable to both, on various levels of abstraction.

The vagueness of every day language tends to conceal such subtle semantic distinctions, but one of the advantages of the abstract approach is that it forces precision upon us.

This is of more general significance than may be immediately obvious. However, if we decided, against the reasoning just given, that a generator, for example *right*, should be allowed to connect to generators of different levels of abstraction we would have to modify the assumptions in 3.4.6. This can certainly be done, and with little effort, but seems unnatural.

The generator *grasp*, of out-arity three, should be read as "grasp something$_3$ between finger$_1$ and finger$_2$". The markers are the bond coordinates. The generator *press*, of out-arity four, should be read "press something_1 against something_2 using finger_3 and finger_4. The remaining generators need no explanation.

4.3.2. The set families $\mathcal{B}_\nu$ are given in Table 4.6, and the partitions according to level and out-arity in Tables 4.7-4.8 respectively.

The index classes are easily calculated; see the discussion in Section 3.5. There are 15 of them given in Table 4.9.

TABLE 4.6

	SET FAMILY $\mathcal{B}_\nu$ FOR $\mathcal{T}_2$
$\mathcal{B}_0$	{A,B}
$\mathcal{B}_1$	{C,D,E,F,G}
$\mathcal{B}_2$	{H,F'}

TABLE 4.7

ν	G_ν^ℓ
0	27,28,29,30,31,32,33,34
1	19,20,21,22,23,24,25,26
2	7,8,9,10,11,12,13,14,15,16,17,18
3	1,2,3,4,5,6

The reader may be interested in recognizing the common characteristics, in every-day language, of generators with the same index α. It is also of interest to compare the index classes with the map μ in Table 4.9 and the relation $\beta \to \beta'$ in Table 4.11 given later in the text.

TABLE 4.8

μ	G_μ^ω
0	27,28,29,30,31,32,33,34
1	3,4,5,6,7,8,9,10,11,12,13,14,15,16,17,18,21,23,24,25,26
2	1,2
3	20
4	22
5	19

TABLE 4.9

α	G^α	$\#(G^\alpha)$
1	{1,2}	2
2	{3,4,5,6}	4
3	{7}	1
4	{8}	1
5	{13,14,15,16}	4
6	{9,10}	2
7	{11,12}	2
8	{17,18}	2
9	{19}	1
10	{20}	1
11	{21}	1
12	{22}	1
13	{23,24,25,26}	4
14	{27,28,29,30,31}	5
15	{32,33,34}	3
		34

The similarities are now many more than in the first example. The group S of similarities is the direct product of full symmetric groups of order 2,4,1,1,4,...; see the last table. Hence

$$\#(S) = 2!4!\ldots \cong 1.6\cdot 10^8. \qquad (4.1)$$

4.3.3. Combining the generators *thumb, index finger, grasp, bolt, brass* we get the regular configuration in Figure 4.2(a). Another one is shown in (b). A more complicated case is given in (c) with $n(c) = 13$. The sub-configuration c' inside the dotted contour is regular, and can be thought of as a macro-generator, see Volume I, p. 32. The whole configuration is of abstraction level three.

Two macro-generators c'' and c''' appear in (d), where two individuals are at work together. This configuration is also of abstraction level 3.

The configuration c_1 in the figure, in (e), is simply related to the one c_2 in (a): they are similar, $c_1 = sc_2$, $s \in S$.

Regular configurations as above, and the resulting images in $\mathcal{T}_2$, describe hand motions of one or two individuals. It would be misleading, however, to say that such images *mean* certain motions in the physical world. To do this we would need *another* formalization of the physical worlds on the level of the natural sciences. In our way of thinking semantics means a correspondence between two regular structures.

As pointed out in Section 2 we do not necessarily assume that the perception of the world of our observer is logically consistent. As a matter of fact the term "logically consistent" requires that second regular structure just mentioned, which may be absent. Without introducing it we should not worry too much if we encounter images in $\mathcal{T}_2$ with individuals with five thumbs, or two individuals sharing an arm.

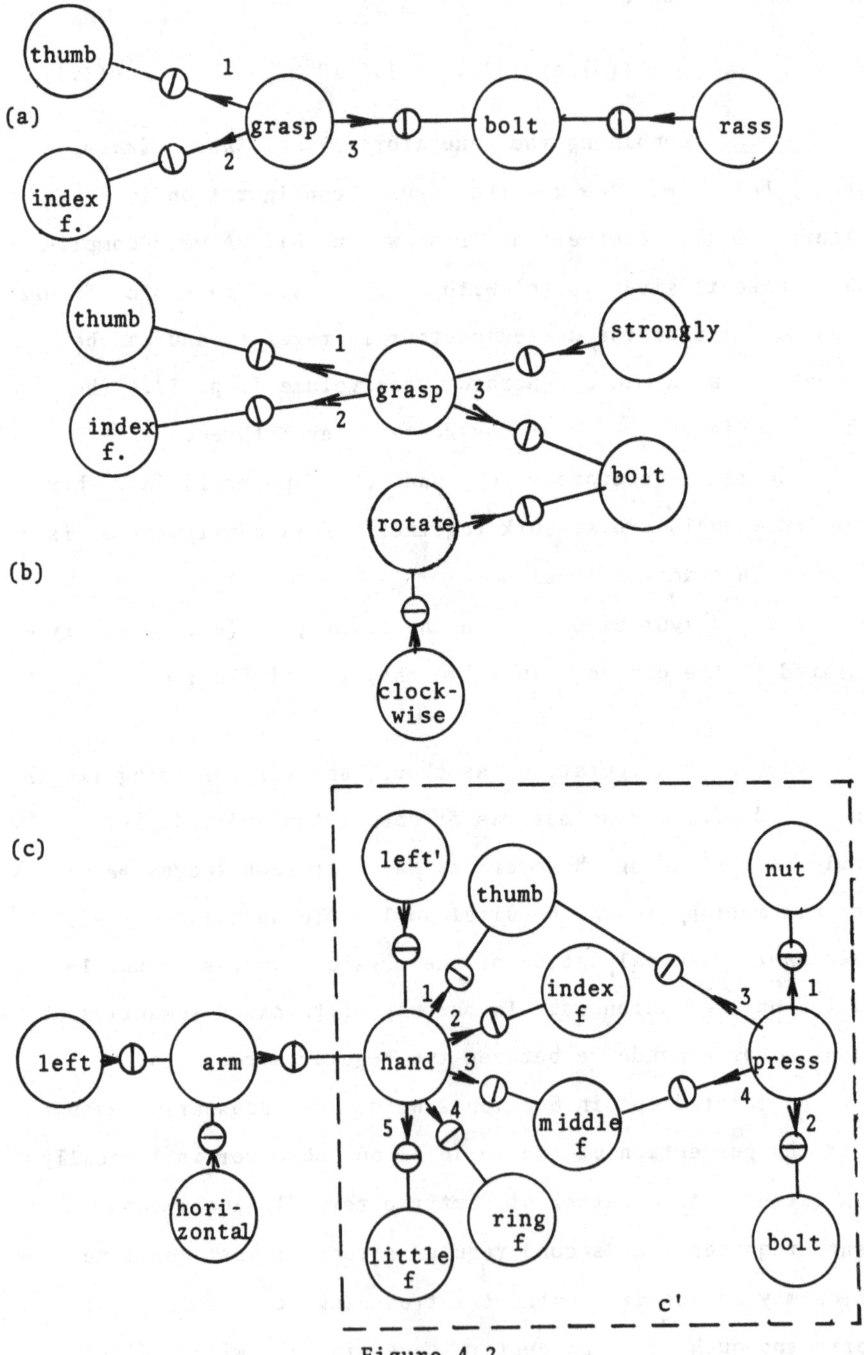

Figure 4.2

CONFIGURATION DIAGRAMS OVER G_2

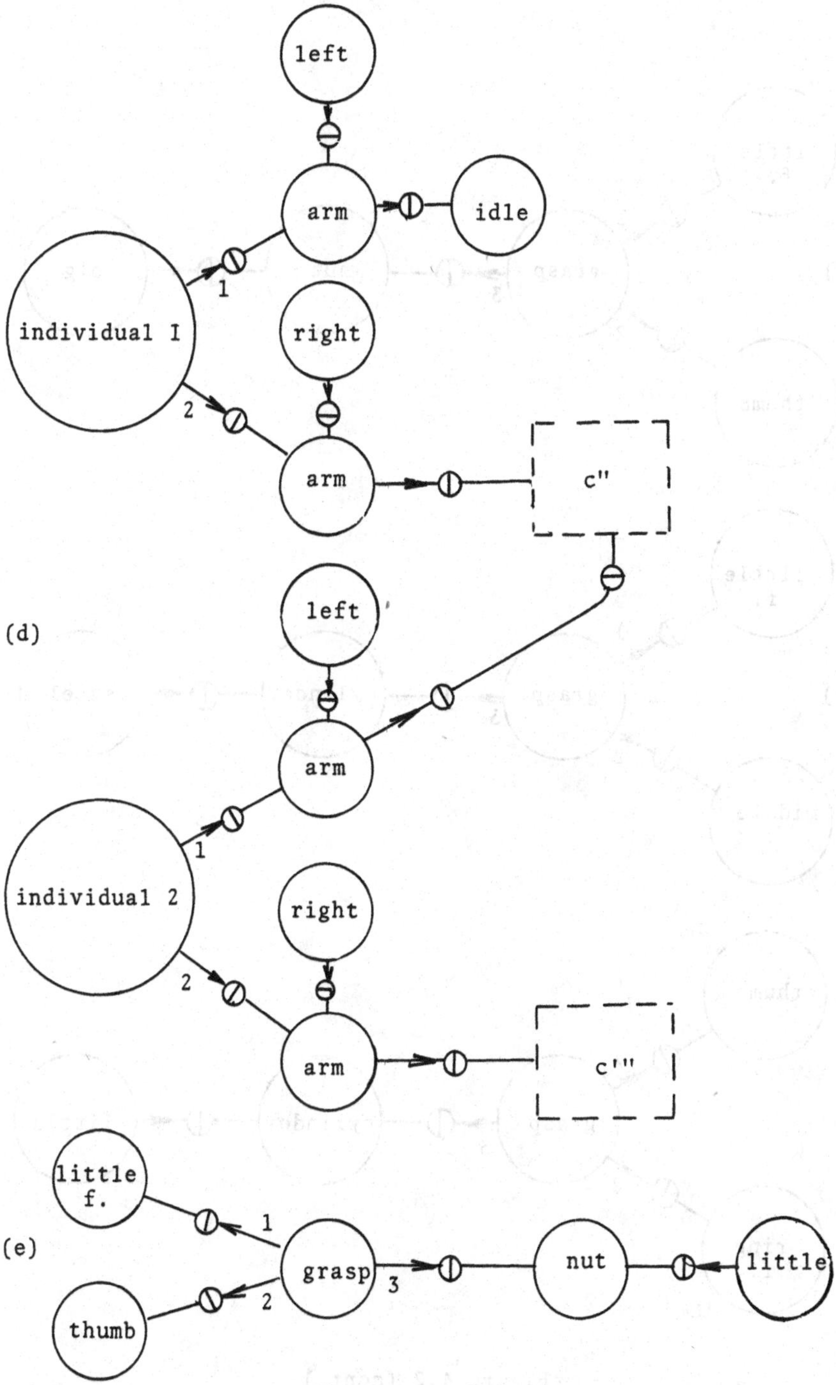

Figure 4.2 (cont.)

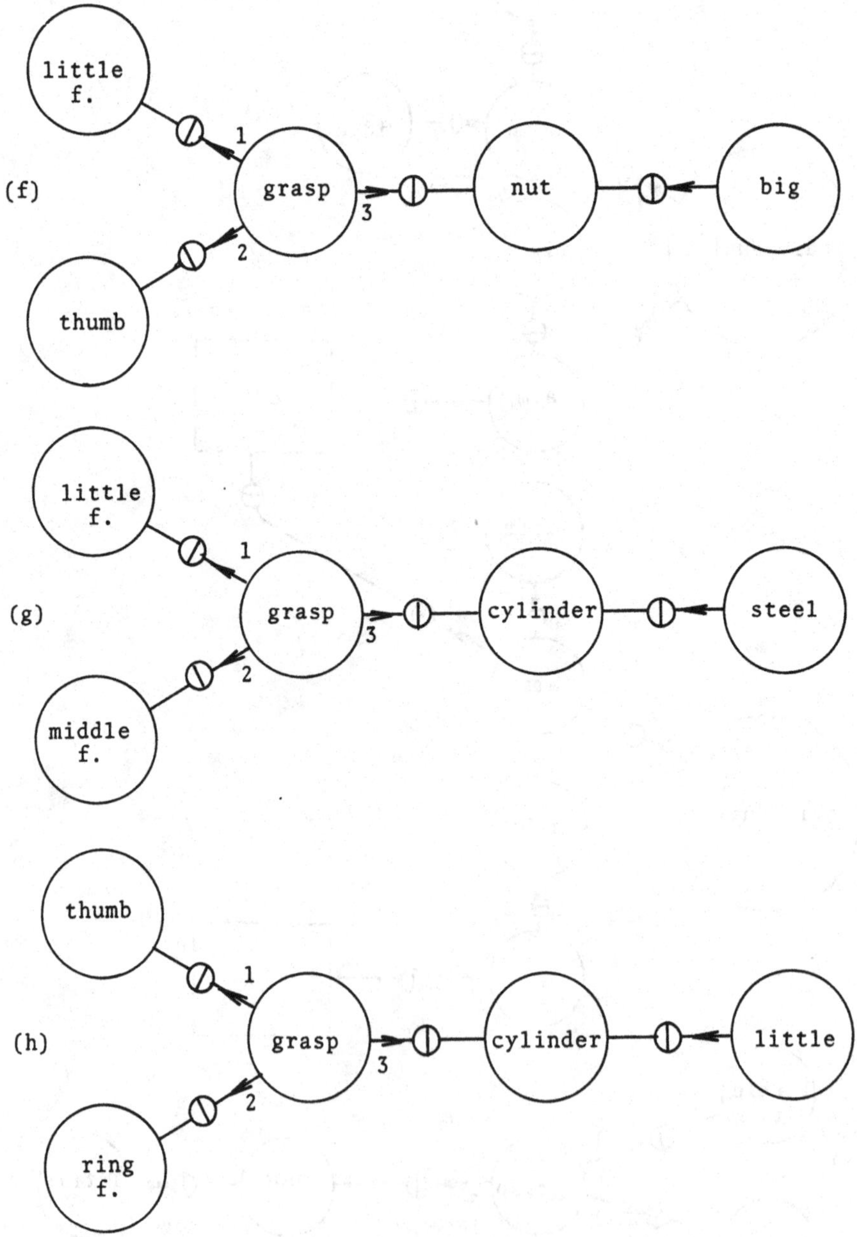

Figure 4.2 (cont.)

If we want to remove such images from the algebra it can be done by labelling in-bonds by markers; at the present stage this does not seem necessary.

The last configurations (f), (g), (h) are similar and belong to the same pattern as the one in (a). This can be checked using Table 4.10 specifying the similarity group S.

4.4. To establish a correspondence between our two configuration spaces (as well as the related image algebras) we consider the following generator map $\mu:G_2 \to G_1$ together with the table of the associated bond value map $\beta \to \beta'$; see Tables 4.10-4.11.

TABLE 4.10

g	μg	g	μg	g	μg
1	a	13	e	25	j
2	a	14	e	26	j
3	b	15	e	27	k
4	b	16	e	28	k
5	b	17	e	29	k
6	b	18	e	30	k
7	c	19	f	31	k
8	d	20	g	32	ℓ
9	e	21	h	33	ℓ
10	e	22	i	34	ℓ
11	e	23	j		
12	e	24	j		

TABLE 4.11

β	β'	β	β'
A	VI	E	II
B	V	F	IV
C	II	F'	III
D	II	G	II
		H	I

Referring to Remark 2 in 3.7.9 we can verify that μ generates a homomorphism $h: \mathscr{C}_2 \to \mathscr{C}_1$. We just have to check that, (a), μ preserves bond-structure, $B^s(\mu g) = B^s(g)$, using Table 4.9 together with Tables 4.5 and 4.1, and that, (b), the bond values behave as required in the quoted Remark 2.

Applying this homomorphism for example to (a) in Figure 4.2, we get the $\mathscr{C}_1$-configuration (a) in Figure 4.1. In the same way the $\mathscr{C}_2$-configuration (b) in Figure 4.2 is carried over into (b) in Figure 4.1.

As typical for homomorphisms, h loses information: a $\mathscr{C}_1$-configuration (or one in $\mathscr{C}_1$) is less informative, although topologically the same, compared to a $\mathscr{C}_2$-configuration (or one in $\mathscr{C}_2$).

9.5. <u>The choice of language type for the study</u>

<u>5.1.1</u>. In accordance with the discussion in Section 2, and with the stated reservations, we shall choose the type of language to be used by the speaker/listener as *finite state*. We can be quite brief when discussing these languages, they are so well known.

<u>5.1.2</u>. The grammar $\mathscr{G}$ will be based on a vocabulary $V = V_T \cup V_N$. Here V_T is the *terminal vocabulary* consisting

of the *words* to be used. We shall sometimes use Greek or Roman letters to denote the elements of V_T, and occasionally common words in English. In the latter case we have to watch out so that we do not forget that *they should be treated abstractly*, not representing a subset of real English.

The *non-terminal vocabulary* V_N consists of *syntactic variables, or states*. They will be denoted by numbers $1,2,3,\ldots F$, where F indicates the final state. We shall use the convention that 1 is the start state.

5.1.3. The *productions* in $\mathcal{G}$ can always be expressed in canonical form as

$$i \underset{x}{\to} j, \quad x \in V_T, \quad i,j \in V_N. \tag{5.1}$$

We can read (5.1) as "the state i goes into j while writing the terminal symbol x". Sometimes it will be convenient to let x in (5.1) indicate a finite string instead, $x \in V_T^*$, but this will not affect the generative power of the grammar.

5.1.4. Just as in Volume II, Chapter 8, we shall assume that the grammar has been reduced to deterministic form so that any productions in $\mathcal{G}$ of the form

$$\begin{cases} i \underset{x}{\to} j \\ i \underset{x}{\to} k \end{cases} \tag{5.2}$$

must coincide, $j = k$. This makes parsing of sentences unambiguous, so that if $x_1x_2x_3\ldots x_n$ is grammatical it has a unique parsing into

$$1 \; x_1 \; i_1 \; x_2 \; i_2 \; x_3 \ldots \; x_n \; F \tag{5.3}$$

In (5.3) we have parsed the sentence into successive productions $1 \underset{x_1}{\rightarrow} i_1, i_1 \underset{x_2}{\rightarrow} i_2, \ldots i_{n-1} \underset{x_n}{\rightarrow} F$.

5.1.5. The set $L \subseteq V_T^*$ of finite strings produced like this constitutes the *language* generated by the language $L = L(\mathcal{G})$.

With the same procedure for producing strings, but not requiring that the state $i_o = 1$ or $i_n = F$, we get *grammatical phrases*

$$i_{o_{x_1}} i_{1_{x_2}} i_{2_{x_3}} \ldots {}_{x_n} i_n \,. \tag{5.4}$$

They need not belong to the language, but can be of linguistic significance anyway.

5.2. Equivalently the language can be represented by a *finite automaton* that we shall often present in diagrammatic form. To clarify the notation consider the finite automaton given by the simple wiring diagram in Figure 5.1.

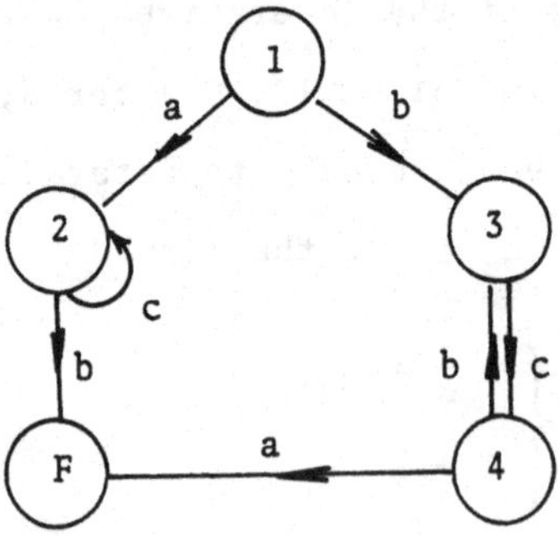

Figure 5.1

The corresponding grammar has

$$\begin{cases} V_T = \{a,b,c\} \\ V_N = \{1,2,3,4,F\} \end{cases} \tag{5.5}$$

and the productions in the table

TABLE 5.1

$1 \underset{a}{\to} 2$	$3 \underset{c}{\to} 4$
$1 \underset{b}{\to} 3$	$4 \underset{b}{\to} 3$
$2 \underset{c}{\to} 2$	$4 \underset{a}{\to} F$
$2 \underset{b}{\to} F$	

It is obviously deterministic. It generates for example sentences parsed as

$$\begin{cases} {}^{1}{}_{a}{}^{2}{}_{c}{}^{2}{}_{c}{}^{2}{}_{b}{}^{F} \\ {}^{1}{}_{b}{}^{3}{}_{c}{}^{4}{}_{b}{}^{3}{}_{c}{}^{4}{}_{a}{}^{F} \\ {}^{1}{}_{b}{}^{3}{}_{c}{}^{4}{}_{a}{}^{F} \end{cases} \tag{5.6}$$

and phrases like

$$\begin{matrix} {}^{2}{}_{c}{}^{2}{}_{c}{}^{2} \\ {}^{3}{}_{c}{}^{4}{}_{b}{}^{3}{}_{c}{}^{4} \end{matrix} \tag{5.7}$$

<u>5.3</u>. Another equivalent way of representing finite state languages is via *regular expressions* from formal logic. Such expressions are built from concatenation, finite repetition (indicated by a star), union, and parentheses to indicate order of execution.

The language generated by the wiring diagram in Figure 5.1, for example, can then be written as

$$L = (ac^*b) \cup (bc(bc)^*a). \tag{5.8}$$

It is clear $\#(L) = +\infty$ if and only if the regular expression contains at least one star. This is the only case of interest for us and will be assumed throughout this paper.

5.4.1. For the following it is of paramount importance that *finite state languages*, as well as many other formal languages, *can be viewed as combinatory regular structures*; see Volume I, Sections 2.4 and 3.2.

The generators will then be represented by the productions of $\mathcal{G}$ (not by words!) They have $\omega_{in} = \omega_{out} = 1$, with the in-bond value given by the state i to be rewritten in (5.1) and the out-bond value as j, the resulting state. Further ρ = EQUAL and Σ = LINEAR.

The identification rule R to be used then identifies two regular configurations (grammatical phrases) if they consist of the same string of terminal symbols and have the same external in-bond value and the same external out-bond value. Remember that the finite automaton was here assumed to be deterministic, then each image consists of a single configuration.

5.4.2. Strictly speaking, this image algebra represents not just L, but all grammatical phrases in L. When we want to limit ourselves to just L, we need two more generators, one g' with $\omega_{in}(g') = 0$, $\omega_{out}(g') = 1$ with out-bond value 1, and another one, g^F, with $\omega_{in}(g^F) = 1$, $\omega_{out}(g^F) = 0$ with the in-bond value F. We shall then let $\mathcal{T}_2$ be the image algebra consisting of all images in $\mathcal{T}$ with the connection type LINEAR with the additional constraint that the out-arity be zero. $\mathcal{T}_2$ will reappear in the next section as the secondary image algebra in semantic relations.

9.6. Semantic maps

6.1.1. We shall now try to formalize in algebraic form the ideas on semantics from Section 2. To begin with we shall do this in a fairly general setting, attempting to bring out clearly the major problems that confront us in our task. Gradually we shall specialize by bringing in restrictions on the semantic maps, and in Section 7 we will examine the detailed structure of some semantic schemes.

6.1.2. In our view *semantics is relative: it relates two or more regular structures to each other.* Consider therefore two image algebras

$$\begin{cases} \mathcal{T}_1 = <\mathcal{L}_1, R_1> \\ \mathcal{T}_2 = <\mathcal{L}_2, R_2>. \end{cases} \tag{6.1}$$

We want to "explain" $\mathcal{T}_2$ in terms of $\mathcal{T}_1$ by relating images from $\mathcal{T}_2$ to some in $\mathcal{T}_1$. To distinguish between them, let us speak of $\mathcal{T}_1$ as the *primary image algebra* and of $\mathcal{T}_2$ as the *secondary image algebra.*

The *semantic map*, $\underline{\text{sem}}: \mathcal{T}_2 \to \mathcal{T}_1$, defined on some subset $\mathcal{T}_2' \subseteq \mathcal{T}_2$, shall be uniquely defined. Otherwise our "explanation" would be ambiguous. This is something we have decided to avoid; see Section 2.4. We shall then say that $\underline{\text{sem}}$ is *adequate for* $\mathcal{T}_2$ *relative to* $\mathcal{T}_1$.

The inverse of $\underline{\text{sem}}$ need not be unique. A given primary image I can correspond to a set (with several elements)

$$\underline{\text{sem}}^{-1}(I) \subseteq \mathcal{T}_2. \tag{6.2}$$

Sometimes it is better to start the analysis of a semantic scheme via this inverse map $\underline{\text{sem}}^{-1}: \mathcal{T}_1 \to 2^{\mathcal{T}_2}$. When the secondary image algebra is a language, $\underline{\text{sem}}^{-1}(I)$ consists of all grammatical utterances that "mean" I, and $\underline{\text{sem}}^{-1}$ expresses the *linguistic strategy* of the speaker; more about this in Section 7.

The following terminology will be used. If sem is defined on the whole of $\mathcal{T}_2$ it is said to be *entire*. If sem: $\mathcal{T}_2 \to \mathcal{T}_1$ is surjective *and* entire it is said to be *perfect*.

6.2.1. In this section we shall always let the primary image algebra be a relation image algebra, as discussed in Section 3. The secondary image algebra shall consist of a finite state language $L(\mathcal{G})$, viewed as a regular structure; see 5.4.

Since we have $\text{card}(\mathcal{T}_1) = \text{card}(\mathcal{T}_2)$, both being denumerably infinite, there is of course no problem with the existence of semantic maps adequate for $\mathcal{T}_2$ relative to $\mathcal{T}_1$. Indeed, there always exist bijective maps $\mathcal{T}_1 \leftrightarrow \mathcal{T}_2$, and infinitely many of them.

6.2.2. The trouble is that a semantic map, even though adequate, even bijective, is of little interest unless it has additional structure. If it is given just as a list of pairs (I_2, I_1), $I_2 \in \mathcal{T}_2$, $I_1 \in \mathcal{T}_1$, and with no more a prioristic knowledge, it could not possibly be learnt from finite experience, nor could it be remembered using a finite memory.

To supply this missing structure we shall exploit the combinatory regularity of the two image algebras.

6.3. Let us approach this topic from a trivial example. Say that G_1 consists entirely of objects. Then any image in $\mathcal{T}_1$ is just a set of generators each of which has out-arity zero, so that they cannot be connected to each other.

To describe such primary images it suffices to introduce the finite state language with $V_T = G_1$, $V_N = \{1,F\}$ and all productions of the form $1 \underset{g}{\to} F$ or $1 \underset{g}{\to} 1$, $g \in G_1$. This language has the regular expression G_1^*.

If $I_2 = g_1, g_2, \ldots g_n$ with $n(g)$ occurrences of the word g, $g \in G_1$, then the semantic map

$$\text{sem}(I_2) = \text{image with } n(g) \text{ generators } g, \quad g \in G_1 \tag{6.3}$$

is obviously adequate. The inverse $\text{sem}^{-1}(I_1)$ gives us all strings over G_1 of length n with $n(g)$ occurrences of g, in arbitrary order; it is not unique.

6.3.2. A reader is certainly entitled to object to this example being too simple-minded: real semantics is infinitely more complicated. And this is just why we picked the example. As soon as we allow connections in the primary image algebra, syntactic constraints will be forced upon us in order to make the semantics adequate.

Consider another example, still quite simple, with the generators in G_1 given in Table 6.1. An arbitrary image in $\mathcal{T}_1$ then consists of, say, r α-generators, to $m_{11}, m_{12}, \ldots, m_{1r}$ of which a γ-generator connects respectively, in addition to s β-generators, to $m_{21}, m_{22}, \ldots m_{2s}$ of which δ-generators are attached; see Figure 6.1 where $r = 3$, $m_{11} = 1$, $m_{12} = 0$, $m_{13} = 2$ and $s = 2$, $m_{21} = 3$, $m_{22} = 0$.

A language suitable to describe such images is easy to construct and we exhibit one in terms of the wiring diagram of its finite automaton in Figure 6.2.

This language will be supplied with a semantic map as follows. Given a grammatical sentence, for each time we pass

TABLE 6.1

number	identifier	level	β_{in}	ω_{out}	β_{out}
1	α	0	A	0	-
2	β	0	B	0	-
3	γ	1	-	1	A
4	δ	1	-	1	B

the branch $2 \underset{a}{\rightarrow} 3$ we add a generator α to the configuration diagram. For each time we pass the branch $3 \underset{g}{\rightarrow} 3$ we add and connect a generator γ to the last α introduced. Each time we pass the branch $4 \underset{b}{\rightarrow} 5$ we add a generator β to the configuration, and for each pass through $5 \underset{d}{\rightarrow} 5$ we add and connect a generator δ to the last β. For other branches in the figure we do not modify the configuration.

This defines $\text{sem}(I_2)$, uniquely on $\mathcal{T}_2$; it is adequate for $\mathcal{T}_2$ relative to $\mathcal{T}_1$. It is also clear that sem is surjective. Given a primary image I_1 let us enumerate its generators of level 0, first the α's and then the β's. After any occurrence of an α we put the γ's attached to it; similarly for the β's and δ's. With this arrangement pass through the diagram in Figure 6.2 writing terminal symbols successively. If no α is present in I_1, we go to 4, writing a y; otherwise to 2 writing an x, and then to 3 writing an a. If the α has one or several γ's attached, loop through $3 \underset{g}{\rightarrow} 3$ the same number of times. If any more α is present go back to 2 and so on; else go to 4 and behave in the same way.

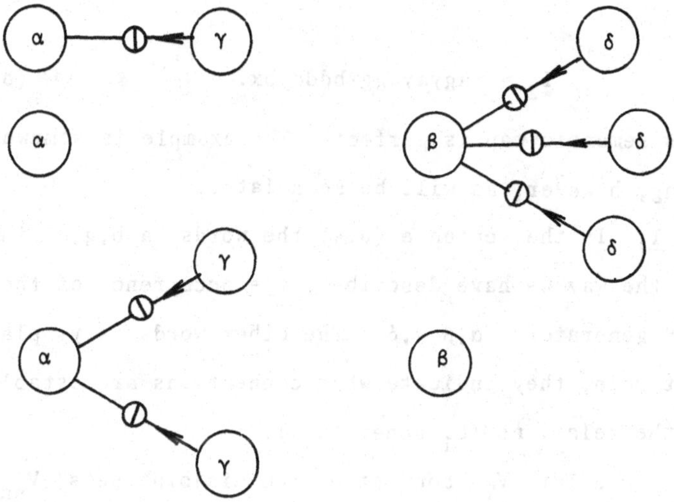

Figure 6.1

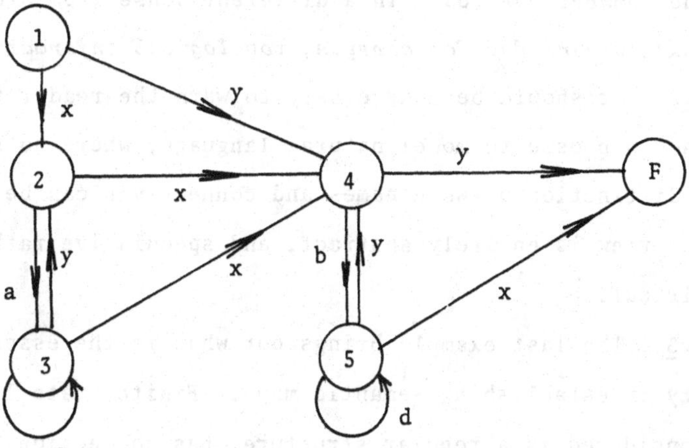

Figure 6.2

Here $V_T = \{a,b,d,g,x,y\}$, $V_N = \{1,2,3,4,5,F\}$ and the productions are the rewriting rules $1 \underset{x}{\to} 2$, and so on, along the branches.

This will produce a grammatical sentence I_2, $I_2 \in \text{sem}^{-1}(I_1)$. For Figure 6.1, for example, we get the sentence

$$I_2 = \text{xagyayaggxbdddybx}. \tag{6.4}$$

This semantic map is perfect. The example is somewhat misleading, however, as will be seen later.

6.4.1. In the sentence (6.4) the words a,b,g,d indicate, in the way we have described, the occurrence of the "related" generators $\alpha,\beta,\gamma,\delta$. The other words x,y play a different role, they indicate what connections are established between the relations (G_1-generators).

One could let V_T consist of two disjoint sets V_{name} and V_{conn}. When we do this, each $g \in G_1$ shall correspond to a set $\underline{name}(g) \subseteq V_{name}$. The inverse $\underline{name}^{-1}$ tells us which generator a name $\in V_{name}$ represents. The words in V_{conn}, the *connectives* (used in a different sense from ordinary syntax), *are needed for carrying topological information.*

6.4.2. It should be unnecessary to warn the reader that this is not supposed to model natural language, where no such clearcut distinction between names and connectives can be made. Our view is entirely abstract, and speculative rather than empirical.

6.4.3. The last example brings out what is the essential difficulty in establishing semantic maps. Finite state language, considered as a regular structure, has connection type Σ_2 = LINEAR; see 5.4.1. Our relation image algebra on the other hand has a much more flexible connection type Σ_1.

Although we shall stay with finite state languages we cannot avoid reminding the reader that with *context free* languages we get a more powerful topology, namely Σ = TREE; see Volume I, Section 2.6. Still more powerful is the connection type for *context-sensitive* languages which allows cycles, just as in our primary image algebra. This should be looked into more carefully in our future work.

6.5.1. The last example contains a clue for the understanding of semantic maps more generally. To make this clear let us return to the wiring diagram in Figure 6.2. For a given grammatical sentence I_2 we start with the empty configuration at state 1. If the first word in I_2 is x we go to 2, keeping the configuration empty. If the next word in I_2 is a, then we go to state 3 and add α to the empty configuration. If the next word in I_2 is g then we connect the generator γ to the generator in our monatomic configuration. We now have a biatomic configuration, this is processed, and we continue until we have reached and used the last word in I_2. Then we have a configuration c from $\mathscr{C}_1$: the corresponding image $R_1(c)$ = sem(I_2).

This is a sequential process, mapping configurations in $\overline{\mathscr{C}}_1$ into others in the same space. Which configuration map will be applied during each step of the process depends upon what branch we are passing through in the wiring diagram.

6.5.2. This leads us to a concept that is fundamental for the following analysis.

Definition 6.1. *By a semantic* (finite-state) *processor from* $\mathscr{C}_2$ *to* $\mathscr{C}_1$ *we shall mean a collection of sets* $C_i \subseteq \overline{\mathscr{C}}_1$, *with* $C_1 = \phi$, $C_F = \mathscr{C}_1$, *with connectors* $\sigma_{ij}(x)\colon C_i \to C_j$; $x \in V_T$, (*where* σ_{ij} *stands for a connector that may or may not contain new generators*)

a) *We start at state* 1 *with* $c = \phi$

b) *A sentence* $x_1 x_2 \ldots x_n \in L(\mathcal{G})$ *is processed left-to-right*

c) *At any transition back to state* 1 c *is made equal to* ϕ *again.*

Remark. The statement in c) means that we start to build a new subconfiguration, starting from the empty one. The new one will not be connected to the one(s) already constructed. Rule c) is less essential than a) and b) and can be left out.

Although we shall study only finite state languages in this paper, the definition has been formulated in such a way that it should be possible to adapt it for more powerful languages.

6.5.3. To gain some intuitive understanding of the role of this definition, let us return to the example in Section 6.3. Introduce the subsets of $\mathcal{C}_1$

$$\begin{cases} C_1 = \phi \\ C_2 = \text{configurations with } r\ \alpha\text{'s},\ r \geq 0, \text{ for each of which may be attached a number of } \gamma\text{'s} \\ C_3 = \text{same as } C_2 \text{ except that } r \geq 1 \\ C_4 = C_2 \\ C_5 = \text{same as } C_2 \text{ but followed by at least one } \beta, \text{ all the } \beta\text{'s may have } \delta\text{'s attached or not} \\ C_F = \mathcal{C}_1. \end{cases} \qquad (6.5)$$

In this example all the C_i-classes consist of regular $\mathcal{C}_1$-configurations, but this need not always be true. More about this later.

The associated configuration maps given by connectors

$\sigma_{ij}(x)$ are defined if $i \underset{x}{\to} j$ is a branch in the diagram

$$\begin{cases} \sigma_{12}(x) = \sigma_{13}(y) = \sigma_{24}(x) = \sigma_{22}(y) = \sigma_{34}(x) = \sigma_{44}(y) \\ \qquad = \sigma_{54}(y) = \sigma_{55}(F) = \text{identity operation} \\ \sigma_{23}(a) = \text{add unconnected } \alpha \text{ to configuration} \\ \sigma_{33}(g) = \text{connect new } \gamma \text{ to last } \alpha \\ \sigma_{45}(b) = \text{add unconnected } \beta \text{ to configuration} \\ \sigma_{55}(d) = \text{connect new } \delta \text{ to last } \beta. \end{cases} \tag{6.6}$$

<u>Remark 1</u>. Since we interpret branching back to state 1 as meaning "begin a new (unconnected) component of the configuration to be calculated", we could have let, for example, the branch $3 \underset{x}{\to} 4$ to 1 instead. Remember that to describe unions of unconnected sub-configurations we need no syntactic information in addition to what is already contained in the sentence to describe the sub-configurations.

<u>Remark 2</u>. In the successive evolution of the c's we may have to refer to generators and bonds, which will be done in terms of the configuration coordinates.

<u>Remark 3</u>. The processor used is related to the concept of tree automata.

<u>6.6.1</u>. Given a semantic processor $\mathscr{C}_2 \to \mathscr{C}_1$ we can extend the configuration maps $\sigma_{ij}(x)$ to be defined for phrases u (in $L(\mathscr{G})$) by putting $\sigma_{ij}(u)$ = undefined if u is not grammatical, and if u is the grammatical phrase (5.4) with $i_o = i, i_n = j$, put

$$\sigma_{ij}(u) = \sigma_{i_{n-1}i_n}(x_n) \cdots \sigma_{i_1 i_o}(x_2)\sigma_{i_o i_1}(x_1). \tag{6.7}$$

Due to associativity (6.7) is well-defined, and since $\mathscr{G}$ is deterministic the string $i_o, i_1, i_2, \ldots i_n$ is unique and hence also (6.7). To the empty string $u = \phi$ we associate $\sigma_{ij}(u)$ = identity.

6.6.2.* With the extended semantic map, the configuration map σ_{1F} represents our semantic map for configurations. We get for the image algebras after R_1-identification has been carried out in $\mathscr{T}_1$

$$\mathscr{T}_2 \to \mathscr{T}_1: \underline{\text{sem}}(I_2) = R_1[\sigma_{1F}(I_1)]. \tag{6.8}$$

6.6.3. We have obtained the semantic map by *sequentially unwrapping the meaning of the given sentence.* This should be compared with the way Wegner (1968) views executing a program as the successive transformation of information structures. In the present case the information structures are configurations from $\mathscr{C}_1$. At each step old bonds may be closed and new generators be added. The out-bonds of the new generators may be left open or be closed immediately. In the example the latter was the case.

The semantic processor still involves operations of too general a nature. In the next section we shall narrow down the choice further.

6.7. Before doing this we mention the following simple observation that serves to bring out more clearly the algebraic structure of the problem of mathematical semantics.

Theorem 1. *The extended semantic processor forms a category.*

Proof: Introduce the objects (in the terminology belonging to categories) C_i and the classes, possibly empty, of morphisms $C_i \to C_j$

$$K_{ij} = \{\sigma_{ij}(u)\ u \in V_T^*\}. \tag{6.9}$$

It is clear that K_{ii} contains the identity map $C_i \to C_i{:}\mathrm{id}_{C_i}$.

The way we have extended the original semantic map in Definition 6.1 to V_T^* it follows directly that

$$\sigma_{ik}(u)\circ\sigma_{k\ell}(v) = \sigma_{i\ell}(uv) \in K_{i\ell} \tag{6.10}$$

where uv stands for the concatenation of the strings u and v. Hence the semantic processor forms a category. Q.E.D.

6.7.1. Consider a configuration $c \in C_i$ with content$(c) = (g_1, g_2, \ldots g_n)$, subscripts are the coordinates of the generators, and bonds $\beta_{k\ell}$, $k = 1,2,\ldots n$, $\ell = 1,2,\ldots\omega_{out}(g_k)+1$. The in-bond of any g_k can be represented by a single value of ℓ, since the values and structural parameters of in-bonds (for one and the same generator) have been assumed to be the same.

Let u be an arbitrary grammatical phrase starting at the state i ending at j, and with the string of arbitrary finite length $x_1x_2\ldots \in V_T^*$. When we apply the corresponding connector $\sigma_{ij}(x_1x_2\ldots)$ to c some of c's bonds will be connected and the rest will not. Denote by $T_i(c)$ the table of the bonds belonging to c that can be connected for some grammatical phrase u starting at i.

Each entry of $T_i(c)$ will consist of three parts. One is the bond coordinate, another consists of bond-structure parameters, and the third is the bond value. During the sequential process we need only keep in memory content(c) and $T_i(c)$.

6.7.2. The memory requirement will therefore depend upon how large the tables content(c) and $T_i(c)$ are. The behavior of $\#[\text{content}(c)]$ is easy to find.

Lemma 1. *We have for* $c' = \sigma_{ij}({}^iu^j)c$

$$\text{content}(c') = \text{content}(c) \cup \text{content}[\sigma_{ij}({}^iu^j)]. \tag{6.11}$$

The proof is immediate, since connectors can only add, not delete generators.

The relation (6.11) implies that

$$\#(c') = \#(c) + \#[\sigma_{ij}({}^iu^j)]. \tag{6.12}$$

The behavior of the size of $T_i(c)$ depends on the particular semantic map and can differ drastically from case to case as will be seen in the next section.

9.7. Special semantic maps

7.1.1. For a given primary image algebra it is easy to construct a scheme that maps any regular configuration into a finite string over a finite vocabulary, in such a way that this string uniquely determines the configuration.

We shall illustrate how this is done via the image algebra in 6.3.2. Choose V_T as consisting of the symbols $\alpha,\beta,\gamma,\delta,I,II$. In other words, we use the elements in G_1 together with two demarcation symbols called I and II.

If content(c) = $(g_1,g_2,\dots g_n)$ we start the string by $g_1g_2\dots g_n II$. If the first out-bond of g_1 goes to g_i we concatenate the string $g_1g_2\dots g_i I$. If the next out-bond goes to g_j we concatenate $g_1g_2\dots g_j I$, and so on. After the last out-bond of g_1 we use the symbol II again, then continue with the out-bonds in g_2, and so on, until the entire configuration has been exhausted. When no out-bonds exist no symbol is used between occurrences of II. We do not use I at the end of the bonds of a generator.

The configuration in Figure 6.1, for example, will be mapped into the string

$$\begin{array}{l}\alpha\gamma\alpha\alpha\gamma\gamma\beta\delta\delta\delta\beta IIII\alpha IIIIII\alpha\gamma\alpha\alpha II\alpha\gamma\alpha\alpha IIII\alpha\gamma\alpha\alpha\gamma\gamma\beta II\\ II\alpha\gamma\alpha\alpha\gamma\gamma\beta II\alpha\gamma\alpha\alpha\gamma\gamma\beta IIII.\end{array} \tag{7.1}$$

The decoding is easy. The substring before the first occurrence of II gives us content(c). The substrings between successive occurrences of II give us the out-bond connections of each generator. Recall that the out-arities are known from G_1, so the references will be unambiguous.

7.1.2. This will give us very long strings, even for simple configurations, such as in Figure 6.1.

We have not specified the syntax of the language. The language is certainly not the entire V_T^*, since most of the elements of this set are not coded representations of elements in $\mathscr{C}_1$. Instead the combinatory regularity $\mathscr{R}_1$ induces syntactic constraints for the coded strings.

We mention parenthetically the reason why we had to refer to generators by strings $g_1 g_2 \cdots g_i$, rather than just by g_i. If the configuration to be talked about contains two identical generators equal to g, say, the latter way of referring to them would be ambiguous. In Figure 7.1 the image in (a)

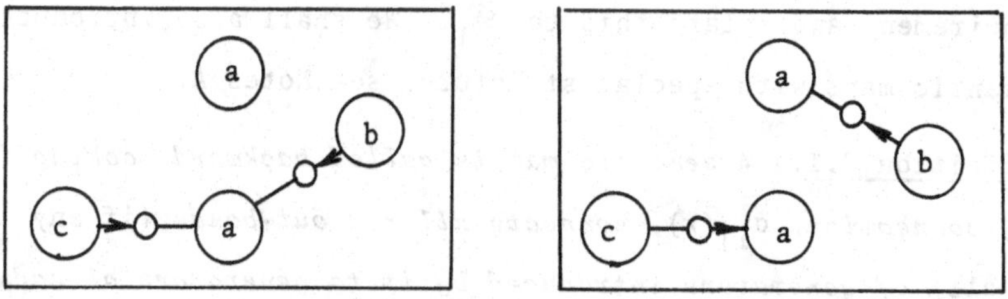

(a) (b)

Figure 7.1

clearly differs from the one in (b), although both are built on the same generator and with connection $c \to a, b \to a$. To specify that b be bonded to a is ambiguous since there are two a's.

This is not always the case, see Figure 7.2. Here it does not matter to which a-generator we connect the out-bond of b, since R_1 will identify the two resulting configurations.

If we could exclude situations like the one in Figure 7.1, we would make our task to construct adequate semantics easier. But that would be to avoid *a difficulty that seems to be intrinsic to the whole topic*, so that we have to face up to the problem in some way.

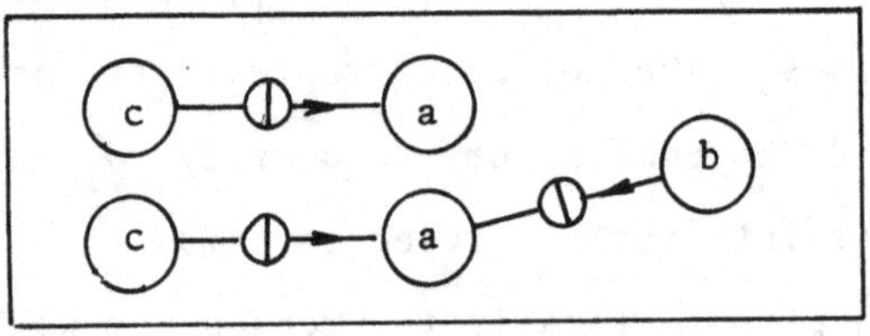

Figure 7.2

7.2.1. Rather than pursuing ad hoc schemes as the one illustrated by (7.1), it is more attractive to start from the other end, with a given semantic map, consider its memory requirement and relate this to $\mathcal{R}_1$. We shall also introduce semantic maps with special structure, see Notes A.

Definition 7.1. *A semantic map is called backward looking if any connection* $\sigma_{ij}(x)$ *connects all new out-bonds* (if any at all) *of generators introduced by it to generators already in* c, $c \in C_i$.

Similarly we could speak of forward looking semantic maps, but this notion will not be more than mentioned in this paper.

7.2.2. As an example of a backward looking semantic map the one in 6.3 can be mentioned. The only connectors $\sigma_{ij}(x)$ that introduce new generators in Figure 6.2 are $\sigma_{23}(a)$, $\sigma_{33}(g)$, $\sigma_{45}(b)$, $\sigma_{55}(d)$ and they all connect the new generators to old ones.

Lemma 1. *If the semantic map is backward looking we have* $\text{sem}({}^1u^i) \in \mathscr{C}_1$ *for any grammatical phrase starting at state* 1.

Proof: Any semantic map, in the sense we use the term, automatically leads to local regularity for $\underline{\text{sem}}({}^1u^i)$. Indeed, $\underline{\text{sem}}({}^1u^i) \in C_i$. But the configurations in C_i belong to $\mathscr{L}_1$ so that all closed bonds satisfy ρ_1. Therefore it is only necessary to verify global regularity. But each connector in the semantic unwrapping of ${}^1u^i$ either does not contain any generator, or if it does, all their out-bonds are connected immediately. Therefore all the out-bonds of $\text{sem}({}^1u^i)$ are closed and the subconfigurations introduced have the connector type Σ_1. In other words $\text{sem}({}^1u^i)$ is $\mathscr{R}_1$-regular.

Q.E.D.

7.2.3. Any grammatical phrase ${}^1u^i$ now means a regular configuration, an important fact that will facilitate the learning of the semantics. The reason for this is that, given a sentence $x_1x_2...x_n \in L$, we can consider each initial phrase $u_k = x_1x_2...x_k$, starting with small values of k, and attempt to learn the meaning of each new branch in the wiring diagram. This makes sense only if, as here, each ${}^1u_k^i$ is meaningful in $\mathscr{C}_1$, not just in $\mathscr{L}_1$ where the configurations do not always imply any meaning to the observer.

7.3.1. To build up a semantic category, see Theorem 6.1, we must construct the connectors $\sigma_{ij}(x)$, but so far we have only seen some simple examples of how this can be done.

To penetrate our problem deeper we shall use the concept of *bonding function which maps syntactic information* (from the sentence in $\mathscr{L}_2$) into *topological information* (for the perceived configuration in $\mathscr{L}_1$). We believe that this concept will be of fundamental importance in further work on mathematical semantics.

We first give the formal definition of a bonding function, and then illustrate its use by examples.

7.3.2. With $B = B_{in}$ = the set of bond values (for G_1) introduce the set

$$D = B \cup (B\times B) \cup (B\times B\times B) \cup \ldots \triangleq D_1 \cup D_2 \cup D_3 \cup \ldots \quad (7.2)$$

and a set Φ of functions ϕ defined on subsets of D. Denote by $D(\phi)$ the domain of such a *bonding function* ϕ, $D(\phi) \subseteq D$.

A bonding function will always be associated with a bond value $\beta \in B$, and we shall assume that for $\delta = (b_1, b_2, \ldots b_n) \in D_n$ the bonding function takes values in the set

$$\Delta_n(\beta) = \{i \mid b_i = \beta\}. \quad (7.3)$$

The purpose of the bonding function is to select one of the bonds of the generators introduced that have the in-bond value β. The set $\Delta_n(\beta)$ can consist of all the integers $1,2,\ldots n$. We shall make sure that no problem arises from the possibility $\Delta_n(\beta) = \phi$ by restricting the domain $D(\phi)$ appropriately.

7.3.3. Recall that the topology of $\mathscr{L}_1$-configurations typically looks as Figure 3.3 with the generators arranged in layers of increasing level of abstraction. This makes it natural to attempt to organize the syntax $\mathcal{G}$ and the syntactic map sem in a similar way. Passing through the wiring diagram we would first handle the objects, level 0, then the properties, level 1, and connect them to the objects, and so on. The connections will be established by bonding functions attached to the branches of the wiring diagram.

Say that we have a branch $i \underset{x}{\to} j$ whose connector operates on level ℓ, $\ell \geq 1$. To this branch we associate at most one generator, say $g \in G_\ell^\ell$ and with $\omega_{out}(g) = \omega$, the out-bond values being $\beta_1, \beta_2, \ldots \beta_\omega$, as well as ω bonding functions $\phi_1, \phi_2, \ldots \phi_\omega$. Here ϕ_r should be associated with the bond value β_r. We allow the degenerate cases when a branch is associated with no generator, only bond functions, or with no generator and no bond function.

Then the connector $\phi_{ij}(x)$ should be formed by connecting the rth out-bond of g to generator number $\phi_r(\delta)$ in the previous level ℓ-1. The vector $\delta = (\beta_1', \beta_2', \ldots \beta_n')$ describes the in-bond values of the subconfiguration consisting of the generators of level ℓ-1, enumerated in the order they have been generated.

In order that this make sense we must ensure that $\delta \in D(\phi)$ which will be done in the following by restricting the selection of any bonding function by what branches precede the current branch in the wiring diagram.

7.4.1. To make the above more intuitive consider the image algebra in 4.2 restricted to generators of levels 0 and 1. Choose L with $V_T = \{\alpha, \beta, \gamma, \delta\}$, and $V_N = \{1, 2, \ldots 10, 11, F\}$ with the wiring diagram in Figure 7.3.

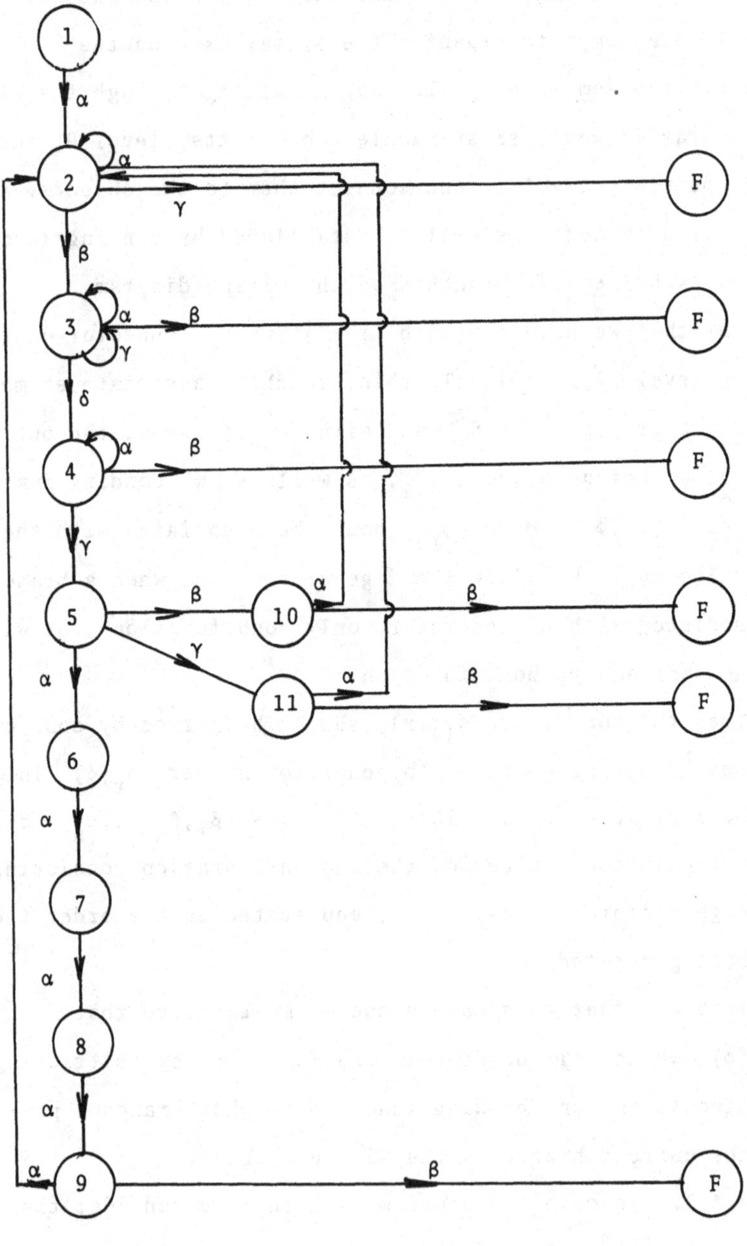

Figure 7.3

TABLE 7.1

branch $i \underset{x}{\to} j$	connector $\sigma_{ij}(x)$
$1 \underset{\alpha}{\to} 2$, $2 \underset{\alpha}{\to} 2$, $2 \underset{\beta}{\to} 3$ $9 \underset{\alpha}{\to} 2$	ϕ_1
$3 \underset{\delta}{\to} 4$, $3 \underset{\delta}{\to} 3$, $3 \underset{\gamma}{\to} 5$ $5 \underset{\alpha}{\to} 6$, $6 \underset{\alpha}{\to} 7$, $7 \underset{\alpha}{\to} 8$	ϕ_2
$8 \underset{\alpha}{\to} 9$, $9 \underset{\beta}{\to} F$	$\phi_3, \phi_4, \phi_5, \phi_6, \phi_7$
$5 \underset{\beta}{\to} 10$	$\phi_8, \phi_9, \phi_{10}$
$3 \underset{\alpha}{\to} 3$	ϕ_{11}
$3 \underset{\gamma}{\to} 3$	ϕ_{12}
$5 \underset{\gamma}{\to} 11$	$\phi_{13}, \phi_{14}, \phi_{15}, \phi_{16}$
all others	no change

To create a semantic map we shall use the bonding functions given in Table 7.2 and interpret the grammatical productions according to Table 7.1.

The "definition of ϕ" listed in the second column of Table 7.2 means the entire action of *all* the bonding functions listed in a row. The role of the individual bonding functions is given only implicitly. For example ϕ_3 means add an f and connect $b_{out,1}$ to the last k, while ϕ_4 means connect

TABLE 7.2

bonding function ϕ	definition of ϕ
ϕ_1	add unconnected ℓ
ϕ_2	add unconnected k
ϕ_i, i = 3,4,5,6,7	add f and connect $b_{out,i-2}$ to $(i-2)^{th}$ of last k
ϕ_i, i = 8,9,10	add g and connect b_{out1},b_{out2} to last two k's and b_{out3} to last ℓ
ϕ_{11}	add h and connect to last ℓ
ϕ_{12}	add j and connect to last ℓ
ϕ_i, i = 13,14,15,16	add i and connect b_{out1} and b_{out2} to last two ℓ's and b_{out3} and b_{out4} to last two k's

$b_{out,2}$ (of the current f-generator) to the next last k.

Consider the sentence

$$I_2 = \alpha\beta\delta\gamma\beta\beta \tag{7.4}$$

with the parsing

$$1_\alpha 2_\beta 3_\delta 4_\gamma 5_\beta 10_\beta F. \tag{7.5}$$

It is grammatical.

To unwrap the meaning of I_2 we get by successively applying the connectors formed by using the bonding functions in the tables:

$1_\alpha 2$

$1_\alpha 2_\beta 3$

$1_\alpha 2_\beta 3_\delta 4$

$1_\alpha 2_\beta 3_\delta 4_\gamma 5$

$1_\alpha 2_\beta 3_\delta 4_\gamma 5_\beta 10$

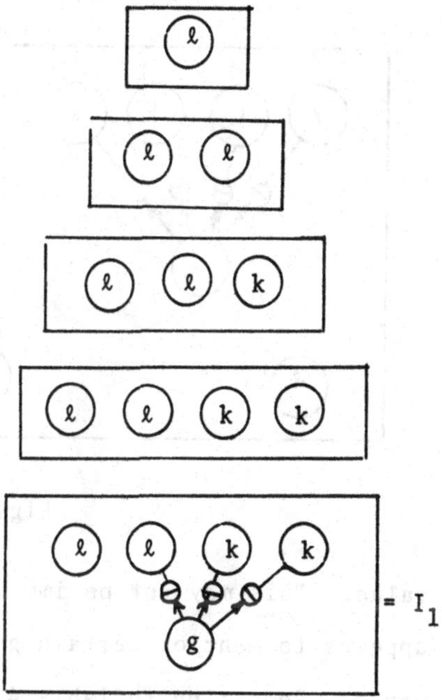

The last transition $10 \to F$ does not change the image.

Now a more complicated example is

$$I_2 = \alpha\beta\delta\gamma\gamma\alpha\beta\delta\gamma\alpha\alpha\alpha\alpha\beta \tag{7.6}$$

parsed into

$$1_\alpha 2_\beta 3_\delta 4_\gamma 5_\gamma 11_\alpha 2_\nu 3_\delta 4_\gamma 5_\alpha 6_\alpha 7_\alpha 8_\alpha 9_\beta F. \tag{7.7}$$

Applying the same unwrapping procedure we see that $\underline{\text{sem}}(I_2) = I_1$ given in Figure 7.4.

Remark 1. The connectors used in the example have two properties that we will meet under more general conditions. Each bonding function connects out-bonds (if any at all) of new generators to in-bonds of old generators; the resulting semantic map is backward looking.

Remark 2. Any bonding function in Table 7.2 is defined in terms of the $1^{st}, 2^{nd}, \ldots$ of the last in-bonds with a given

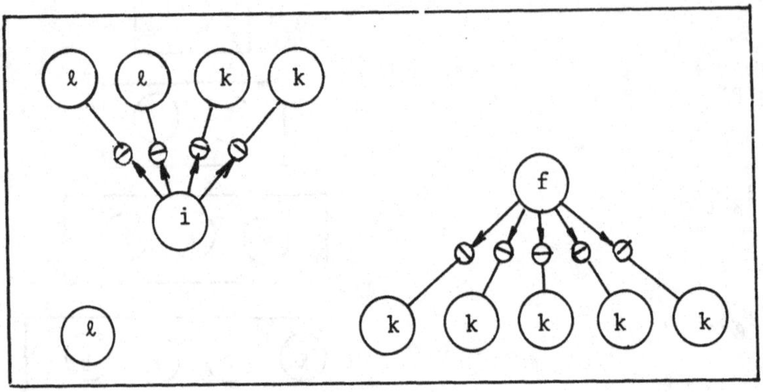 $= I_1$

Figure 7.4

value. This may not be immediately obvious since Table 7.2 appears to mention certain generators rather than their in-bonds. Referring to Table 4.1, the last two rows, we see however that this amounts to the same thing in the present example. In other, more general cases, this distinction must be kept in mind. Such bonding functions, depending only upon the order in which generators and bonds have been introduced will be said to employ *ordered reference*.

7.4.2. Now return to the wiring diagram in Figure 7.3. A state can be identified with the set of strings ${}^1u^i$ leading from 1 to i. Actually, Nerode's theorem tells us that if we use as states the congruence classes over V_T^* we get the minimal wiring diagram.

A semantic map, given in terms of such bonding functions that were mentioned in the last two remarks, depends crucially upon the numbers

$$N_i(\beta) = \min\#\{g\text{'s introduced by any } {}^1u^i \text{ with } \beta_{in}(g) = \beta, \quad \beta \in B,\ i \in V_N\}. \tag{7.8}$$

In (7.8) the minimum is taken over all phrases starting in 1 and ending in i.

In our example we have

$$\begin{cases} N_1(\beta) = 0, \text{ all } \beta \\ N_2(I) = N_2(II) = N_2(III) = N_2(IV) = N_2(V) = 0, \; N_2(VI) = 1 \\ N_3(I) = N_2(II) = N_2(III) = N_2(IV) = N_2(V) = 0, \; N_2(VI) = 2 \\ \cdots \end{cases} \tag{7.9}$$

as can be verified going back to Table 4.1, fourth column.

The numbers $N_i(\beta)$ tell us how much topological information we have built up at state i expressed in terms of a lower bound for the number of potential in-bonds.

7.4.3. A related set of numbers are the *lags* $\lambda_\beta(\phi)$ of a bonding function ϕ employing ordered references. It means

$$\lambda_\beta(\phi) = \max\{\text{number of steps backwards of references to } \beta\text{-values for } \phi\} \tag{7.10}$$

In the example we have

$$\begin{cases} \lambda_\beta(\phi_1) = \lambda_\beta(\phi_2) = 0 \\ \lambda_V(\phi_i) = i-2, \; \lambda_\beta(\phi_i) = 0; \; i = 3,4,5,6,7 \\ \lambda_V(\phi_i) = i-7, \; \lambda_\beta(\phi_i) = 0 \text{ all other } \beta; \; i = 8,9 \\ \lambda_{VI}(\phi_{10}) = 3, \; \lambda_\beta(\phi_{10}) = 0 \text{ all other } \beta; \\ \cdots \end{cases} \tag{7.11}$$

The lag tells us how far back we have to remember potential in-bonds of generators that have already been unwrapped.

7.5. Leaving the example, consider now the connectors constructed as above. Does it lead to a semantic category as in Theorem 6.1? An answer is given by

Theorem 1. *Consider a backward looking strategy and assume that for any branch* $i \underset{x}{\to} j$ *in the wiring diagram any associated bonding function* ϕ *with bond value* β *satisfies*

$$\lambda_\beta(\phi) \leq N_i(\beta). \tag{7.12}$$

Then the construction leads to an entire semantic map.

Proof: We construct the connectors $\sigma_{ij}(x)$ directly by executing the commands in the bonding functions ϕ belonging to the branch $i \underset{x}{\to} j$. Each time we have zero or one generator whose out-bonds have to be connected. The bonding functions do this without ambiguity since only one generator is concerned as far as out-bonds go.

With the aid of the $\sigma_{ij}(x)$ connectors we can now build up the classes C_i, starting with the empty configuration at state 1 and connecting more generators or closing bonds as commanded by the $\sigma_{ij}(x)$. We have to make sure that these classes are subsets of $\mathscr{C}_1$; see Definition 6.1.

This is so; in the present case we can even assert that $C_i \subseteq \mathscr{L}_1 \subseteq \mathscr{C}_1$: all the configurations that we unwrap sequentially are regular. As for global regularity this follows from what was said in the proof of Lemma 7.1.

Local regularity does not follow quite as directly. Indeed, it could happen when we build up the classes C_i that the value of a connector, when applied to the current configuration, is not defined. But such a connector is made up by bonding functions, each ϕ of which only refers backwards a certain number of steps in the order of reference. If fewer generators with the relevant in-bonds had been introduced so far the procedure would fail. Condition (7.12) insures that this cannot happen: we have access to the required number of relevant in-bonds in our list of potential ones. Therefore ϕ is always defined, the bonds can be closed without violating the bond relation ρ, and the new configuration will be regular. Q.E.D.

7.6. It is obvious how a *forward looking strategy* would be organized. This will not be examined in detail here, but we shall have occasion to study strategies looking both forward, for some bonding functions, and backward for others. Theorem 7.1 will then have to be modified.

Since lags can then be both positive and negative we also need a function given as the maximum of the absolute value of the negative lags involved; we will use both $\lambda_\beta^+(\phi)$ as before and the new $\lambda_\beta^-(\phi)$.

We also need an analogue of the numbers $N_i(\beta)$ in (7.8) and introduce

$$M_i(\beta) = \min \#\{g' \text{ introduced by any } {}^i{}_uF \text{ with } \beta_{in}(g) = \beta\};\ \beta \in B,\ j \in V_N. \tag{7.13}$$

There will now be two conditions

$$\begin{cases} \lambda_\beta^+(\phi) \leq N_i(\beta) \\ \lambda_\beta^-(\phi) \leq M_i(\beta) \end{cases} \tag{7.14}$$

in order that our construction yield a semantic category.

Note that we can no longer assert that $C_i \subseteq \mathscr{L}_1$, only that $C_i \subseteq \bar{\mathscr{L}}_1$.

9.8. Learning semantics

Assuming the semantic map to be fixed but unknown, of backward looking type, and expressed as in the previous section,

$$\text{sem} = \{\sigma_{ij}(x)\} \tag{8.1}$$

how can it be learnt from finite experience? More precisely how can sem be estimated when we have observed a sequence of *image couples*

$$[I_1(t), I_2(t)];\ t=1,2,\ldots N;\ I_1(t) \in \mathcal{T}_1;\ I_2(t) \in \mathcal{T}_2 \tag{8.2}$$

where $\text{sem}[I_2(t)] = I_1(t)$. The sequence represents our finite experience and we have left out syntactically correct but meaningless sentences that may have been encountered.

Each $I_1(t)$ can be expressed by listing the generators and connections in one of the configurations that make up the image. Similarly each sentence $I_2(t)$ can be expressed by its passing into a sequence of productions $i \underset{x}{\to} j$. We are looking for a map from the set of productions to the set of generator - connectors.

The set of productions is given already and finite. The set of primitive connectors can be infinite. We shall bound it by assuming the maximum lag to be finite. We then are considering mappings between two finite sets. Consider two finite sets X and Y with $\#(X) = n$ and $\#(Y) = m$. An unknown map $\alpha: X \to Y$ should be estimated when we have access to the observed set couples, $x(t) \subseteq X$, $y(t) \subseteq Y$,

$$\begin{cases} x(1), y(1) = \alpha x(1) \\ x(2), y(2) = \alpha x(a) \\ \vdots \\ x(N), y(N) = \alpha x(N) \end{cases} \tag{8.3}$$

Of course αx stands for the set of all y-values obtained from the point-to-point map α applied to all elements of x.

We have not said anything about the way the $x(t)$-sequence has been generated. In the following it will be assumed that it is obtained as an i.i.d. sample with respect to a given probability measure over 2^X. The measure could be for example one of the syntax controlled measures for context free languages studied in Volume I, or some other measure of interest.

It is important to realize that our problem has been reduced to estimating a *function, here denoted by* α, *when only sets can be observed.* This important problem does not seem to have been studied in the statistical literature, so that we have to start by developing methods for doing this.

Before doing this let us just mention the following possible extensions of our problem. When we have the parsing of an image for $\mathscr{T}_2$ it consists of a string of productions. We abstract the information of set type from this string: the set of productions that occur in the string. Of course this destroys information. The string (1,2,2,7,14,11,7,2) is reduced to the set {1,2,7,11,14}.

We could preserve more information if we also counted frequencies of occurences in the string. In the example we would not just say that "2" occurred, but that it occurred twice, and so on. In other words we would be treating $x(t)$ in (8.3) as a multi-set, not just a set, and similarly for $y(t)$.

Still better would be to treat the string as ordered, then we would lose no information. Similarly treat the generator-connector occurences in $\mathcal{L}_1(\mathcal{R})$ as a POSET.

We are not ready to deal with these two more informative versions of the problem and hope that other researchers will do this.

Returning to the first version, how can we estimate α? It will be instructive to do this first assuming that

$$\alpha \text{ is bijective (so that } n = m). \tag{8.4}$$

We shall later see what happens under more general conditions. We can immediately make a simple statement

Lemma 1. *The observations imply that*

$$\alpha(\delta_i) \in D_{i,N} = \left[\bigcap_{x_i(t)=1} y(t)\right] \cap \left[\bigcap_{x_i(t)=0} y^c(t)\right]; \ \forall i; \tag{8.5}$$

where δ_i *consists of the* i^{th} *element of* X *and* $x(x_1(t), x_2(t), \ldots x_n(t))$ *is written as an indicator function. This is the strongest possible statement.*

Proof: Consider the set

$$E_{i,N} = \left[\bigcap_{x_i(t)=1} x(t)\right] \cap \left[\bigcap_{x_i(t)=0} x^c(t)\right]. \tag{8.6}$$

It is clear that $\delta_i \in E_{i,N}$ since both factors in (8.6) have the ith element in it for all t. But then $\alpha(\delta_i) \in \alpha(E_{i,N})$, and using the fact that α is *homomorphic* for set operations since it is bijective, we get $\alpha(E_{i,N}) = D_{i,N}$.

We cannot strengthen (8.5). Indeed, let the map α satisfy (8.5). If $\#(D_{i,N}) = 1$ then the value $\alpha(\delta_i)$ is completely determined. If $\#(D_{i,N}) > 1$ then $D_{i,N}$ contains two distinct elements, say k and ℓ, and with $\alpha(\delta_i) = k$. We now claim that two $E_{i,N}$ sets are either equal or disjoint.

We can write the intersection of $E_{i,N}$ and $E_{j,N}$ as

$$E_{i,N} \cap E_{j,N} = \left[\bigcap_A x(t)\right] \cap \left[\bigcap_B x^c(t)\right] \tag{8.7}$$

where

$$\begin{cases} A = \{t \mid x_i(t)=1 \text{ or } x_j(t)=1\} \subseteq \{1,2,\ldots n\} \\ B = \{t \mid x_i(t)=0 \text{ or } x_j(t)=0\} \subseteq \{1,2,\ldots n\} \end{cases} \tag{8.8}$$

It is clear that if A and B have an element in common then the intersection in (8.7) is empty. But if A and B exclude each other, since $A \cup B = \{1,2,\ldots n\} = n$, we must have $A = B^c$. This means that

$$[x_i(t)=1 \text{ or } x_j(t)=1] \longleftrightarrow [x_i(t)=1 \text{ and } x_j(t)=1]. \tag{8.9}$$

Then we must have, for a given t, either $x_i(t) = x_j(t) = 0$, or one of them is $= 1$. In the letter case (8.9) implies that both are equal to 1. Hence $x_i(t) = x_j(t)$, $\forall t$. Recalling the definition (8.6) we see that $E_{i,N} = E_{j,N}$. Hence $\{E_{i,N}\}$ constitutes a partition of X. Introduce a new map α' which is equal to α except that it permutes k and ℓ. Note that α maps the sets $E_{i,N}$ to $D_{i,N}$ so that the construction is possible. But α' is consistent with the observations; hence (8.5) is the strongest possible statement based on the observations. Q.E.D.

Remark 1. The proof uses the homorphic property several times, and this is based on the assumption that α is bijective.

Given a sequence of sets we shall say that it is *complete* if the resulting $E_{i,N} = \delta_i$, $\forall i$.

Theorem 1. *As* $N \to \infty$ *we will learn* α *consistently (with probability* 1) *if and only if* support (P) *forms a complete set.*

Proof: Assume that support (P) is complete. Then, with probability one, the infinite i.i.d. sample will contain each of the elements in the support. Then Lemma 1 gives us α unambiguously.

On the other hand in order that $E_{i,N} \to \delta_i$, $\forall i$, as $N \to \infty$ with probability one (which is needed according to the indirect part of the lemma) we can argue as follows. As N increases we will get more and more factors in the intersection on the right side of (8.6). The sequence of $E_{i,N}$ sets is monotonic and hence has a limit. This limit is a.c., with $S = \text{support}(P)$,

$$E_{i,\infty} = \Big[\bigcap_{\substack{x_i=1 \\ x\in S}} x\Big] \cap \Big[\bigcap_{\substack{x_i=0 \\ x\in S}} x^c\Big]. \tag{8.10}$$

But $E_{i,\infty} = \delta_i$, $\forall i$, if and only if S forms a complete set, as asserted. Q.E.D.

Given a set S we can give an algorithm for determining whether it is complete or not. This could be done similarly by Gram-Schmidt orthogonalization (sequentially), or directly.

One case in which P is obviously complete is when $\text{support}(P) = 2^X$: all sets occur with positive probability.

We now turn to the somewhat more complicated question: what is the speed of learning? To ensure completeness say that we are in the situation just mentioned when $\text{support}(P) = X$. For example, all sets x could be given the same probability 2^{-n}: the uniform distribution over X. An answer is given by the following result; see Notes A.

Theorem 2. *If* $\text{support}(P) = 2^X$ *then the probability that we will have learnt* α *completely after* N *trials satisfies*

$$\lim_{N\to\infty} \frac{1-P(\text{complete learning})}{\nu\mu^N} = 1 \tag{8.11}$$

where

$$\mu = \max_{i<j} P(x_i = x_j) \tag{8.12}$$

and ν *is the number of pairs* $i < j$ *realizing the maximum.*

<u>Proof</u>: To determine the probability that $x(1), x(2), \ldots x(N)$ is complete, $E_{i,N} = \delta_i$, $\forall i$, introduce for $i < j$ the event F_{ij} with indicator function

$$\prod_{x_i(t)=1} x_j(t) \cdot \prod_{x_i(t)=0} [1-x_j(t)]. \tag{8.13}$$

Then the event F = "the system is complete" can be expressed as

$$F = \bigcap_{i<j} F_{ij}^c; \quad F^c = \bigcup_{i<j} F_{ij}. \tag{8.14}$$

But then

$$P(F^c) = \sum_{i<j} P(F_{ij}) - \sum_{\substack{i<j \\ k<\ell}} P(F_{ij} \cap F_{k\ell}) + - \ldots \tag{8.15}$$

where we have ordered all pairs (i,j), $i < j$, and consider only $(i,j) < (k,\ell)$ etc. in the above sums.

We have, $i < j$, using the i.i.d. property

$$P(F_{ij}) = E\left\{\prod_{x_i(t)=1} x_j(t) \prod_{x_i(t)=0} [1-x_j(t)]\right\} = c_{ij}^N \tag{8.16}$$

where

$$\begin{aligned} c_{ij} &= P\{x_i=1 \text{ and } x_j=1 \text{ or } x_i=0 \text{ and } x_j=0\} \\ &= P\{x_i=x_j\}. \end{aligned} \tag{8.17}$$

Hence the terms in the first sum of (8.15) that matter asyptotically give together the contribution $\nu\mu^N$, and it remains to show that the rest is negligible.

Consider a term in (8.15) of the form

$$P(F_{ij} \cap F_{k\ell}) \tag{8.18}$$

where $(i,j) \neq (k,\ell)$. First, assume that all four subscripts are different. Then the probability (8.18) is

$$E\left\{\prod_{x_i(t)=1} x_j(t) \prod_{x_i(t)=0} [1-x_j(t)] \prod_{x_k(t)=1} x_\ell(t) \prod_{x_k(t)=0} [1-x_\ell(t)]\right\}$$
$$= c^N_{ij;k\ell} \tag{8.19}$$

where

$$\begin{aligned} c_{ij,k\ell} &= P\{x_i = x_j \text{ and } x_k = x_\ell\} \\ &= c_{ij} P\{A|B\} \end{aligned} \tag{8.20}$$

with

$$\begin{cases} A = \{x_k = x_\ell\} \\ B = \{x_i = x_j\}. \end{cases} \tag{8.21}$$

If $P(A|B) = 1$ we have $P(A \cap B) = P(B)$ and $P[B \cap (A \cap B)^c] = 0$ so that $P[B^c \cup (A \cap B)] = 1$. But

$$(x_i \neq x_j) \vee [(x_k = x_\ell) \wedge (x_i = x_j)] \tag{8.22}$$

is not a tautology; it is false, for example when $x_i = x_j = 0$, $x_k = 0$, $x_\ell = 1$. Hence $P(A|B) < 1$ and

$$c_{ij,k\ell} < c_{ij} \tag{8.23}$$

and the corresponding terms in (8.15) are asymptotically negligible.

Second, assume that $i = k$ differs from j and from $\ell \neq j$. Then (8.19) should be replaced by

$$E\left\{\prod_{x_i(t)=1} x_j(t) x_\ell(t) \prod_{x_i(t)=0} [1-x_j(t)][1-x_\ell(t)]\right\} = d^N_{i,j\ell} \tag{8.24}$$

where

$$d_{i,j\ell} = P\{x_i = x_j = x_\ell\} = P(C|B) c_{ij} \tag{8.25}$$

with

$$C = \{x_\ell = x_j\}. \tag{8.26}$$

The same reasoning as above leads now to the Boolean function

$$(x_i \neq x_j) \vee [(x_i = x_j) \wedge (x_\ell = x_j)] \tag{8.27}$$

which is false, for example when $x_i = x_j = 0$, $x_\ell = 1$.

The remaining terms are treated in the same way. Q.E.D.

Remark 1. Given a level $\varepsilon > 0$ we can expect to have a complete system of observations if N is at least (note that $\ell n\, \mu < 0$!)

$$N_\varepsilon = \frac{\ell n\, \varepsilon - \ell n\, \nu}{\ell n\, \mu}. \tag{8.28}$$

In the special case of P uniform over $X = 2^m$ we have $P(x_i = x_j) = 1/2$ for all $i < j$ so that $\mu = 1/2$ and $\nu = n(n-1)/2$. With base 2 log's in (8.28) we get

$$N_\varepsilon = \log_2 \frac{n(n-1)}{2} - \log_2 \varepsilon \sim 2 \log_2 n \text{ for large } n \tag{8.29}$$

where the factor 2 seems counter-intuitive! (Compare with the lower bound $\log_2 n$.)

In a small mathematical experiment P was made uniform and sampling over 2^X was continued until a complete system was obtained. An APL program was written to test for completeness. The following sample sizes T were observed

n	T	n	T
4	8	16	7
4	7	16	7
4	8	16	13
4	4	32	9
8	5	32	7
8	8	32	13
		64	12

The behavior predicted by (8.29) seems to apply fairly well for the larger values of n.

Remark 2. The leading term $\nu\mu^N$ is not stable. Indeed we can change P by an arbitrary small amount such that μ is not changed but ν changes by ± 1.

Remark 3. The expression in (8.11) makes it natural to define the *asymptotic learning rate* by

$$R = \lim_{N\to\infty} P^{-1/N} \quad \text{(not complete learning).} \tag{8.30}$$

In the present case $R = 1/\mu$. The worst cases occur when μ in (8.12) is large so that there is some pair (i,j) for which $P(x_i=x_j)$ is close to 1. Fast learning occurs if all $P(x_i=x_j)$ are small, and the question arises what the smallest value of $\mu = \mu(P)$ is when n is fixed. This is less obvious then the opposite extreme case, and is answered by

Theorem 3. *For given* n *we have* (with brackets indicating integral part) *and the minimum taken over all* P

$$\min \mu(P) = \frac{\left[\frac{(n-1)^2}{2}\right]}{n(n-1)} . \tag{8.31}$$

Proof: Introduce the indicator function

$$d_i \triangleq d_i(x) = 1_{x_i=1}(x) \tag{8.32}$$

and note that

$$r_{ij} = P(x_i=x_j) = E[1-(d_i-d_j)^2] \tag{8.33}$$

so that

$$A_n = \sum_{i\neq j} r_{ij} = E(B_n); \quad B_n = \sum_{i\neq j} [1-(d_i-d_j)^2] . \tag{8.34}$$

Expanding the square in (8.34) and noting that $d_i^2 = d_i$ we can write

$$
\begin{aligned}
B_n &= \sum_{i\neq j} [1-d_i^2-d_j^2 + 2d_id_j] \\
&= n(n-1)-2(n-1)\sum_i d_i^2 + 2\sum_i d_i \sum_{j\neq i} d_j \\
&= n(n-1)-2(n-1)\sum_i d_i + 2\sum_i d_i\Big(\sum_j d_j - d_i\Big) \\
&= n(n-1)-2n\,D + 2D^2
\end{aligned}
\qquad (8.35)
$$

with $D = \Sigma d_i$. This quadratic polynomial attains its minimum for $D_o = n/2$.

Now D is a stochastic variable taking as possible values $0,1,2,\ldots n$. If n is even, $n = 2m$, then the inequality follows by direct calculation (again with brackets indicating integral parts)

$$B_n \geq \frac{n^2}{2} - n = 2m^2-2m = [\frac{n^2}{2} - n + \frac{1}{2}] = [\frac{(n-1)^2}{2}] . \qquad (8.36)$$

If n is odd, $n = 2m+1$, we get instead, since D can take integral values only,

$$B_n \geq \frac{n^2}{2} - n + \frac{1}{2} = \frac{(n-1)^2}{2} . \qquad (8.37)$$

Hence (8.36) and (8.37) yield

$$A_n = E(B_n) \geq [\frac{(n-1)^2}{2}] . \qquad (8.38)$$

But

$$A_n = \sum_{i\neq j} r_{ij} \leq n(n-1) \max_{r\neq j} r_{ij} = n(n-1)\mu \qquad (8.39)$$

so that, as claimed,

$$\mu \geq \frac{[\frac{(n-1)^2}{2}]}{n(n-1)} . \qquad (8.40)$$

It is clear from the method of the proof that in order that equality hold in (8.40) we must have, for even n,

$$P[D(x) = \frac{n}{2}] = 1 \qquad (8.41)$$

and, for odd n,

$$P[D(x) = \frac{n-1}{2} \text{ or } \frac{n+1}{2}] = 1. \qquad (8.42)$$

To show that the bound (8.40) is achieved pick a vector $\xi = (1,1,1,\ldots 0,0)$ where the number of 1's is $n/2$ if n is even, and $(n+1)/2$ if n is odd. Apply all permutations to ξ and define P by having the same probability $1/n!$ for each one. Then P is symmetric so that all r_{ij}, $i \neq j$, are equal. But this common value r must then satisfy, since $D(x)$ satisfies (8.41) or (8.42),

$$A_n = n(n-1)r = [\frac{(n-1)^2}{2}] \qquad (8.43)$$

so that the bound (8.40) is achieved. Q.E.D.

The estimate of α given in (8.5) will be called the *intersection algorithm* when we want to distinguish it from others to be introduced later. Theorem 1 tells us that if the support of P forms a complete set the intersection algorithm is guaranteed to converge to the correct map. Furthermore, we know the asymptotic speed of convergence given by Theorem 2, and the asymptotic learning rate equals $1/\mu$ with μ defined in Eq. (8.12). This implies that unless P is very skewed over 2^X we can expect to find the map α quickly. It is only when P is quite skew that we need to expect trouble. An extreme case occurs when P attributes probability zero to many subsets of X so that the support is not complete: Then we cannot hope to get convergence to the true α.

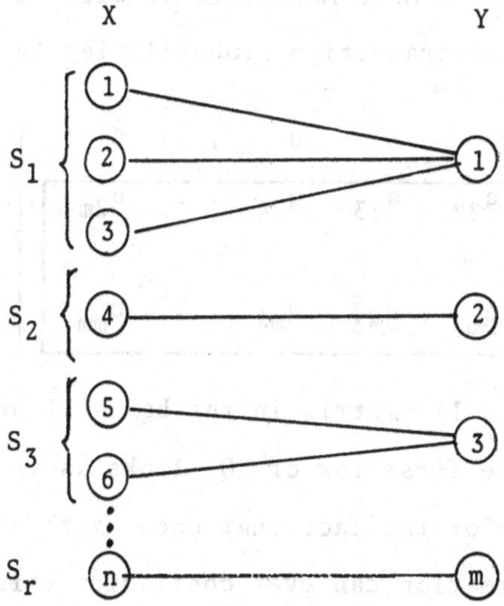

Figure 8.1

An estimate of quite different type has been suggested by S. Geman. This *random algorithm* starts at iteration $t = 1$ with an arbitrary assignment to α, call it $\alpha^*(1)$. When we observe the set couple $[x(t), y(t)]$ we do not change any component of $\alpha(t)$ that is consistent with our information. A component that is not consistent is changed at random, with equal probability to the consistent possibilities.

We shall study the asymptotic learning behavior of the random algorithm. We shall no longer assume that it is bijective - only surjective. The α-map therefore looks typically as in Figure 8.1 after a relabelling of the x- and y-values.

Let the set S_k that is mapped by α into $y = k$ be of size n_k. If all $n_k = 1$, α is bijective. Consider one of the components of $\alpha^*(t)$, for example the first one. Denote it by $a(t)$ so that we would like to assert that $a(t) \to 1$ in the case illustrated in the figure. It is easy to verify that $a(t)$ forms a Markov chain with m states and some matrix Q of transition probabilities looking as

$$Q = \left\{\begin{array}{c|ccccc|} 1 & 0 & 0 & 0 & \cdots & 0 \\ \cline{2-6} q_{21} & q_{22} & q_{23} & q_{24} & \cdots & q_{2m} \\ \vdots & & & & & \\ q_{m1} & q_{m2} & q_{m3} & q_{m4} & \cdots & q_{mm} \\ \cline{2-6} \end{array}\right\} \tag{8.44}$$

where the $(m-1) \times (m-1)$ matrix in the box will be denoted R. The fact that the first row of Q looks as in (8.44) is a direct consequence of the fact that once $a(t)$ has attained the value 1 no information can ever contradict this since it is the true value; therefore $a(t)$ will not change any further.

The probability at iteration t that the first component of the map has not been correctly identified is then

$$P[a(t) \neq 1] = p^T Q^t e = \sum_{k=2}^{m} p_k \sum_{\ell=2}^{m} Q_{k\ell}^t \tag{8.45}$$

where $Q_{k\ell}^t$ stands for the (k,ℓ) element of the tth power of Q, p is the m-vector of initial probabilities, and e is the vector $(0,1,1,\ldots 1)$. Note that only (k,ℓ) values belonging to the matrix R appear in the sum in (8.45).

To get an asymptotic bound for (8.45), consider q_{i1} for $2 \leq i \leq m$. This means the conditional probability

$$q_{i1} = P[a(t+1) = 1 | a(t) = i] \tag{8.46}$$

and in order that the event $a(t+1) = 1$ in (8.46) takes place when $a(t) = i$ we must have

$$\left\{\begin{array}{l} i \notin y(t) = \alpha x(t) \\ \text{and happen to pick 1 at random from the set } \alpha x \end{array}\right. \tag{8.47}$$

To find the probability that (8.47) occurs see Figure 8.1 and let for example $i = 2$. To get information leading us to change $a(t) = 2$ to $a(t+1) = 1$ we must have an occurence of $x = 1$ but no occurences of $x = 4$ (x in S_2). We will then pick 1 with probability $(1+e_3+e_4+\dots)^{-1}$ if e_k stands for the indicator function of at least one $x \in S_k$ occurring. This is so since we then pick the new value with equal probabilities from a set of size $1+e_3+e_4+\dots$. Combining these probabilities we get

$$q_{i1} = \frac{1}{2}\left(\frac{1}{2}\right)^{n_i} E\left(\frac{1}{1+e_3+e_4+\dots}\right) . \tag{8.48}$$

Using the inequality between the harmonic and arithmetic means (8.48) leads to the inequality

$$q_{i1} \geq \frac{2^{-1-n_i}}{E(1+e_3+e_4+\dots)} \geq \frac{2^{-1-n_i}}{m-1} . \tag{8.49}$$

It is easy to sharpen the inequality but we shall not do this here. Instead we note that (8.49) implies

$$\|R\| \leq \max_{2\leq i\leq m} \sum_{j=2}^{m} p_{ij} = 1 - \min_{2\leq i\leq m} q_{i1} \leq 1 - \frac{2^{-1-\nu}}{m-1} \tag{8.50}$$

with $\nu = \max n_i$, leading to the crude bound,

$$\|R\| \leq 1 - \frac{1}{2^{\nu+1}(m-1)} . \tag{8.51}$$

Going back to (8.45) we see that the error probability for the first component is bounded geometrically as the number of iterations increases

$$P\left[a(t) \text{ takes the wrong value}\right] = 0\left[\left(1 - \frac{1}{2^{\nu+1}(m-1)}\right)^t\right]. \tag{8.52}$$

As an example, if α is again bijective, $\nu = 1$, we see that for large $m = n$ it is enough to let t be considerably larger than n in order to be sure about the convergence for any given component of the estimate map $\alpha^*(t)$. While the random algorithm learns more slowly than the intersection algorithm, its behavior is quite acceptable for learning bijective maps. If α is highly non-injective, however, so that ν is large, the behavior deteriorates.

In terms of the semantic map lack of injectivity means that several productions in the language correspond to the same bonding function in $\mathscr{L}_1(\mathscr{R})$. If $\mathscr{L}_1(\mathscr{R})$ involves generators with high abstraction levels we have found (only empirically) that it often happens that many productions mean the same bonding function: α is far from bijective; see Notes B. This will then lead to slower abduction of the semantic map.

9.9. Abduction of semantic maps

The two algorithms studied in the previous section can now be applied to estimate sem. We can of course not expect that the map between the semantic category and that of connectors be bijective. It may be thought that the intersection algorithm is still optimal in the sense that no stronger statement can be made than the one in Lemma 8.1. This was so in the bijective case. But one can show by examples that this is not so in general. Nevertheless it is believed that the intersection algorithm performs moderately well even in the non-bijective case.

Computer experiments have been performed by P. Flanagan (1980) in order to study how the semantic map can be learned. The results will be reported elsewhere.

OUTLOOK

Work on the theory of patterns that has been presented in these three volumes was begun more than ten years ago. During this period the theory has matured and deepened resulting in a lifting of the horizon: what originally appeared as a narrow and well delimited research area is now seen to be of general scope and comprises a broad spectrum of questions, methods, and answers.

The basic ideas in the theory as it stands today are surprisingly unchanged when compared to the early formulation of our research strategy. At the same time, however, the accumulation of results for concrete cases has brought out the unity of the underlying principles. In spite of many gaps in the theory we can see a unified theory emerging from the multitude of special cases.

To complete this theory, and to apply it practically, is a challenging task that will require a sustained research effort over years to come.

APPENDIX

For the reader's convenience a selection of Wittgenstein's aphorisms is given from *Tractatus* and with the original numbering.

1. The world is everything that is the case.

1.1 The world is the totality of facts, not of things.

1.11 The world is determined by the facts, and by these being *all* the facts.

1.12 For the totality of facts determines both what is the case, and also all that is not the case.

1.2 The world divides into facts.

2.01 An atomic fact is a combination of objects (entities, things).

2.021 Objects form the substance of the world. Therefore they cannot be compound.

2.0272 The configuration of the objects forms the atomic fact.

2.03 In the atomic fact objects hang one in another, like the links of a chain.

2.032 The way in which objects hang together in the atomic fact is the structure of the atomic fact.

2.12 The picture is a model of reality.

2.13 To the objects correspond in the picture the elements of the picture.

2.141 The picture is a fact.

2.15 That the elements of the picture are combined with one another in a definite way, represents that the things are so combined with one another.

2.21 The picture agrees with reality or not; it is right or wrong, true or false.

2.22 The picture represents what it represents, independently of its truth or falsehood, through the form of representation.

2.221 What the picture represents is its sense.

2.222 In the agreement or disagreement of its sense with reality, its truth or falsity consists.

2.224 It cannot be discovered from the picture alone whether it is true or false.

3.14 The propositional sign consists in the fact that its elements, the words, are combined in it in a definite way.

The propositional sign is a fact.

3.2 In propositions thoughts can be so expressed that to the objects of the thoughts correspond the elements of the propositional sign.

3.201 These elements I call "simple signs" and the proposition "completely analyzed".

3.202 The simple signs employed in propositions are called names.

3.203 The name means the object. The object is its meaning.

3.21 To the configuration of the simple signs in the propositional sign corresponds the configuration of the objects in the state of affairs.

3.22 In the proposition the name represents the object.

3.25 There is one and only one complete analysis of the proposition.

4.021 The proposition is a picture of reality, for I know the state of affairs presented by it, if I understand the proposition, without its sense having been explained to me.

4.026 The meanings of the simple signs (the words) must be explained to us, if we are to understand them.

4.0311 One name stands for one thing, and another for another thing, and they are connected together. And so the whole, like a living picture, presents the atomic fact.

NOTES

<u>1.1.A</u>. The detailed development of a theory of inference for non-traditional sample spaces will be presented in the author's forthcoming book "Abstract Inference". This development was motivated mainly by questions arising in pattern inference.

<u>1.2.A</u>. Figure 2.1 is from S. Grenander (1929).

<u>B</u>. Some mathematical results in this direction can be found in Volume I, pp. 280-311. See also Chapter 5 in the present volume.

<u>C</u>. See Volume I, Section 3.7 for a particular instance of this sort.

<u>D</u>. Taxonomy - the science of classifying objects into meaningful groups - has a long history. In Foucault (1973) the following old Chinese taxonomy is mentioned:

Animals are divided into:

a) belonging to the Emperor
b) embalmed
c) same
d) sucking pigs
e) sirens
f) fabulous
g) stray dogs
h) included in the present classification
i) frenzied
j) innumerable
k) drawn with a very fine camel hair brush
l) et cetera
m) having just broken the water pitcher
n) that from a long way off look like flies.

European biological taxonomy in the 18th Century is not as delightful, but perhaps more scientific.

2.1.A. Bonds need not be directed. When they are we speak of directed regularity, otherwise of symmetric regularity. This will be elaborated in Chapter 3.

2.2.A. If the regularity is symmetric we do not qualify statements by "in-bonds to out-bonds".

B. The function A is called an acceptance function.

C. We shall often let trees have directed edges contrary to common usage.

2.4.A. More details can be found in Volume II, Chapter 2.

B. See Volume I, Chapter 4.

3.1.A. The question whether two generators should be considered identical or not when they are related by a bond group transformation is similar to the one where pairs of points in space are related by a coordinate transformation. Such a pair $P_1(x_1,y_1,z_1)$, $P_2(x_2,y_2,z_2)$ can be viewed as P_1 belonging to one $\mathbb{R}^3$ space and P_2 to another copy of $\mathbb{R}^3$; the two copies related to each other by a coordinate transformation. Alternatively one can view P_1 and P_2 as the same point in a single space expressed in two different coordinate systems. In Section 3.1 we have chosen the second of these viewpoints, but the other one would be just as valid.

Chapter 3 is based on Grenander 1977(a,c,d,e,f).

Throughout Volume 3 the similarities will form a group.

3.3.A. Since connectors can be restricted by generator index values it is often natural to have several connectors with the same topological function but associated with different sets of generator indices. For example "left-right concatenation" can be specialized to $\alpha_1 \to \alpha_2$ bond couples, or to $\alpha_3 \to \alpha_4$ bond couples, and so on.

3.3.B. The set of connectors can be given *indirectly*, from $\mathscr{R}$, as those σ that appear as $c = \sigma(c_1,c_2)$, c_1,c_2, $c \in \mathscr{L}(\mathscr{R})$, or *directly* as an explicitly defined set $\{\sigma\}$. Which is the case will be clear from context.

C. This connector differs from earlier ones in that it depends also upon bond values.

3.4.A. In the older algebraic literature a homomorphism was often assumed to be surjective but the modern usage of the term does not do this. It is of course only a matter of convenience which alternative to choose; we have decided on the second.

3.4.B. In the proof of the indirect part of Theorem 1 we used the fact that $h[\{g\}]$ must have the same bond structure as g. This is a consequence of the assumption in the statement of the theorem that $h\sigma = \sigma$, all σ: the topology is left unchanged when we apply h to any regular configuration. In particular the topology of a generator, i.e. its bond structure, is left unchanged since it forms a regular configuration when $\mathscr{L}$ is monatomic.

3.4.C. It was believed originally that (iii) holds only under some additional assumption, for example $\mathscr{R}$ = FREE but this does not seem to be required.

3.7.* In the text of this section we use the notion of sub-image algebra $\mathscr{T}'$ of $\mathscr{T}$: the set $\mathscr{T}'$ should be a subset of $\mathscr{T}$, for any $s \in S$ and $I \in \mathscr{T}' \Rightarrow sI \in \mathscr{T}'$, and if $I_1,I_2 \in \mathscr{T}'$ and $I = \sigma(I_1,I_2)\mathscr{T} \Rightarrow I \in \mathscr{T}'$.

** We make the following observation in the case of monotonic Σ. Say that $I_1,I_2,I_3 \in \mathscr{T}$ together with

$I = \sigma(I_1,I_2,I_3)$ and write this combination as $\sigma'(\sigma''(I_1,I_2),I_3)$. Form $I' = \sigma''(I_1,I_2)$ by deleting I_3 and its bonds from I. Then $I' \in \mathcal{T}$ since its global regularity holds because of Σ being monotonic, and its local regularity is guaranteed since all the bond relations in I' remain true. Hence we can write $I = \sigma'(I',I_3)$ with an image $I' \in \mathcal{T}$.

*** The discussion on p. 96 of Volume I is not complete in that it gives the impression that only local regularity matters (the bond relation holds). The expression $I = \sigma(I_1,I_2)$ means, however, that $c_1,c_2,c = \sigma(c_1,c_2)$ exist such that $[c_1]_R = I_1$, $[c_2]_R = I_2$, $[c]_R = I$ where all the three configurations are regular, both locally and globally.

3.7.A. The statement that the empty configuration ϕ belongs to $\mathcal{L}(\mathcal{R})$ or does not is a property of the connection type. The role of the empty configuration is similar to that of the empty string in formal languages. It is a matter of convenience whether to allow ϕ or not as a member of $\mathcal{L}(\mathcal{R})$. Note that if Σ is monotonic then ϕ is a regular configuration - a more precise terminology would be to speak of a partial unit w.r.t. a given connector.

3.7.B. The notion of congruence relation, which is clear for a universal algebra, is more sensitive if some of the basic operations of the algebra are only partial which is the case of greatest interest for us. The reader is referred to Gratzer (1968), pp. 79-99, knowledge of which is assumed below. If we consider a basic operation σ connecting c_1 and c_2 into a regular configuration c, and c_1' and c_2' into a regular configuration c' then we have asked that

$c_i R c_i'$, i = 1,2,; ⇒ cRc'. Hence our notion of identification rule corresponds to "congruence", not "strong congruence". For the basic operations $c \to sc$ this distinction is unnecessary, since c is regular if and only if sc is regular, when the similarities form a group, which has been assumed throughout this volume.

C. In this volume we have relaxed the concept of homomorphism to allow maps between spaces with different global regularity. In the same spirit it would seem natural to generalize Definition 1.1 of Chapter 3 of Volume I by omitting, or at least weakening, condition (ii). We have not studied this possibility.

D. Definition 1 extends the definition given in Volume I in that it allows the connectors to change when the mapping is applied.

3.8.A. The case of fixed global regularity was the only one allowed in the earlier definitions of homomorphisms used in Volume I and in Grenander (1977a).

B. In the earlier definition h was said to be an isomorphism if it was homomorphic and bijective. We now also require that h^{-1} be homomorphic, which forced us to use stronger assumptions to make the conclusion of Theorem 2 valid.

C. It should be noticed that here we require the identification to be strong.

3.9.A. The question often arises whether to allow ϕ as a regular configuration, and then $e = [\phi]_R$ as an image. If Σ is monotonic this will be the case, but in general it is a matter of convenience which alternative one chooses.

B. It may seem reasonable to modify Definition 1 in such a way that we rule out combination $\sigma(I_1,I_2)$ where one of the I_ν is a conditional unit. It is not clear at this time what are the consequences of such a modification.

C. The definition in (16) appears reasonable but it could be questioned if $\mathcal{T}$ has conditional units $\neq$ e. See also remark after the proof of Theorem 2.

4.1.A. If the topology on $\mathcal{T}_\infty$ can be described by a metric, and if $\mathcal{T}_\infty$ is not complete, we can extend $\mathcal{T}_\infty$ to a metric complete space $\mathcal{T}$, and the problem arises what properties we can state about polynomials on $\mathcal{T}$ in analogy to the Theorem above. We have not studied this question yet.

5.1.A. We remind the reader that a probability measure P_C on $\mathcal{C}(\mathcal{R})$ induces another P_I on $\mathcal{T}$ by the relation

$$P_I(E) = P_C(c \mid [c]_R \subset E)$$

where E is any Borel set in $\mathcal{T}$. A typical version of this in a discrete case can be found in Eq. (10.66), Chapter 2, Grenander (1976).

B. The first serious study of controlled probabilities was given in Grenander (1967b).

C. The constant Z plays a role similar to that of a partition function in statistical mechanics.

D. One conditioning, that will not be discussed explicitly in Chapter 5 but will occur occasionally in implicit form is the following one.

Fix $n = \#(c)$ and content(c), enumerate the bond couples by a double subscript (k,ℓ) as usual. Consider a subset BC of (k,ℓ)'s such that to each k there is at most one ℓ

for which $(k,\ell) \in BC$, and the same for each ℓ. We allow only bond couples in BC to be closed, but they are also allowed to be left open.

5.2.A. The results in Theorems 1 and 2 are due to a collective effort by several members of the Research Seminar on Pattern Theory.

B. The proof of Lemma 3 is a simplified version of a result in Thrift (1977). The simplification was given by S. Geman. Theorem 3 can also be derived using the methods presented in Suomela (1976).

5.4.A. Scheffé's theorem says that if $f_n \to f$ a.c. and if

$$\int_{\mathbb{R}^n} f_n(x)m(dx) = \int_{\mathbb{R}^n} f(x)m(dx) = 1$$

then the associated measures P_n tend weakly to P where

$$\begin{cases} \dfrac{P_n(dx)}{m(dx)} = f_n(x) \\ \dfrac{P(dx)}{m(dx)} = f(x) \end{cases}$$

See Billingsley (1968).

5.6.A. A tempting idea is to base the treatment of infinite configurations on non-standard analysis, but this possibility has not been explored.

B. Related results can be found in Krée-Tortrat (1973).

5.7.A. A subset R of T is called cofinal if for any $x \in T$ there exists an element $y \in R$ such that $y \geq x$.

B. By k copies of σ we mean the union of σ k times.

C. If E is a closed subset of $\mathbb{R}^n$ and $1/2x + 1/2y \in E$ for any $x, y \in E$ then it follows that E is convex. To see this consider the line segment from x to y, map it linearly onto the interval [0,1] on the real line. Any point on the line segment that corresponds to a finite binary expansion on [0,1] then belongs to E by repeated "halving". Closure of E then implies convexity.

5.8.A. The notion of marginal in a configuration space was introduced in Grenander (1978c).

5.15.A. Theorems 1 - 4 are from Grenander (1977b) and (1978a). The argument in (15.33) - (15.39) is due to C.-R. Hwang.

B. Relation (112) must be known but we have not been able to find it in the literature.

6.1.A. The literature on hypothesis formation in artificial intelligence is vast. A standard reference is Nilsson (1971) where many other references can be found. An unconventional approach is given in Hájek-Havránik (1978).

6.2.A. In Ristow's work referred to in 6.1 "looping" generators were also included. This led to a more general type of global regularity than we use in this chapter, namely POSET structure of blocks. The blocks are allowed to include looping operations, but in a very restrictive way in order to make outputs well defined.

6.3.A. One could also introduce generators $C(\beta)$ meaning continuous distributions on the set β; then β would be the out-bond value.

B. It is of course true that avoiding the linear algebra generators will make G simpler, but the price we pay for this is loss of flexibility. If these generators had been included many configuration diagrams could be drastically simplified. This is typical of the general pattern synthesis decision; whether to choose a large repertoire for G, or to accept the fact that the configuration diagrams become cumbersome.

C. Two similar generators need not be equal since they can be realized by two different functions of ω, the element in the probability space on which their randomness is defined.

6.4.A. It may seem unnecessary to let $X_n(\beta)$ have the variable attribute β. This is needed, however, if we want to have control over the range set of operations that follow a "copy" generator in the partial order of $\sigma \in$ POSET.

6.5.A. We may add certain images to our generator space, treat them as macrogenerators. For example the generator N with its first in-bond going to $\mathbb{R}$ and the second to $\mathbb{R}^+$ means *all* normal distributions and is of course useful. It has arity one, only a single out-bond equal to $\mathbb{R}$.

6.6.A. The isomorphism established in the theorem seems related to the identification rule that identifies configurations with the same out-bonds and the same moments of order one and two (for fixed inputs). This idea has not been pursued; it should be compared with Theorem 3.8.1.

7.1.A. The results of Chapter 7 are from Grenander (1978c).

7.3.A. The program CONNECT was written by P. Flanagan.

8.1.A. A reader who wishes to learn about modern tendencies in quantitative taxonomy is advised to consult Sokal-Sneath (1963) and Jardine-Sibson (1971).

9.2.A. The following interpretation is strongly influenced by Wedberg (1966), Chapter V, and von Wright (1957), pp. 134-154.

B. While studying probability measures induced by syntactic constraints, Grenander (1967), the problem of inference was left open. We could prove that the maximum likelihood led to consistent estimates but this result was of limited usefulness and not a natural procedure in this context.

9.3.A. Whether to accept this very stringent definition of equality or not amounts to deciding whether the coordinates are absolute or relative. In Chapter 9 they will be treated as absolute for the configurations but in other cases it is more natural to consider them as relative. In the latter case we identify condigurations by the R-rule used in 9.3.8.

When we use absolute coordinates we can think of the coordinates as structural parameters. Compare with the distinction graph-labelled graph.

B. It should be remarked that the statement that a configuration is irreducible does not mean the same as that it is irreducible modulo R. See Volume I, p. 99.

C. The identification rule in 3.8.1 is of general scope: it makes the absolute coordinates relative.

D. In this image algebra generators can be identified for configurations making up any given image.

9.7.A. A weakening of Definition 7.1 of some interest is to ask only that those new connections established by any $\sigma_{ij}(x)$ go to earlier generators, but not that all new out-bonds are immediately closed. Then Lemma 7.1 does not apply.

<u>9.8.A</u>. The theorem is valid for bijective maps only but it is believed that the method of proof can be applied also for non-bijective cases.

<u>B</u>. The statement that high abstraction levels seem to lead to lack of injectivity is based only on the examination of some special cases. It should not be believed in uncritically.

BIBLIOGRAPHY

P. Billingsley (1968): Convergence of probability measures, John Wiley & Sons, New York.

I. Chase (1974): Models of hierarchy formation in animal societies, Behav. Sci., vol. 19, pp. 374-382.

H. Cramér (1945): Mathematical methods of statistics, Princeton Univ. Press, Princeton.

K. Culik, II and A. Lindenmeyer (1976): Parallel graph generating and graph recurrence systems for multicellular development, Int. J. General Systems, 3, pp. 53-66.

John Dalton (1808): A New System of Chemical Philosophy, R. Bickerstaff, London.

P. J. Davis (1979): Circulant Matrices, John Wiley & Sons, New York.

J. E. Dennis and J. F. Traub (1976): The algebraic theory of matrix polynomials, SIAM J. of Numerical Analysis, 13, pp. 831-845.

M. Eden (1961): On the formalization of handwriting, in Symp. Appl. Math. Proc. 12, Am. Math. Soc., Providence, R. I.

G. Fant (1973): Speech Sounds and Features, M.I.T. Press, Cambridge, Mass.

P. Flanagan (1980): Abduction of semantics, unpublished manuscript, Div. Appl. Math., Brown University.

M. Foucault (1973): The Order of Things, Vintage Books, New York.

J. G. Frazer (1951): The Golden Bough, Macmillan Co., New York.

R. Frazer, W. Duncan and A. Collar (1952): Elementary matrices, Cambridge University Press, Cambridge.

K. S. Fu (1974): Syntactic Methods in Pattern Recognition, Academic Press, New York.

B. V. Gnedenko and A. N. Kolmogorov (1954): Limit distributions for sums of independent random variables, Cambridge, Mass.

G. Grätzer (1968): Universal Algebra, Van Nostrand Co., Princeton.

S. Grenander (1929): Astronomiens grunder, A. Bonniers, Stockholm.

U. Grenander (1963): Probabilities on algebraic structures, Almqvist & Wiksells, Stockholm.

U. Grenander (1967a): Foundations of pattern analysis, Graphics/IBM-2 also (1969) in Quart. Appl. Math., 27, pp. 1-55.

U. Grenander (1967b): Syntax-controlled probabilities, R.P.A. No. 2, Brown University.

U. Grenander (1970): A unified approach to pattern analysis, Advances in Computers, vol. 10 (ed. W. Frieberger) Academic Press, N. Y., pp. 175-216.

U. Grenander (1977): An affinity logic for taxonomic patterns, Reports in Pattern Analysis No. 53, Brown University.

U. Grenander (1977a): Algebraic aspects of regular structures, R.P.A. No. 54, Brown University.

U. Grenander (1977b): Is there a law of large numbers in combinatory pattern theory? R.P.A. No. 57, Brown University.

U. Grenander (1977c): A representation theorem for image algebras in terms of <EQUAL,LINEAR>-regularity, R.P.A. No. 60, Brown University.

U. Grenander (1977d): On the concept of homomorphisms in pattern theory, R.P.A. No. 61, Brown University.

U. Grenander (1977e): A representation theorem for image algebras in terms of <INCLUSION,LINEAR>-regularity, R.P.A. No. 63, Brown University.

U. Grenander (1977f): A representation theorem for image algebras in terms of <EQUAL,TREE>-regularity, R.P.A. No. 64, Brown University.

U. Grenander (1978a): Another law of large numbers for symmetric <EQUAL,LINEAR>-regularity, R.P.A. No. 65, Brown University.

U. Grenander (1978b): On mathematical semantics: a pattern theoretic view, R.P.A. No. 71, Brown University.

U. Grenander (1978c): Synthesis of social patterns of domination, R.P.A. No. 74, Brown University.

P. Hájek and T. Havránek (1978): Mechanizing hypothesis formation, Springer-Verlag, Berlin, Heidelberg, New York.

E. J. Hannan (1970): Multiple time series, John Wiley & Sons, New York.

F. Harary (1969): Graph theory, Addison-Wesley, Reading, Mass.

H. Helson and D. Lowdenschlager (1958): Prediction theory and Fourier series in several variables, Acta Math., 99, pp. 165-202.

G. Holton (1973): Thematic Origins of Scientific Thought, Harvard Univ. Press, Cambridge, Mass.

D. H. Hubel and T. N. Wiesel (1965): Receptive fields and functional architecture in two non-striate visual areas (18 and 19) of the cat, J. Neurophysiol. 28, pp. 229-289.

D. Hume (1951): A Treatise of Human Nature, reprinted from the original edition, Oxford University Press, Amen House, London.

C. R. Hwang (1978): Frozen patterns and minimal energy states, R.P.A. No. 68, Brown University.

N. Jardine and R. Sibson (1971): Mathematical Taxonomy, John Wiley & Sons, New York.

J. J. Katz and J. A. Fodor (1963): The structure of language as semantic theory, Language 39, pp. 170-210.

J. L. Kelley (1955): General topology, Springer-Verlag, New York, Heidelberg, Berlin.

P. Krée and A. Tortrat (1973): Désintégration d'une loi gaussienne dans une somme vectorielle, C. R. Acad. Sci., Paris, 277, Series A, pp. 695-697.

P. Lancaster (1966): Lambda matrices and vibrating systems, Pergamon Press, New York.

H. G. Landau (1951): On dominance relations and the structure of animal societies, Bull. Math. Biophysics, 13, pp. 245-262.

H. G. Landau (1965): Development of structure in a society with a dominance relation when new members are added successively, Bull. Math. Biophysics 27, pp. 151-160.

A. J. Larkin (1969): Work Study, McGraw-Hill Book Co., New York.

R. Ledley (1964): High-speed automatic analysis of biomedical pictures, Science, 146, pp. 216-223.

J. Y. Lettvin, H. R. Maturana, W. S. McCulloch and W. Pitts (1959): What the frog's eye tells the frog's brain, Proc. Inst. Radio Engrs. 47, pp. 1940-1951.

C. Lévi-Strauss (1955): The structural study of myth. J. of Am. Folklore, LXXVIII, pp. 428-444.

A. M. Lieberman, F. S. Cooper, D. P. Sankweiler and M. Studdert-Kenedy (1967): Perception of the Speech Code, Psych. Rev. 74, pp. 431-461.

A. Lindenmayer (1975): Developmental systems and languages in their biological contact, in Developmental Systems and Languages, ed. G. T. Herman and G. Rosenberg, North-Holland Pub. Co., Amsterdam.

A. L. Loeb (1976): Space Structures, Their Harmony and Counterpoint, Addison-Wesley Pub. Co., Reading, Mass.

F. Lorrain (1975): Réseaux sociaux et classifications sociales, Essai sur l'algèbre et la géometrie des structures sociales, Actualité Sci. et Ind., No. 1368, Hermann, Paris.

A. L. Mackay (1974): Generalized structural geometry, Acta Cryst. A30, pp. 291-305.

A. I. Mal'cev (1973): Algebraic Systems, Springer-Verlag, Berlin, Heidelberg, New York.

S. Marcus (1967): Algebraic Linguistics: Analytical Models, Academic Press, New York.

M. Marcus and H. Minc (1964): A survey of matrix theory and matrix inequalities, Allyn and Bacon, New Jersey.

P. Martin-Löf (1966): The definition of random sequences, Inf. and Control, 9, pp. 602-619.

W. F. Miller and A. C. Shaw (1968): Linguistics in picture processing - a survey, Proc. AFIPS Fall Joint Comp. Conf., pp. 279-290.

J. W. Milnor and J. D. Stasheff (1974): Characteristic classes, Princeton University Press, Princeton.

R. Narasimhan (1964): Labelling schematic and syntactic description of pictures, Inf. Contr. 7, pp. 151-179.

N. Nilsson (1971): Problem solving methods in artificial intelligence, McGraw-Hill, New York.

O. Peterson, E. Barsamian and M. Eden (1966): A study of diagnostic performance, J. Med. Ed. 41, pp. 797-803.

P. M. Prenter (1975): Splines and Variational Methods, John Wiley & Sons, New York.

J. Prewitt (1972): Parametric and nonparametric recognition by computer: An application to Lenkocyte image processing, Adv. Comp. 12, pp. 285-414.

V. Propp (1971): Morphology of the Folktale, English translation by L. Scott 2nd ed., Univ. Texas Press, Austin.

R. H. Robins (1967): A Short History of Linguistics, Indiana Univ. Press, Bloomington, Indiana.

E. A. Robinson (1967): Multichannel time series analysis with digital computer programs, Holden-Day, Calif.

W. Rudin (1973): Functional analysis, McGraw-Hill, New Jersey.

E. J. Sandewall (1971): Representing natural language information in predicate calculus, in Machine Intelligence (eds. Meltzer and Mitchie), 5, Edinburgh Univ. Press, Edinburgh.

C. J. Schneer (1960): The Search for Order, Harper & Brothers, New York.

H. Schubert (1968): Topology, Allyn & Bacon, Boston.

I. E. Segal and R. H. Kunge (1978): Integrals and Operators, 2nd ed., Springer-Verlag, Berlin, Heidelberg, New York.

R. F. Simmons (1973): Semantic networks: Their computation and use for understanding English sentences, in Computer Models of Thought and Language, edited by Roger C. Schank and K. M. Colby, pp. 63-113.

F. Sinic (1971): Sequence - covering and countably bi-quotient mappings, Gen. Topology and its Appl., 1, pp. 143-154.

R. R. Sokal and P. H. A. Sneath (1963): Principles of Numerical Taxonomy, Freeman, San Francisco, London.

R. J. Solomonoff (1964): A formal theory of inductive inference, Part I, Inf. and Control 7, pp. 1-22.

E. M. Stein (1970): Singular integrals and differentiability properties of functions, Princeton Univ. Press, Princeton.

P. Suomela (1976): Construction of nearest neighbor systems, Ann. Ac. Sci. Fennicae, A., pp. 1-57.

R. Tarjan (1972): Depth-first search and linear graph algorithms, SIAM J. Computing, I, pp. 146-160.

D'Arcy Wentworth Thompson (1961): On Growth and Form (Abr. Ed.) Cambridge Univ. Press, Cambridge.

P. Thrift (1977): Conditioning for topologically controlled random variables, R.P.A. No. 53, Brown University.

P. Thrift (1979): Autoregression in homogeneous Gaussian configurations, R.P.A. No. 77, Brown University.

G. H. von Wright (1957): Logik, filosofi och språk, Bonniers, Stockholm.

A. Wedberg (1966): Filosofiens historia, 3, Bonniers, Stockholm.

P. Wegner (1968): Programming languages, information structures, and machine organization,

H. Weyl (1939): On the volume of tubes, Am. J. of Math. 61, pp. 461-472.

H. Weyl (1952): Symmetry, Princeton Univ. Press, Princeton, New Jersey.

P. Whittle (1963): Prediction and Regulation, The English Universities Press.

T. Winograd (1972): Understanding Natural Language, Academic Press, New York.

L. Wittgenstein (1955): Tractatus Logico-Philosophicus. Sixth Edition, London.

W. A. Wood (1970): Transition network grammars for natural language analysis, Comm. Assoc. Comp. Mach. 13, pp. 591-606.

INDEX

Applied Mathematical Sciences

1. John: **Partial Differential Equations**, 3rd ed.
2. Sirovich: **Techniques of Asymptotic Analysis.**
3. Hale: **Theory of Functional Differential Equations**, 2nd ed. (cloth)
4. Percus: **Combinatorial Methods.**
5. von Mises/Friedrichs: **Fluid Dynamics.**
6. Freiberger/Grenander: **A Course in Computational Probability and Statistics**, Rev. Ptg.
7. Pipkin. **Lectures on Viscoelasticity Theory.**
8. Giacaglia: **Perturbation Methods in Non-Linear Systems.**
9. Friedrichs: **Spectral Theory of Operators in Hilbert Space.**
10. Stroud: **Numerical Quadrature and Solutions of Ordinary Differential Equations.**
11. Wolovich: **Linear Multivariable Systems.**
12. Berkovitz: **Optimal Control Theory.**
13. Bluman/Cole: **Similarity Methods for Differential Equations.**
14. Yoshizawa: **Stability Theory and the Existence of Periodic Solutions and Almost Periodic Solutions.**
15. Braun: **Differential Equations and Their Applications**, 2nd ed. (cloth)
16. Lefschetz: **Applications of Algebraic Topology.**
17. Collatz/Wetterling: **Optimization Problems.**
18. Grenander: **Pattern Synthesis. Lectures in Pattern Theory**, vol. I.
19. Marsden/McCracken: **The Hopf Bifurcation and Its Applications.**
20. Driver: **Ordinary and Delay Differential Equations.**
21. Courant/Friedrichs: **Supersonic Flow and Shock Waves.** (cloth)
22. Rouche/Habets/Laloy: **Stability Theory by Liapunov's Direct Method.**
23. Lamperti: **Stochastic Processes. A Survey of the Mathematical Theory.**
24. Grenander: **Pattern Analysis. Lectures in Pattern Theory**, vol. II.
25. Davies: **Integral Transforms and Their Applications.**
26. Kushner/Clark: **Stochastic Approximation Methods for Constrained and Unconstrained Systems.**
27. de Boor: **A Practical Guide to Splines.**
28. Keilson: **Markov Chain Models—Rarity and Exponentiality.**
29. de Veubeke: **A Course in Elasticity.**
30. Śniatycki: **Geometric Quantization and Quantum Mechanics.**
31. Reid: **Sturmian Theory for Ordinary Differential Equations.**
32. Meis/Marcowitz: **Numerical Solution of Partial Differential Equations.**
33. Grenander: **Lectures in Pattern Theory**, vol. III.
34. Cole/Kevorkian: **Pertubation Methods in Applied Mathematics.** (cloth)